# Basic
# Mathematics
# for
# Technical
# Occupations

# PRENTICE-HALL SERIES
# IN TECHNICAL MATHEMATICS

## Frank L. Juszli, *Editor*

APPLIED MATHEMATICS FOR ELECTRONICS
*Westlake and Noden*

APPLIED MATHEMATICS FOR ENGINEERING AND SCIENCE
*Shere and Love*

APPLIED MATHEMATICS FOR TECHNICIANS
*Moore, Griffin, and Polhamus*

BASIC MATHEMATICS FOR TECHNICAL OCCUPATIONS
*Bohuslov*

CALCULUS WITH ANALYTIC GEOMETRY
*Niles and Haborak*

CONTEMPORARY TECHNICAL MATHEMATICS
*Paul and Shaevel*

CONTEMPORARY TECHNICAL MATHEMATICS WITH CALCULUS
*Paul and Shaevel*

ELEMENTARY TECHNICAL MATHEMATICS, 2nd ed.
*Juszli and Rodgers*

ELEMENTARY TECHNICAL MATHEMATICS WITH CALCULUS
*Juszli and Rodgers*

ESSENTIALS OF TECHNICAL MATHEMATICS
*Paul and Shaevel*

MATHEMATICS FOR APPLIED ENGINEERING
*Cairns*

MATHEMATICS FOR INDUSTRIAL TECHNICIANS
*Pachucki*

MATHEMATICS FOR TECHNICIANS
*Tronaas*

PRECALCULUS MATHEMATICS FOR TECHNICAL STUDENTS
*Mavrommatis and Reichmeider*

TECHNICAL MATHEMATICS WITH CALCULUS
*Placek*

# Basic Mathematics for Technical Occupations

RONALD L. BOHUSLOV
*College of Alameda*

PRENTICE-HALL, INC., *Englewood Cliffs, New Jersey*

*Library of Congress Cataloging in Publication Data*

BOHUSLOV, RONALD.
  Basic mathematics for technical occupations.

  1. Mathematics—1961–  I.  Title.
QA39.2.B63      513′.1      75-25879
ISBN 0–13–063396–8

*To Esther Fern*

PRENTICE-HALL INTERNATIONAL, INC., *London*
PRENTICE-HALL OF AUSTRALIA PTY, LTD., *Sydney*
PRENTICE-HALL OF CANADA, LTD., *Toronto*
PRENTICE-HALL OF INDIA PRIVATE LIMITED, *New Delhi*
PRENTICE-HALL OF JAPAN, INC., *Tokyo*
PRENTICE-HALL OF SOUTHEAST ASIA (PTE.) LTD., *Singapore*

# Contents

# Preface

*Basic Mathematics for Technical Occupations* contains the mathematics necessary to understand and solve problems in technical and trade vocations. An awareness of numbers and a desire to achieve mathematical skills are the only prerequisites for mastery of the material.

The computations and techniques are designed to motivate understanding and appreciation of the principles used in technical mathematics. Each topic is developed through realistic examples, discussions, illustrations, and pertinent problems using the latest data and information. The problems are from the major occupational areas and are designed to encourage the student to develop problem-solving techniques as well as manipulative skills.

Content and arrangement of chapters give the instructor maximum flexibility in choosing the course content to meet the needs of individual classes. Most chapters are limited to two or three major topics, and are organized around a principle or idea in such a manner that the student is able to comprehend a complete concept or procedure in one or two class periods. Practice Tests enable the student to evaluate his or her progress.

The text contains:

> Approximately:   400 worked examples
>                     6,000 problems
>                     350 illustrations
> Tables of Interrelation of Units of Measurement
> (U. S. Customary and Metric)
> Answers to odd-numbered problems
> Answers to Practice Tests
> List of mathematical symbols

Chapters 1 through 5 contain the important concepts of place value and the number line, rounding off, significant digits, measurements, and the arithmetic of the whole numbers. To avoid cumbersome computations while studying these ideas, only positive whole numbers are used. In chapters 6 through 14 these concepts are applied to fractions, decimals, percent, basic geometry, basic trigonometry, and basic algebra.

Although this book is specifically designed for students in technical occupations, it also meets in a very real way the needs of general mathematics students. Arithmetic processes are reviewed, and concepts ranging from whole numbers through basic algebra are developed and explored. The use of current, realistic situations enables the nontechnical student to relate mathematics to practical values and concrete ideas.

In appreciation of assistance and cooperation, I express my gratitude to
The staff of Prentice Hall, especially Cary F. Baker, Margaret McAbee, and Steve Lux;
The reviewers, especially Edwin Conry, Orange Coast College;
The College of Alameda, especially the Transportation Division;
My students, especially Jim Thomsen.

R. L. B.

# To the Student

Are you, or do you plan to become an auto mechanic, diesel mechanic, auto-body repairman, sheetmetal worker, machine shop worker, laboratory assistant, carpenter, or other similar technician or tradesman? If so, this book will help you learn the mathematical skills that will put you into positions of high responsibility and top income.

Do you need to understand mathematics? Do you have difficulty grasping abstract mathematical concepts? If so, this book will enable you to learn the mathematics necessary for your own specific purposes.

You do not need to have had previous experience in mathematics to use this text. In chapters 1 through 5 you will review or learn basic arithmetic. In chapters 6 through 14 you will use the skills you developed in the first five chapters to work with fractions, decimals, percent, basic geometry, basic trigonometry, and basic algebra. Each topic is clearly developed, and you should have no difficulty mastering the material.

To get maximum results from your efforts, establish a few study practices, such as doing your work at a certain time and place, and follow these general guidelines:

Read the explanations thoroughly.
Work through the examples.
When necessary, memorize rules, definitions, and formulas.
Do the Practice Tests. (Your results on these tests will show you your mastery of the material, as well as areas of weakness that need to be strengthened by more study.)

If you are studying alone, choose the material in the text that is applicable to your needs. If you are in school, your instructor will design the course of study to get the maximum results for the class.

R. L. B.

# Mathematical Symbols and Their Meaning

| Symbol | Meaning | Example |
|--------|---------|---------|
| $+$ | Add (plus) | $4 + 5 = 9$ |
| $+$ | Positive | $+8$ |
| $-$ | Subtract (minus) | $7 - 3 = 4$ |
| $-$ | Negative | $-9$ |
| $\times$ or $\cdot$ | Multiply (times) | $4 \times 5 = 20,\ 4 \cdot 5 = 20$ |
| $\div$ or $-$ or $\overline{)}$ or $/$ | Divide | $6 \div 3 = \dfrac{6}{3} = 3\overline{)6} = 6/3$ |
| ( ) parentheses } [ ] brackets } | Grouping | $2(3 + 6) = 2(9) = 18$ <br> $2[3 + 6] = 2[9] = 18$ |
| $=$ | Equals | $4 = 3 + 1$ |
| $:$ | Is to (used between ratios) | $9 : 5$ |
| $\neq$ | Does not equal | $4 \neq 3 - 1$ |
| $<$ | Is less than | $6 < 17$ |
| $>$ | Is greater than | $15 > 10$ |
| $\sqrt{\phantom{x}}$ | Positive square root | $\sqrt{64} = 8$ |
| $\sqrt[n]{\phantom{x}}$ | Real $n$th root | $\sqrt[3]{8} = 2,\ \sqrt[4]{16} = 2$ |
| $\angle$ | Angle | $\angle A = 36°$ |
| $°$ | Degree symbol | $56°F$ (temperature) <br> $45°$ (angle) |
| $\%$ | Percent | $15\%$ |
| $\pi$ | Pi, the ratio of the circumference of a circle to its diameter | $3.1415926536$ |

# 1

# Whole Numbers, Number Representation, Place Value, and the Number Line

Today's modern technology would not exist without man's ability to express his observations and carry out his ideas by means of numbers. Eons ago man had little need for numbers, fractions, decimals, or rules of algebra. He was not concerned with the price per pound of dinosaur steak, the family budget, or the square foot area of his cave. But his enterprising nature soon made him want some way to keep track of his possessions. He wanted to be able to count the number of people in his group, the tools he made, and maybe the number of sabre-tooth tigers he killed. Perhaps at first he did this by counting on his fingers. When he wanted to count more than ten things, he met his first big problem in counting.

According to history, some *counter*, such as a stone or a stick, was used to mark the multiples of ten. Figures 1-1, 1-2, 1-3, and 1-4 show how the counter developed during the centuries and how it is used in today's business and technical work.

Many early civilizations had special signs or symbols for the numbers 1, 5, 10, 100, and so on. Table 1-1 shows some of the early number systems, as well as our modern Hindu-Arabic system.

Our arithmetic would be very difficult if we had to use the symbols from an earlier system. Man is constantly improving his method of displaying

**Figure 1-1**   *Counters in the sand*

**Figure 1-2**   *The abacus*

**Figure 1-3**   *Manual calculator*

**Figure 1-4**   *Modern accounting machines and computers*

TABLE 1-1   NUMBER SYMBOLS

| Hindu-Arabic | (one) 1 | (five) 5 | (ten) 10 | (hundred) 100 |
|---|---|---|---|---|
| Roman | I | V | X | C |
| Egyptian | \| | \|\|\|  \|\| | ∩ | ⟨ |
| Babylonian | ▼ | ▼▼▼ ▼▼ | < | ◀ |
| Hebrew | א | ה | ן | ק |

numbers. Perhaps some day our system may look as strange to other people as the Egyptian or Babylonian numbering system looks to us now.

The method of using counters is not new. In the Orient, the abacus, or counting board, has been used for centuries. Figure 1-5 shows a simple counting board. Each column has 10 counters that look like buttons. The column on the far right is used to record single items. When ten single items are counted, one counter is brought down in the second column from right. This is called the 10's column. Then the single item count starts over again in the l's column. If the count goes so high that ten counters are brought down in the 10's column, then one counter is moved down in the third column from the right. This is the 100's column, and it is used to record that ten 10s have been counted. Then the 10's column is cleared by moving the counters to the top of the board. Then the column may be used to record 10s again.

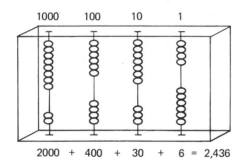

**Figure 1-5**  *Counting board*

Look at Fig. 1-5. The number 2,436 is the total shown on the counting board. Note that a single counter in the 10's column takes the place of ten counters in the 1's column. A single counter in the 100's column means that ten items have been counted in the 10's column, and so on through the 1,000's column.

If numbers larger then 10,000 are to be counted, add more columns to the counting board. The next column placed to the left would be for 10,000. Then the next column to the left would be for 100,000, and so on.

The mileage register on the speedometer of an automobile is a mechanical version of the counting board. This register, called an odometer, is shown in Fig. 1-6. If the odometer is removed from the dashboard, it looks like a set of wheels set side by side (See Fig. 1-7). Each wheel has ten numerals from zero (0) to nine (9) spaced around the wheel. The first wheel on the right is used to record single miles. The second wheel from the right records every ten miles, the third wheel records the hundreds of miles, the

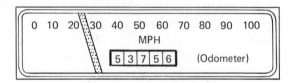

**Figure 1-6**   *Odometer*

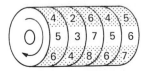

**Figure 1-7**   *Odometer wheels*

next wheel record thousands, and the last wheel records the tens of thousands of miles. Here, as on the counting boards, each number place to the left is ten times the number place on the right. Both the counting board and the odometer use the *decimal* system of counting. This means that they count by groups of ten. Look at the odometer reading in Fig. 1-7. There are 5 ten-thousands, 3 thousands, 7 hundreds, 5 tens, and 6 ones. The odometer reading is:

$$5 \times 10,000 + 3 \times 1,000 + 7 \times 100 + 5 \times 10 + 6 \times 1$$

which is read as 53,756 miles.

Most modern counting devices are based on the same principles as the counting board and odometer, but there are some variations. The following material will help you recognize and understand other forms of counting devices. Do not assume that all counting devices work on the decimal, or base 10, system. There are some exceptions. Figure 1-8 and 1-9 show two exceptions. Digital clocks record time in digital (numeral) form.

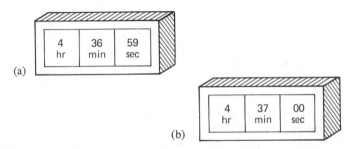

**Figure 1-8**   *Digital clock: (a) A time reading; (b) 1 second later*

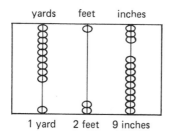

**Figure 1-9**  *Inches-feet-yards counting board*

Since the minute and hour are based on 60 units rather than 100 units, the minute counter on a clock changes every 60 seconds, and the hour counter changes every 60 minutes. Figure 1-8 shows what happens in 1 second.

If you construct a counting board to record time in the way a digital clock does, the first column on the right will have the usual 10 counters. The second column needs only 6 counters (one for each 10 seconds). The third column will have 10 counters. The fourth column again needs 6 (one for each 10 minutes). The left column will have either 12 or 24, depending on the method for recording the hours in a day. It is easy to see that a digital clock is a sophisticated version of a counting board.

Here is a second variation from the familiar decimal system—a counting board that records *yards-feet-inches*. If you measure a table top and find that it is 69 inches long, how do you express this measurement in *yards-feet-inches*? To make the conversion, use the counting board shown in Fig. 1-9. The first column at the right has 12 counters, since there are 12 inches in 1 foot. The second column has only 3 counters, since there are 3 feet in 1 yard. The third column has 10 counters which makes it possible to record large measurements. The counting board shows 1 yd. 2 ft. 9 in. because:

$$
\begin{array}{r}
1 \text{ yard} = 36 \text{ inches} \\
2 \text{ feet} = 24 \text{ inches} \\
9 \text{ inches} = \phantom{0}9 \text{ inches} \\
\hline
1 \text{ yd. } 2 \text{ ft. } 9 \text{ in.} = 69 \text{ inches}
\end{array}
$$

(See tables of conversion facts—US customary system pages 465-469)

As you see, counters and counting boards serve us in our need to know how many objects we have, how far we have gone, or what time of day it is. Counting boards, whether in the form of the ancient oriental abacus or the modern high-speed computer, will be with us for a long time. Your ability to read counting boards, gauges, dials, and other counting devices correctly is an important part of your future in business and technology.

Study the problems that follow. Find out how well you can *count* on a counting board.

## PROBLEMS: Sec. 1 Whole Numbers

1. What number is represented by each of the counting boards shown?

**Example:**

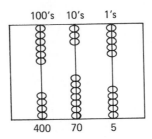

**Figure 1-10**

*Answer* _____ 475 _____

(a)

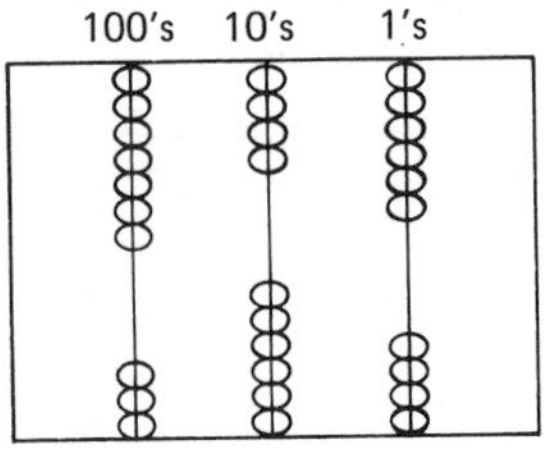

**Figure 1-11**

(b)

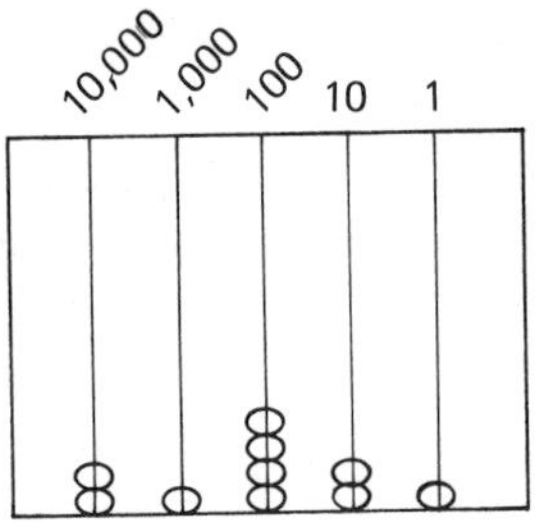

**Figure 1-12**

(c)

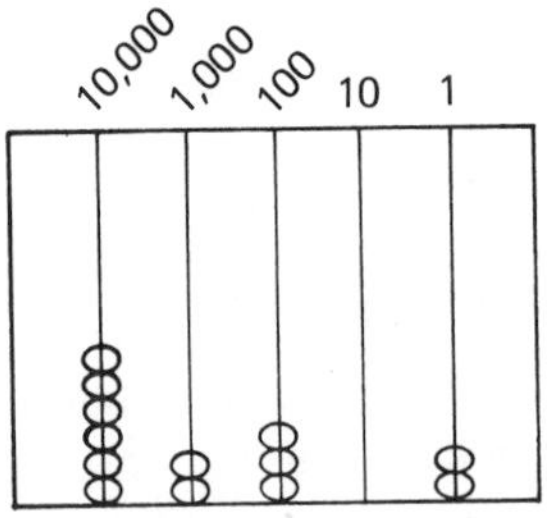

**Figure 1-13**

(d)

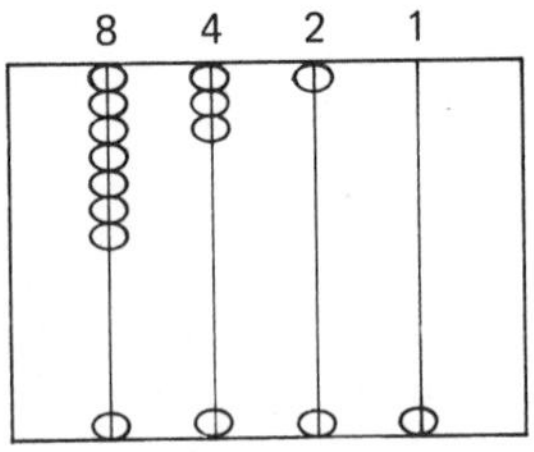

**Figure 1-14**

NOTE: Some of the following problems may require the use of tables that appear at the end of the text.

**2.** Each of the following figures shows a system of counting units. For each problem (a) to (d) write down on paper, the bottom (total) number of units in each column.

**Example:**

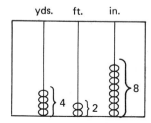

**Figure 1-15** *Linear measure (Hindu-Arabic)*

*Answer*   4 yds., 2 ft., 8 in.

(a)                                              (b)

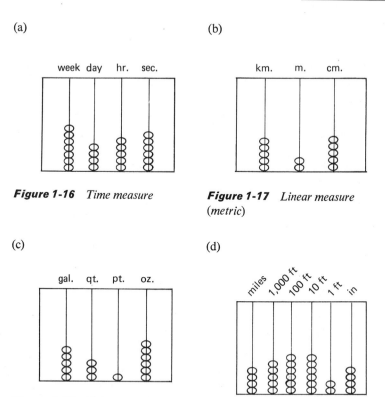

**Figure 1-16** *Time measure*

**Figure 1-17** *Linear measure (metric)*

(c)                                              (d)

**Figure 1-18** *Volume measure (liquid)*

**Figure 1-19** *Distance measure*

## SEC. 2   NUMBER REPRESENTATION AND PLACE VALUE

Our number system uses ten symbols:

$$0, 1, 2, 3, 4, 5, 6, 7, 8, 9$$

These symbols are called *digits*.

These digits and the numbers 10, 100, 1,000, 10,000, and so on, give us a method of representing any number. Using these digits, the number three thousand, four hundred eighty-seven is written as 3,487. The use of the digits 3, 4, 8, and 7, is a compact way of stating a number. Using these digits and appropriate multiples of ten, the number 3,487 is stated in *expanded* form as:

$$3 \times 1,000 + 4 \times 100 + 8 \times 10 + 7$$

The correct number of thousands, hundreds, tens, and units appears clearly in this form.

You know that the digits 0, 1, 2, 3, 4, 5, 6, 7, 8, 9 are used to represent numbers, and that each digit in a number has both a *face* value and a *place* value. For example, the number *nine hundred twenty-seven* is written as 927. Here the 9 has a *face* value of nine and a *place* value of one hundred, since it is in the third place from the right, which is the 100's place. The 9 in this place tells you there are nine hundreds in the number 927. The 2 in 927 has a *face* value of two and a *place* value of ten, because it is in the second place from the right, which is the 10's place. The 2 in this place shows that there are two tens. The 7 in 927 has a *face* value of seven and a *place* value of one. The 7 in this place shows you that there are seven ones in the number.

Figure 1-20 shows a *place-value* chart. Each pocket in this chart represents a *place* in which one of the digits 0, 1, 2, 3, 4, 5, 6, 7, 8, 9, may appear. In this figure the number 23,458 is represented.

When you read a number such as 345, be careful to read it, "three

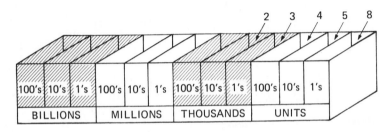

***Figure 1-20***   *Place-value chart*

hundred forty-five" (NOT "three hundred *and* forty-five"). The word *and* is used to locate the decimal point when writing and reading numbers.

Note that the pockets are in groups of three. There are three pockets at the right for units. The next three pockets are for thousands, and the next three pockets are for millions, and so on. These groups of three are called *periods*. In this chart, four periods are shown (units, thousands, millions, billions). When a large number is written, *each period is separated by a comma (,)*.

> For instance: The number *three million, four hundred fifty-six thousand, seven hundred thirty-eight* is written:
> 3,456,738

## PROBLEMS: Sec. 2 Number Representation and Place Value

**1.** Write each of these numbers in expanded form.

> **Example:** $5,867 = 5,000 + 800 + 60 + 7$
> $= 5 \times 1,000 + 8 \times 100 + 6 \times 10 + 7$

| | | |
|---|---|---|
| (a) 389 | (b) 576 | (c) 765 |
| (d) 1,652 | (e) 1,500 | (f) 17,488 |
| (g) 45,890 | (h) 40,785 | (i) 16,005 |
| (j) 345,884 | (k) 970,750 | (l) 1,657,754 |
| (m) 5,800,650 | (n) 34,650,895 | |

**2.** Write each of these expanded numbers in standard number form.

> **Example:** $2 \times 10,000 + 8 \times 1,000 + 5 \times 100 + 6 \times 10 + 4 = 28,564$

(a) $4 \times 1,000 + 5 \times 100 + 6$
(b) $8 \times 1,000 + 1 \times 100 + 9$
(c) $9 \times 10,000 + 3 \times 1,000 + 0 \times 100 + 3 \times 10 + 0$
(d) $5,000 + 600 + 80 + 5$
(e) $4 \times 10,000 + 6 \times 1,000 + 8 \times 100 + 6 \times 10 + 5$
(f) $5 \times 10,000 + 6 \times 10 + 9$

**3.** What is the *place* value of the digit 5 in each of these numbers?

> **Example:** In 3,452 the 5 has a *place* value of 10 because it is in the 10's place.

(a) 5,987. The 5 has a place value of _____.
(b) 3,500,874. The 5 has a place value of _____.
(c) 345. The 5 has a place value of _____.
(d) 24,500. The 5 has a place value of _____.

**4.** Write the correct number for each word statement.

**Example:** Two thousand, five hundred sixty-four <u>2,564</u>.

(a) Two hundred seventy-seven _____.

(b) Three thousand, thirty _____.

(c) Four million, six hundred forty-eight thousand, nine hundred eighty-five _____.

(d) Fifty-seven hundred _____.

(e) Seven hundred thousand, seven hundred seven _____.

(f) Four hundred fifty-six thousand, six _____.

(g) Three thousand, five hundred sixty-five _____.

(h) Four hundred thousand, four hundred four _____.

**5.** A practical example of a counting board that is used widely is the ordinary gas, electric, or water meter. These meters have a set of circular dials that indicate the amount used, and they express that amount as a multiple of ten. In the problems given, find the reading on the meter.

**Example:** The meter shown in Fig. 1-21 measures the number of cubic feet of gas used. The pointer on each circular dial shows a multiple of ten cubic feet used.

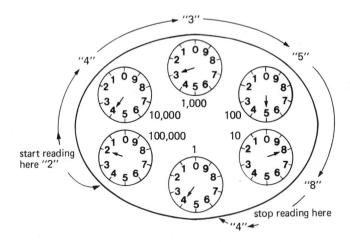

**Figure 1-21**  *Gas meter read in cubic feet*

These meter dials are read in a clockwise direction, starting at lower left.

*Solution:* Start at the 100,000 dial and read 2. Read clockwise to the 10,000 dial and read 4. Go to the 1,000 dial and read 3.

You read the 100, 10, and 1 (unit) dials 5, 8, and 4, in that order.

*Answer:* The complete meter reading in expanded form is:

$2 \times 100{,}000 + 4 \times 10{,}000 + 3 \times 1{,}000 + 5 \times 100 + 8 \times 10 + 4 \times 1$

or a total of 243,584 cubic feet of gas.

(a) Read the gas meter (Fig. 1-22) in cubic feet.

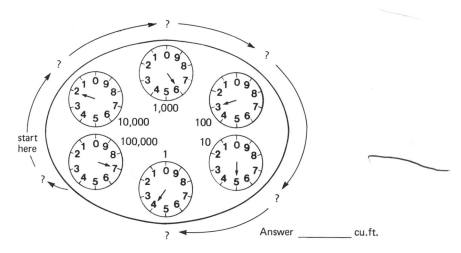

**Figure 1-22**

(b) Read the gas meter (Fig. 1-23) in cubic feet.

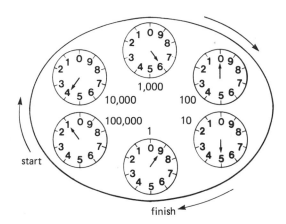

**Figure 1-23**

(c)  Read the electric meter (Fig. 1-24) in kilowatt hours.

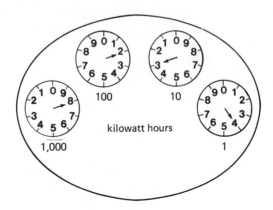

*Figure 1-24*

(d)  Read the water meter (Fig. 1-25) in gallons.

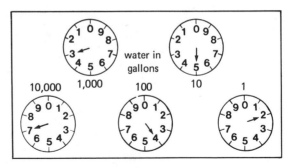

*Figure 1-25*

## SEC. 3   THE NUMBER LINE

It has been said that "One picture is worth a thousand words." The use of pictures to explain an idea is helpful in mathematics. Perhaps you remember from a mathematics course how to show certain facts and information by use of line graphs, pie graphs, or bar graphs. In this section you will learn to show sets of numbers by drawing a number line. Study the pictures carefully. The methods they show are important in the chapters on rounding off numbers, adding and subtracting whole numbers and fractions, and the arithmetic of signed numbers.

Now you will study the method of placing zero and the positive numbers along a line. The result is called the *number line*. A number line looks

Figure 1-26  *Number line*

very much like a ruler or yardstick with markings at regular intervals. Just as a ruler starts at zero, your number line starts at zero. The number line shown in Fig. 1-26 has an arrowhead at the right. This arrow suggests that the line extends without end.

Start your number line at zero. Place the positive whole number one (1) anywhere to the right of zero. When you have chosen where to put the 1, divide the rest of the line into segments equal to the length of the segment from zero to one.

At the right end of each line segment, write the 2, next the 3, then the 4, 5, 6, . . . in order. This is a number line. The length and the number of line segments drawn will depend on the length of the first line segment and on how many numbers you wish to show. Figure 1-27 shows three different choices for number lines.

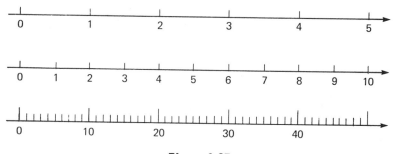

Figure 1-27

To *graph* a certain number on a number line means to find the one point where that number is placed. In Fig. 1-27, find the number 3, the number 4, and the number 30. The number that goes with the point is called the *coordinate*.

When a number coordinate is graphed, it is often shown by placing a large dot ( ● ) at the point.

For instance:  Graph the numbers 3, 5, and 7 on the number line. See the graph in Fig. 1-28.

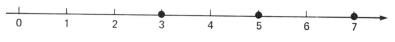

Figure 1-28

For instance: See Fig. 1-29 and name the coordinate number that should appear at point *a*. What number should be placed at point *b*? at point *c*? at point *d*?

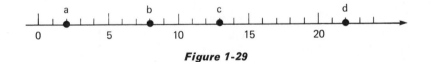

**Figure 1-29**

The coordinate of point *a* is 2. The coordinate of *b* is 8. The coordinate of *c* is 13. The coordinate of *d* is 22.

In Chapter 2 the number line is used to study the methods of rounding off numbers.

## PROBLEMS: Sec. 3  The Number Line

1. Graph each set of numbers on a separate number line.
   (a) The numbers 4, 6, and 8.
   (b) The odd numbers between 4 and 12.
   (c) The even numbers between 21 and 29.
   (d) The whole numbers between 5 and 12.

2. Graph each set of numbers on a separate number line.
   (a) The numbers that are less than 8.
   (b) The whole numbers that are greater than 5, but less than 11.
   (c) The whole numbers that are greater than or equal to 4, but less than 7.
   (d) The whole numbers that are less than or equal to 15.

3. Graph the number that is midway between 8 and 18.

4. In Fig. 1-30, name the coordinate number for each point marked *a*, *b*, *c*, and *d*.

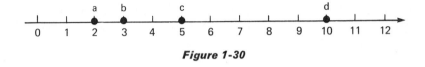

**Figure 1-30**

5. In Fig. 1-31 name the coordinate number for each point marked *r*, *s*, *t*, and *u*.

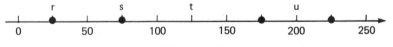

**Figure 1-31**

### COMMENTS

Developing the tools and methods of counting is a slow process. The symbols used for counting change as technology changes. The use of counters, counting boards, and the graphing of numbers on a number line makes counting easier. These tools and methods are useful today. Scientists are working on new ways to express numbers. With modern technology and electronic computers we may soon be using new number systems in our everyday work.

## Practice Test for Chapter 1

Take this practice test without looking at the text. These 14 problems cover the basic ideas that are most important for you. You should finish the test in less than one hour. If you miss any of the problems, restudy the chapter.

1. What number is shown on this counting board?

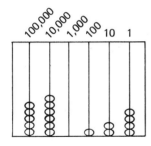

**Figure 1-32**                    *Answer* : _____

2. Draw counters on the next counting board so that the total number 3,465 is shown.

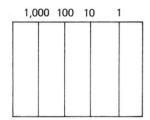

**Figure 1-33**

3. In the next figure, write the correct unit of time at the top of each column.

Then draw counters in each column to show 2 years, 4 months, 3 days, 12 hours.

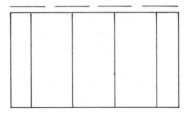

*Figure 1-34*

4. Write in expanded form.
   2,456,654 = _____

5. Express 4,000,040 in expanded form.

   _____

6. In the number 356,097, the digit _____ is in the 1's place, the digit _____ is in the 1,000's place, and the digit _____ is in the 10,000's place.

7. Use the digits 1, 3, 4, 5 *only once* and write the largest possible number with 4 in the 10's place. _____

8. Use the digits 1, 2, 5, 7 only once and write the smallest possible whole number with 2 in the 100's place. _____

9. Write the word name for the number 24,509. _____

10. Write the word name for the number 456,597. _____

11. Write the number for *thirty-five thousand, four hundred, sixty-nine.*

   *Answer:* _____

12. Write the number for *eight hundred fifty-three thousand, fifty-eight.*

   *Answer:* _____

13. Read the gas meter in Fig. 1-35.

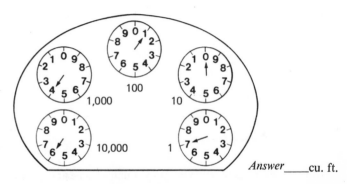

*Answer* ____ cu. ft.

*Figure 1-35*

**14.** Draw the pointers in the gas meter in Fig. 1-36 to indicate 897,906 cubic feet.

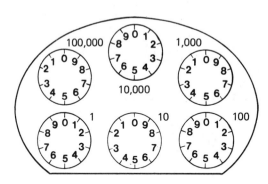

**Figure 1-36**

# 2

# Approximating, Rounding off, and Significant Digits

## SEC. 1 APPROXIMATING AND ROUNDING OFF

How often have you been asked or heard questions like these?

"How old are you?"
"What time is it?"
"How far is it from San Francisco to New York?"
"How much do you weigh?"
"What is the horsepower of your car?"

These questions, and others like them, usually get answers such as:

"I'm nineteen."
"It's about twelve-thirty."
"I'd say about thirty-five hundred miles."
"I weigh one hundred sixty-eight pounds."
"It's rated at 325 horsepower."

The lack of precision in these questions and answers shows that we commonly use an approximate number, or an estimate of number values. In modern technology the approximation or rounding off of numbers is an easy way to make a quick count.

To estimate values, such as finding the number of hours it will take to paint a house, or the number of gallons of paint needed to paint it, it is common practice to estimate the numbers to the nearest ten, hundred, thousand, ten-thousand, etc. This method of estimating is called rounding off a number. An approximate number is a number that has been rounded off and is correct to the closest unit, ten, hundred, or thousand, etc.

**Example 1:** Estimate the number 58 to the nearest 10.

*Solution:* The number 58, to the nearest ten, is 60 because 60 is the *multiple of ten* nearest to the number 58. Here a number line helps explain the process of approximation.

*Answer:* The number line in Fig. 2-1 shows that 58 is closer to 60 than it is to 50. In this figure, 60 is the multiple of ten nearest (closest) to the number 58.

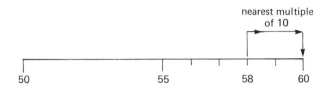

**Figure 2-1** *Number scale*

**Example 2:** Round off the number 456 to the nearest hundred. Is the number closer to 400 or 500?

*Solution:* Use the number line in Fig. 2-2 to check you approximation. The number line shows that the two closest hundreds are 400 on the left and 500 on the right. Since 456 is to the right of 450, it is closer to 500.

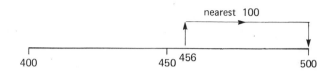

**Figure 2-2** *Number scale*

*Answer:* Since 500 is the hundred nearest to 456, round off the value of 456 *to the nearest hundred*, which is 500.

When you want to round off a number to a certain place value, look at the first digit to the right of the place value in the number you are rounding off. When the digit to the right is 0, 1, 2, 3, or 4, then the number you are rounding off is rounded down. When the digit to the right is 6, 7, 8, or 9, then the number you are rounding off is rounded up to the next higher number.

**Example 3:** (a) Write the number 236. Round off this number to the nearest ten.

*Solution:* The digit 3 is in the 10's place. The digit to its right is 6. When the digit to the right of the place value in the number you are rounding off is a 6, 7, 8, or 9, round up the number to the next higher value.

*Answer:* Round 236 up to 240 to the nearest 10.

**Example 3:** (b) Round off the number 236 to the nearest hundred.

*Solution:* The 2 is in the 100's place. The first digit to its right is 3. When the digit to the right of the place value in the number you are rounding off is 0, 1, 2, 3, or 4, round the number down.

*Answer:* Round 236 down to 200 to the nearest 100.

**Example 4:** Round off 1,849 to (a) nearest ten; (b) nearest hundred.

*Solution:* (a) Since 1,849 has a 4 in the 10's place and a 9 in the unit's place, the number is rounded up.

*Answer:* 1,849 is rounded off to 1,850 to the nearest 10.

*Solution:* (b) Since 1,849 has an 8 in the 100's place and a 4 in the 10's place (which is the first digit to the right of the specified place) the number is rounded down.

*Answer:* 1,849 is rounded off to 1,800 to the nearest 100.

From these examples, a rule for rounding off can be made:

### Rule For Rounding Off—Place Value

*To round off a number to a specified place value, find the digit holding that place value. If the first digit to its right is 0, 1, 2, 3, 4, the number is **rounded down**. If the first digit to the right is 6, 7, 8, 9, the number is **rounded up** to the next higher value.*

When rounding off a number to a specified place value, *all digits* to the left of that place value remain the same. *All digits* to the right of that place value must be *replaced* by zeros.

**Example 5:** Round off the number 365,748 to the nearest thousand.

*Solution:* The digit 5 is in the 1,000's place. The digits 3 and 6 to the left of the 5 stay the same. The digit 5 is replaced

by a 6 because the first digit to its right is a 7. The digits
7 4 8 to the right of the 1,000's place are replaced by
zeros.

*Answer:* 365,748 is rounded off to 366,000 to the nearest 1,000.

When the first digit to the right of a specified place value is a 5, the
number may be rounded up or down. This will depend on the digits that
follow the 5. If there are any non-zero digits to the right of the 5 (that is
1 through 9), then the number is rounded up. It is rounded up because it is
more than halfway to the next higher value. Here is an example:

**Example 6:** Round off 3,506 to the nearest 1,000.

*Solution:* The digit in the 1,000's place is a 3. The digits to the right
of the 1,000's place are 5, 0, and 6. Since there is a
nonzero digit at the right of the 5 (the 6 is a nonzero
digit), round 3,506 up to the next higher value, which is
4,000.

*Answer:* The number line in Fig. 2-3 shows that 3,506 is more than
halfway to 4,000.

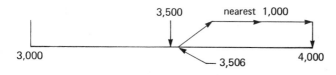

**Figure 2-3**

If the first digit to the right of a specified place value is a 5 followed
by zeros only, use the following rule to round off the number:

**Midway Rule:** *When the given number is midway between the values
on left and right, the digit in the specified place value
is rounded off to the nearest EVEN digit.*

**Example 7:** Round off the number 34,500 to the nearest thousand.

*Solution:* The number 34,500 is **midway** between 34,000 on the left
and 35,000 on the right (Fig. 2-4). Use the midway rule

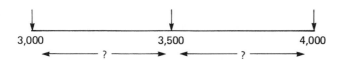

**Figure 2-4** *Which thousand is the closest?*

to round 34,500 down to 34,000 which is the nearest EVEN thousand.

*Answer:* 34,500 is rounded off to 34,000 to the nearest thousand.

**Example 8:** Round off 15,500 to the nearest thousand.

*Solution:* The number 15,500 is **midway** between 15,000 on the left and 16,000 on the right. Use the midway rule to round 15,500 up to 16,000, which is the nearest EVEN thousand.

*Answer:* 15,500 is rounded off to 16,000 to the nearest thousand.

In Table 2-1, the numbers are rounded off to specified place values. The methods you learned in this section are used to get the results. Study the entries in this table.

**TABLE 2-1**

| Given number | Nearest 10 | Nearest 100 | Nearest 1,000 | Nearest 10,000 |
|---|---|---|---|---|
| 375,465 | 375,460 | 375,500 | 375,000 | 380,000 |
| 565,555 | 565,560 | 565,600 | 566,000 | 570,000 |
| 354,500 | 354,500 | 354,500 | 354,000 | 350,000 |
| 465,500 | 465,500 | 465,500 | 466,000 | 470,000 |

## PROBLEMS: Sec. 1 Approximating and Rounding Off

1. Round off each of these numbers to the nearest ten.

   **Example:** When rounded off to the nearest ten, 368 is 370 because 368 is closer to 370 than to 360.

   (a)    479        348          653          932          46
          78         184          579
   (b)  4,566      2,194        4,659        5,096        4,007
        6,702
   (c) 14,753     16,977       20,076       15,006       19,728
        22,801
   (d) 234,657    695,788    1,789,456    25,899,707

2. Round off each of these numbers to the nearest hundred.

   **Example:** When rounded off to the nearest hundred, 4,687 is 4,700 because 4,687 is closer to 4,700 than to 4,600.

   (a)     589        456          938          792
   (b)   1,680      3,754        4,089        7,443
   (c)  14,759     34,067       71,506       29,665
   (d) 565,756    648,904    1,345,067    23,500,078

3.  Round off each of the numbers to the nearest thousand.
    | | | | |
    |---|---|---|---|
    | (a) | 2,340 | 3,568 | 4,789 | 6,377 |
    | (b) | 54,780 | 45,403 | 21,900 | 65,508 |
    | (c) | 145,785 | 235,700 | 560,750 | 642,400 |
    | (d) | 1,476,800 | 2,565,450 | 16,759,560 | 12,121,121 |

4.  Fill in the blanks in Table 2-2 by rounding off the numbers to the specified values. The first row of the table is filled in as an example.

**TABLE 2-2**

| Given number | Round off to the nearest | | |
|---|---|---|---|
| | 10 | 100 | 1,000 |
| 3,575 | 3,580 | 3,600 | 4,000 |
| (a)  2,176 | _____ | _____ | _____ |
| (b) 15,782 | _____ | _____ | _____ |
| (c) 13,667 | _____ | _____ | _____ |
| (d)  6,096 | _____ | _____ | _____ |
| (e) 25,047 | _____ | _____ | _____ |
| (f) 235,704 | _____ | _____ | _____ |
| (g) 501,090 | _____ | _____ | _____ |

5.  Round off the numbers to the specified place value. Use the rule for rounding off numbers that are *midway* between two values.

    **Example:** Round off 5,500 to the nearest thousand.

    *Answer:* When rounded off to the nearest thousand, 5,500 is 6,000 because 6,000 is the nearest EVEN thousand.

    (a) 450 to nearest 100
    (b) 1,500 to nearest 1,000
    (c) 2,500 to nearest 1,000
    (d) 9,500 to nearest 1,000
    (e) 7,450 to nearest 100
    (f) 16,500 to nearest 1,000
    (g) 3,465 to nearest 10
    (h) 5,550 to nearest 100
    (i) 1,500,000 to nearest million
    (j) 4,500,000 to nearest million

6.  Round off each number in this statement to the nearest thousand: "In one week the number of automobile engines produced by GM was 23,458; by Chrysler-Dodge 21,676; by Ford 22,509; and by American Motors 16,478."

7.  Round off each number in this statement to the nearest hundred thousand: "Today the New York Stock Exchange handled 14,156,359 shares of stock. This compares with 15,456,098 shares yesterday, and 11,080,876 shares on this date a year ago."

8.  When a certain whole number was rounded off to the nearest thousand, the result was 7,000. Before rounding off, the greatest value this number could have been was 7,499. What is the *least* value this number could have been?

9. Each of these numbers is the result of rounding off to the nearest 1,000. Find the greatest and least value each number could have been before rounding off.

> **Example:** The result of rounding off a number to the nearest 1,000 is 8,000. The *greatest* value before rounding off was 8,500. The *least* value before rounding off was 7,500.

| Given number | Greatest value before rounding off | Least value before rounding off |
|---|---|---|
| (a)  5,000 | _____ | _____ |
| (b)  16,000 | _____ | _____ |
| (c)  45,000 | _____ | _____ |
| (d)  40,000 | _____ | _____ |
| (e)  165,000 | _____ | _____ |
| (f)  1,470,000 | _____ | _____ |

10. Last year a carpenter for the Jerry Bilt Construction Co., earned $19,489.00.
    (a) What were his earnings to the nearest $10?
    (b) What were his earnings to the nearest $100?
    (c) What were his earnings to the nearest $1,000?

11. A plumber estimates that he will need 450 feet of copper pipe for a certain job. If his estimate is to the nearest 10 feet, what are the greatest and least amounts the plumber measured to obtain the estimate?

12. One acre of ground is equal to 43,560 square feet.
    (a) Round off to nearest 100 square feet.
    (b) Round off to nearest 1,000 square feet.

13. A nautical mile is equal to 6,076 feet to the nearest foot.
    (a) Round off this value to the nearest 10 feet.
    (b) Round off this value to the nearest 100 feet.

14. One cubic yard equals 46,656 cubic inches. Round off to the nearest 1,000 cubic inches.

15. The areas of the three largest oceans of the world are:
    Pacific Ocean: 63,801,700 square miles
    Atlantic Ocean: 31,830,800 square miles
    Indian Ocean: 28,356,300 square miles
    (a) What is the area of the Pacific Ocean to the nearest 100,000 square miles?
    (b) What is the area of the Atlantic Ocean to the nearest 10,000 square miles?
    (c) What is the area of the Indian Ocean to the nearest 1,000,000 square miles?

16. Within the continental United States the five tallest mountains are:
    Mount Whitney:    14,495 feet
    Elbert:    14,431 feet
    Massive:    14,418 feet

Rainier:    14,408 feet
Harvard:    14,399 feet

What is the elevation of each of these mountains to the nearest (a) 10 feet?
(b) 100 feet?

**17.** The distance from the earth to the moon is 238,857 miles.
(a) What is the distance to the nearest 1,000 miles?
(b) What is the distance to the nearest 10,000 miles?

**18.** In January the sun and earth are 91,450,000 miles apart, and in July they are 94,560,000 miles apart. What are these distances to the nearest 100,000 miles?

**19.** The total area of the United States is 3,615,122 square miles. Of this area, 3,536,855 square miles is land, and 78,267 square miles is inland water.
(a) What is the total area of the U.S. to the nearest 100,000 square miles?
(b) To the nearest 1,000 square miles, how much of the U.S. is inland water?
(c) What is the land area of the U.S. to the nearest 10,000 square miles?

**20.** The major automobile manufacturers reported that passenger car production for one year was: American Motors, 235,669; Chrysler-Dodge, 1,313,306; Ford, 2,176,335; and GM, 4,853,015.
(a) Round off each of these numbers to the nearest 100,000 cars.
(b) How many cars did GM produce to the nearest million?

# SEC. 2    SIGNIFICANT DIGITS

When a number is rounded off there is a loss of some information. After it is rounded off, it is correct only to the nearest 10 100 1,000 or whatever value has been specified. When a number is rounded off, the discarded digits to the right of the specified value are replaced by zeros. The number of places in the answer must be the same as in the original number. These replacement zeros are place holders. The digits to the left of the specified place value (including the digit in the rounded off place) are the *important ones*. They show approximately how many things are being considered. These digits are called *significant digits*. For instance, when you round off 45,875 to the nearest 100 (which is 45,900), the 7 and the 5 are replaced by zeros, and the 8 is changed to a 9 to show the nearest hundred. The three nonzero digits, 4, 5, and 9, are the *significant digits*. The two zeros to the right of the 9 are only place holders.

If after a number is rounded off and the digit zero is in the specified place value, then the zero becomes a significant digit. Because of this it is necessary to have a general rule showing the difference between a significant zero and a placeholder zero. To show this difference, engineers, scientists, and technicians have agreed to place a bar (-) over the zero when it is a significant digit in the specified place position.

For instance, when rounded off to the nearest hundred, the number 21,013 is 21,$\bar{0}$00. The zero in the hundreds place *IS significant*. When this same number is rounded off to the nearest thousand, it is again 21,000. This time the zero in the hundreds place is *NOT significant.*

> **Example 9:** The number 21,013 is written as 21,0̄00 when it is rounded off to the nearest hundred. [Note the bar (-) over the zero in the hundreds place.]

The bar is used when rounding off numbers such as 34,998 to the nearest ten. The result is 35,0̄00. The bar over the zero in the 10's place means that the zero is significant. The number 35,0̄00 has four significant digits (3, 5, 0, 0̄).

Based on these observations and examples, here are the rules for writing and reading the significant digits in a number.

### Rules for Determining Significant Digits

> 1. *All nonzero digits in a number are significant.*
> 2. *Zeros between significant digits are significant.*
> 3. *A zero with a bar over it ($\bar{0}$) is a significant digit.*
> 4. *Zeros used in any other way are NOT significant.*

Here are four important examples.

> **Example 10:** (a) The number 45,300 has three significant digits. The zeros in the tens and units place are place holders and are not significant (Rule 4).

> **Example 10:** (b) The number 45,067 has five significant digits. The zero in the hundred's place is significant because it is between significant digits. This zero means that there are no hundreds in the number (Rule 2).

> **Example 10:** (c) The number 40,050 has four significant digits. The zero at the right of the 5 is not significant (application of Rule 4). But the two zeros—in the hundred's place and in the thousand's place—are significant (Rule 2).

> **Example 10:** (d) The number 67,0̄00 has four significant digits. The 6 and the 7 are significant (Rule 1). The zero in the hundred's place is significant (Rule 2). The zero in the ten's place is significant (Rule 3). The zero in the unit's place is not significant (Rule 4).

The use of significant digits is another way of rounding off a number.

### Rule for Rounding Off—Significant Digits

*To round off a number to a specified number of significant digits, count the digits from LEFT to RIGHT. Stop at the digit in the desired position and round off using the rule for place value.*

**Example 11:** (a) Round off the number 4,576 to one significant digit.

*Solution:* Start at the left and count one digit. Stop at the 4. Look at the first digit to the right which is a 5. Round off using the rule for place value.

*Answer:* 4,576 is 5,000 to one significant digit.

**Example 11:** (b) Round off the number 4,576 to two significant digits.

*Solution:* Start at the left and count two digits. Stop at the 5. Look at the first digit to the right which is a 7. Round off using the rule for place value.

*Answer:* 4,576 is 4,600 to two significant digits.

Here is a comparison of the two methods for rounding off a number.

| To round off a number to a specified number of significant digits, count the digits from LEFT to RIGHT. | To round off a number to a specified place value, count the place positions from RIGHT to LEFT. |
|---|---|
| For Significant Digits (Count LEFT to RIGHT) | For Place Value (Count RIGHT to LEFT) |
| GIVEN NUMBER | |

**Example 12:** (a) Round off 195,483 to three significant digits.

*Solution:* Start at the digit 1 at the LEFT and count three digits. Stop at the digit 5. Look at the next digit which is a 4, and round off using the rule for place value.

*Answer:* 195,483 is 195,000 to three significant digits.

**Example 12:** (b) Round off 195,483 to the nearest 100.

*Solution:* Start at the unit's place and count place values until you get to the 100's place which has the digit 4. Round off using the rule for place value.

*Answer:* 195,483 is 195,500 to the nearest 100.

## PROBLEMS: Sec. 2 Significant Digits

**1.** Fill in the blanks in this table. The first row is filled in as an example.

| Given number | Number of significant digits | Round off to nearest 100 |
|---|---|---|
| (a)    2,468 | 4 | 2,500 |
| (b)  340,359 | | |
| (c)   36,109 | | |
| (d)      599 | | |
| (e)   35,501 | | |
| (f) 1,908,876 | | |

**2.** In this table round off the numbers to 2, 3, or 4 singificant digits. The first row is filled in as an example.

| Given number | Round off to the indicated number of significant digits | | |
|---|---|---|---|
| | (2 digits) | (3 digits) | (4 digits) |
| 23,456 | 23,000 | 23,500 | 23,460 |
| (a)    134,890 | | | |
| (b)  1,345,809 | | | |
| (c) 10,675,345 | | | |
| (d) 34,009,098 | | | |
| (e)      3,456 | | | |

**3.** Use the four *Rules for Determining Significant Digits*, and fill in this table. In the second column write the number of significant digits for each measure. In the third column write the place value of the last significant digit for each measure.

**Example:** 47,500 pounds. Number of Significant digits = 3
Place value of last significant digit = 100

NOTE: Counting LEFT to RIGHT, the last significant digit is 5, which is in the 100's place.

| Given measure | Number of significant digits | Place value of last significant digit |
|---|---|---|
| (a) 235 grams | | |
| (b) 4,565 meters | | |
| (c) 150 centimeters | | |
| (d) 4,1$\bar{0}$0 miles | | |
| (e) 14,000 square feet | | |
| (f) 2,00$\bar{0}$,000 acres | | |
| (g) 8,$\bar{0}$00 kilometers | | |
| (h) 25 cubic centimeters | | |

4. For each value (i) round off to the number of significant digits specified. (ii) round off to the place value specified.

> **Example:** 56,785 kilograms. Round off to:
>       (i) *three significant digits.* 56,800 kg.
>       (ii) *nearest thousand.*     57,000 kg.

  (a) Given 2,598 meters. Round off to:
     (i) two significant digits.
     (ii) nearest hundred.
  (b) 456 feet. Round off to:
     (i) one significant digit.
     (ii) nearest ten.
  (c) 2,450,000 pounds. Round off to:
     (i) two significant digits.
     (ii) nearest million.
  (d) 3,010 square inches. Round off to:
     (i) two significant digits.
     (ii) nearest hundred.

5. The capacity of an underground storage tank is 55,782 gallons. What is the capacity to two significant digits?

6. A surveyor estimates the distance across a river is 1,460 feet. What is the distance to two significant digits?

7. The trucking industry reports that it employs 8,894,000 persons and has an annual payroll of $61,355,349,000.
  (a) Round off to two significant digits the number of people who are employed by the trucking industry.
  (b) What is the annual payroll to five significant digits?

8. From Miami, FL., to Boston, MA., is a distance of 1,539 miles.
  (a) Round off the distance to two significant digits.
  (b) Round off the distance to three significant digits.

9. Recent statistics show that 5,199 persons graduated with Bachelor degrees in Trade or Industrial training in one year.
  (a) Round off the number of graduates to two significant digits.
  (b) Round off the number of graduates to the nearest hundred.

10. The average yearly income for a person with a high-school education is $10,080. The average yearly income for a person with two years of college is $12,111.
  (a) Round off the income of a high school graduate to two significant digits.
  (b) Round off the income of a two-year college graduate to three significant digits.

11. The odds against getting a royal flush in a poker game are 649,739 to 1. Express these odds to two significant digits.

12. The odds against getting all 13 cards of one suit in a bridge game are 158,755,357,992 to 1. Express these odds to three significant digits.

13. An acre of ground contains 43,560 square feet. If 1 inch of rain falls on 1 acre of land, the ground will be covered with 6,272,640 cubic inches of water. Round off the number of cubic inches of water to three significant digits.

14. One inch of rain falling on one acre of land produces 27,192 gallons of water. This is how many gallons to one significant digit?

15. California has 213,280 Japanese, 170,131 Chinese, and 138,859 Filipino citizens. Round off each of these population figures to three significant digits.

16. California's total land area is 100,314,000 acres. This area is made up of:
    Cropland in farms: 10,235,000 acres
    Pasture and range in farms: 17,074,000 acres
    Other range land: 9,226,000 acres
    Commercial forest land: 17,317,000 acres
    Noncommercial land (parks, etc.): 25,224,000 acres
    Other land (cities, towns, highways, etc.): 21,224,000 acres
    (a) Round off each value to nearest hundred thousand.
    (b) Round off each value to four significant digits.
    (c) Round off each value to nearest million.
    (d) Round off each value to nearest ten million.

## COMMENTS

This chapter covers the methods and rules used to find approximate numbers. Rounding off to a specific place value and finding the number of significant digits are two methods used to obtain an approximation. Either method must be used with care. You must decide whether or not the rounded off value will seriously affect the result when it is used in computations. In Chapter 3 you will need a great deal of the material you learned in this chapter. Approximations may seem difficult at first. But if you carefully study the examples and work the problems, you will be able to make estimates with ease and confidence.

## Practice Test for Chapter 2

1. Round off each number to the nearest place value specified.

| Given number | nearest 10 | Round off to nearest 100 | nearest 1,000 |
|---|---|---|---|
| (a)   23,465 | _____ | _____ | _____ |
| (b)    1,101 | _____ | _____ | _____ |
| (c)    2,505 | _____ | _____ | _____ |
| (d) 236,650,009 | _____ | _____ | _____ |

2. Find the *number* of significant digits in each of these given values. Round off each value to *two* significant digits.

| Given values | Number of significant digits | Round off to two significant digits |
|---|---|---|
| (a)    23,549 | _____ | _____ |
| (b) 2,435,096 | _____ | _____ |
| (c)    345,700 | _____ | _____ |
| (d)         254 | _____ | _____ |

3. Draw a number line and locate the number 876. On the scale, mark (a) the nearest 1,000; (b) the nearest 100; and (c) the nearest 10.

4. Each of these values is the result of rounding off. What is the *least* value for the number *before* rounding off? What is the *largest* value *before* rounding off?

| Given value | Least value | Largest value |
|---|---|---|
| (a) 45,000 pounds | _____ | _____ |
| (b) 230 meters | _____ | _____ |
| (c) 1,600 kilograms | _____ | _____ |
| (d) 2,000,$\bar{0}$00 miles | _____ | _____ |

5. For the three measurements (a), (b), and (c), round off to the *number* of significant digits specified. Then round off to the *place* value specified.
   (a) 198 grams. Round off to:
      two significant digits.    _____
      nearest hundred.    _____
   (b) 2,488 feet. Round off to:
      one significant digit.    _____
      nearest ten.    _____
   (c) 234,575 miles. Round off to:
      five significant digits.    _____
      nearest ten thousand.    _____

6. The estimated distance from the earth to the sun is 93,000,000 miles. This number has already been rounded off to two significant digits. What is the *least* distance to the sun? _____ What is the *greatest* distance to the sun? _____

7. In (a) through (d) round off to the number of significant digits specified. Mark a significant zero with an over bar when it is the last significant digit.
   (a) 2,000,000    three significant digits    _____
   (b) 3,400,000    five significant digits    _____
   (c)      3,000    two significant digits    _____
   (d)         100    three significant digits    _____

8. According to the Bureau of Labor Statistics, there are 27,184,000 blue-collar workers in the U.S. What is this number to two significant digits?

_____

**9.** In the 1972 elections Nixon received 45,767,218 votes, and McGovern received 28,357,668 votes.

(a) Round off each number to the nearest million.

_____ ; _____

(b) Round off each number to three significant digits.

_____ ; _____

**10.** Here are the heights of some of the tallest buildings in the United States. Round them off to the nearest 100 feet.

(a) World Trade Center      1,350 feet _____
(b) Empire State Building    1,472 feet _____
(c) Chrysler Building        1,046 feet _____
(d) Sears Building           1,450 feet _____

# 3

# Measurement

Measurement came into being because people wanted to compare one thing to another. First, people counted on their fingers. Then they counted by comparing a number of stones, sticks, or marks in the dirt with something else, such as the number of animals in a flock. Later they used the counting board to simplify this counting process. These early attempts to use units of measurement were of the simplest type. Having no basic unit of measurement, they used parts of the body such as the hand or foot or the length of a person's stride (Fig. 3-1).

Here are some of the ways the human body was used to establish units of measurement:

A *cubit* is the length of the forearm from the elbow to the end of the middle finger (Fig. 3-2). Its accepted values are 18 inches or 45.72 centimeters.

A *span* is the distance from the end of the thumb to the end of the little finger of a spread hand (Fig. 3-3). Its accepted values are 9 inches or 22.86 centimeters.

A *yard* was decreed by King Henry I of England to be the distance from the tip of his nose to the tip of his middle finger (Fig. 3-4). Its accepted values are 3 feet or 91.44 centimeters.

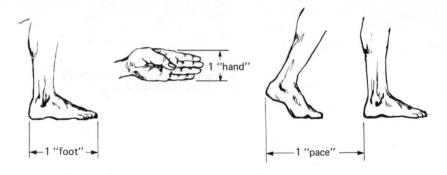

Figure 3-1

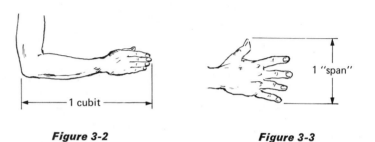

Figure 3-2                    Figure 3-3

These first basic units of measure came from practical, economic, and basic physical situations. They were not based on logical or scientific facts. They were not standardized. But they were convenient. Each person carried with him his own set of measuring instruments. But these units of measure were not the same for all people. Their hands and feet were not the same size, nor was the distance from the tip of the nose to the tip of

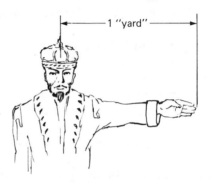

Figure 3-4

the middle finger. Because of these differences it was necessary to agree upon some standard value for these measurements. The merchants, especially, needed to have these units of measure standardized. The demand for measurements that were the same for everyone resulted in the two systems of measurement which you will study in this chapter. One system is the *U. S. Customary System* which uses the inch-pound-second as its basic units. The other system is the *Metric System* which uses the centimeter-gram-second as its basic units. The *Metric System* is now called the *International System* and it is usually shortened to SI.

Measurement is an important part of our daily lives. We use it in our homes, in school, in our jobs, and in our entertainment. Stop for a moment and ask yourself how many times in the last few days you have used the idea of measurement. Here are a few questions for you:

How many *gallons* of gas did it take to fill your car?

Did you buy a *quart* of oil?

Did you buy a *pound* of bacon?

What was the *temperature* at 5: 30 A.M. today?

What is your *age*?

What is the *cubic inch* displacement of your car's engine?

Do you have a $\frac{7}{16}$ inch socket in your tool kit?

These and questions like them use the idea of measurement. In this chapter you will learn about measurements and some do's and don'ts about methods of calculation with measurements.

The *basic principle of measurement* is expressed by the following:

> *When something is measured, a* number *is given to the particular* property *of the thing being measured. The property may be length, width, volume, temperature, power, etc. The association of the number and the property is made through a unit of measure such as feet, inches, cubic centimeters, degrees Fahrenheit, horsepower, and so on.*

In the United States the U. S. Customary System is used for everyday practical measurements. The metric system (International or SI System) is often used in engineering, science, and technical work. It is important for you to be thoroughly familiar with both the metric and the U. S. Customary System of measurement.

Let's look at some of the important relationships between these two systems of measurement. In the metric system the basic unit of length is the *meter*. It has a scientific basis. The standard meter is marked on a bar of platinum iridium of "X"-shaped cross section. This bar is kept at the International Bureau of Weights and Measures near Paris, France. Originally, the meter was meant to be one ten-millionths of the distance from the North Pole to the equator. This distance was to be measured along a meridian (longitude) as shown in Fig. 3-5. The original meter has now

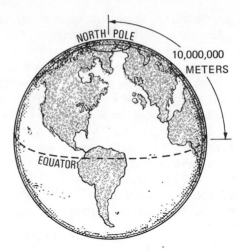

*Figure 3-5*

been proved to be in error by about 0.023 %. The wavelength of an isotope of krypton is now used as a basic measure for the meter. The actual distance from the North Pole to the equator is 10,001,997 meters.

In the U. S. Customary System the basic unit of length is the *inch*. Other units, such as the foot, yard, etc., are given in terms of the inch. A few of these relations are:

    1 foot = 12 inches
    1 yard = 36 inches
    1 mile = 5,280 feet = 63,360 inches

In the metric system the basic unit is the *meter*, and the other units of the system are expressed in terms of it. A few of these relations are:

    1 meter = 100 centimeters
    1 kilometer = 1,000 meters = 100,000 centimeters

A table of relations for these two systems and the conversions between them is given on page 465.

Because the metric system is a decimal system, it is widely used in engineering, science, and technology. Its groups are multiples of 10. Conversions within this system are made by either multiplying or dividing the units given by 10 100 1,000 and so on. For example, to change meters to centimeters, multiply by 100. To change meters to kilometers, divide by 1,000.

Although the metric and customary systems had been used for hundreds of years in the United States, it was not until 1866 that Congress passed a law that established the conversions between the two systems. Today most measurements are in the familiar U. S. Customary System.

But, as you know, more and more measurements are being made in the metric system. In a few years the metric system will be the standard system of measurement throughout the world.

Study the following table which gives some conversions between the U. S. customary and the SI metric system.

SHORT TABLE OF CONVERSION

(length measure only)
U.S. Customary
Metric


1 inch

2.54 cm.


1 foot

30.48 cm.


1 yard

91.44 centimeter (cm.)


You are familiar with the common yardstick. It is equal to 36 inches in length. In any laboratory and most shops you will also see the meterstick. The meterstick is equal to 100 centimeters in length. When you compare a yardstick with a meterstick, you will find that the meterstick is about three inches longer. (See Fig. 3-6.) Use the table of conversion factors to see that  100 centimeters = 39.37 inches

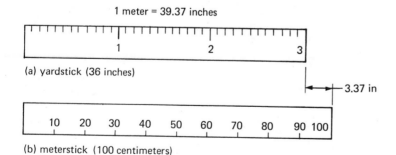

**Figure 3-6**  (a) Yardstick (36 inches); (b) Meterstick (39.37 inches)

To change U. S. customary to SI metric units, or to change SI metric to U. S. customary units, use the conversion table on page 465. Some of the values in the table may not seem important now, but they are necessary for you to master other sections of this text.

The use of an instrument to measure some property of an object gives only an approximate value because no measuring tool is without some small error. Consider measuring the width of this page. How do you proceed? You might measure it with a ruler, a yardstick, a steel tape, or a meterstick. If several people measure the width of this page, they may not agree on the value.

Measurement is only an approximation of the exact value.

There are many reasons for variations in values. A few of these are:

- A wooden ruler marked in sixteenths of an inch is not as accurate as a steel ruler that is marked in sixty-fourths of an inch.
- Not everyone reads the same instrument in the same way.
- Not all measuring instruments are of the same quality.
- Acuteness of eyesight affects the final result.

When a unit of measure such as inch or centimeter is compared *directly* to the object being measured, the measurement is called a *direct measurement*. Most measurements are direct measurements. A ruler, a micrometer, a pressure gauge, a thermometer, and a voltmeter are instruments that give direct measurements.

There are other measurements that are not obtained directly. These measurements are called *indirect measurements*. Examples of indirect measurements are:

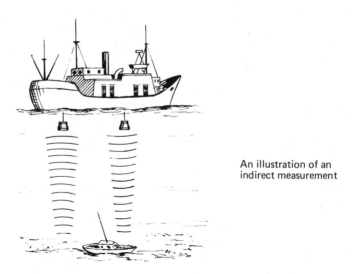

An illustration of an indirect measurement

**Figure 3-7**    *Determining the depth of the ocean by using sonar*

- the diameter of the moon
- the depth of the ocean, as in Fig. 3-7
- the temperature of the sun
- the diameter of an atom of hydrogen
- the age of a dinosaur bone
- the height of a tree, as in Fig. 3-8

An illustration
of an indirect
measurement

**Figure 3-8** *Determining the height of the tree by using a transit and certain ratios*

## PROBLEMS: Sec. 1 Measurement

1.  There are units of measurement in these statements. Read each statement carefully. Underline the word (or words) that give the unit involved.

    **Example:** One *horsepower* is equal to 746 *watts*. (Horsepower and watts are both units of measurement.)

    (a) The average gas tank holds about 16 gallons.
    (b) One gallon of water weighs 8.33 pounds.
    (c) Area is often expressed in square feet.
    (d) My brother's age is twenty-nine years.
    (e) The two common systems for stating the temperature are degrees Fahrenheit and degrees Celsius.
    (f) One atmosphere is about 15 pounds per square inch.
    (g) The speed limit on most highways is 55 miles per hour, or about 90 kilometers per hour.
    (h) A ton of feathers weighs more than a ton of gold.
    (i) A large tree produces several thousand board-feet of lumber.

2.  Illustrations (a) through (d) show common, everyday objects. For each object list five or more properties that can be measured.

    **Example:** A box of soda crackers (Fig. 3-9)

    Properties that can be measured are:
    *length* of box
    *width* of box
    *height* of box

**Figure 3-9**

*volume*
*surface area*
*weight* of box (empty) or (full)
*thickness* of material used in manufacture
*strength* of material. (There are more properties that are not mentioned.)

(a)  A common eight-penny nail
(b)  A piece of lumber
(c)  A golf ball
(d)  An engine valve

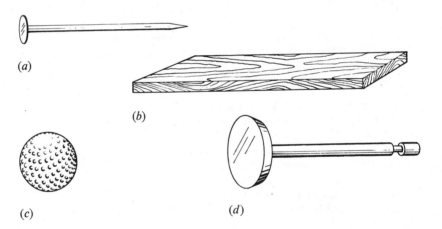

3.  Find the equal value of each of the given measurements. Use the tables at the end of the text to find the value you want.

   **Example:** A cement contractor estimates it will take 3 cubic yards of concrete for the foundation of a new house. How many cubic feet is this?

   *Solution:* From the table find:
   _____ cubic yards × 27 = _____ cubic feet
   The known value of 3 cubic yards is put into the blank on the

left. It is then multiplied by the conversion factor of 27. The result is:

$$3 \text{ cubic yards} \times 27 = 81 \text{ cubic feet}$$

*Answer:* The contractor needs 81 cubic feet of concrete.

(a) Mr. Holmes bought 10 square yards of carpet for an upstairs bedroom. How many square feet did he buy?

(b) In a cross-country race, Don Thompson, a track star, ran 15 miles. How many yards did he run?

(c) A fuel tank has a holding capacity of 100 cubic feet. Approximately how many U. S. gallons will it hold?

(d) Mr. Britt, a physicist, drives to and from school each day, a total of 10 kilometers. How many miles does he drive each day?

(e) The E-2-S rocket burns 3 tons of fuel per second in the lift-off stage. How many pounds of fuel are used in the first 5 seconds?

(f) Mr. Jowise wants to determine the cubic-inch displacement of his foreign-made sports car. Upon reading the owner's manual, he finds the displacement is given in liters.

The owner's manual states:
Displacement: 4 *liters*

What is the engine displacement in *cubic inches*?

(g) Mr. Penning, who believes in physical fitness, went on a diet and lost 3 kilograms in one week. How many pounds did he lose?

(h) While driving in Mexico last summer, Ms. Quasada saw a traffic sign like the one in Fig. 3-10. She knew it meant the speed limit. What is the (approximate) maximum speed in *miles per hour*?

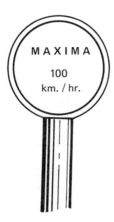

**Figure 3-10**

**4.** In problems (a) though (f) fill in the blanks with the correct word or number. Try to determine the correct answer without referring to the text.

**Example:** In the United States we use two systems of measurement. They are: <u>U. S. Customary</u> and the <u>International Standard</u> (SI) or <u>Metric</u>.

(a) Various parts of the body are used as basic units of measurement. Three of these parts are: _____ , _____ , _____ .

(b) Of the two systems of measurement, the _____ system has a logical and scientific basis.

(c) To measure something means to give a _____ to the _____ of the object being measured.

(d) The main advantage of metric measurement is the fact that conversions are obtained by multiplying or dividing the given number by a multiple of ten. The metric system is a _____ system.

(e) In 1866 the United States Congress approved the conversion factors between the two systems of measurement. Two of these conversions are:
One inch = _____ centimeters
One meter = _____ inches

(f) When you measure with an instrument, the result is only an _____ of the exact value.

## SEC. 2    PRECISION IN MEASUREMENT

There are no truly accurate measurements because perfect measuring instruments have not yet been made. Thus, when you take or use measurements, you must know how *reliable* they are. The reliability of a measurement is expressed in terms of *precision*.

Look at the line segment CD in Fig. 3-11. If you measure this line with

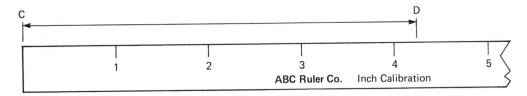

*Figure 3-11*

a ruler that has subdivisions in inches only, the measurement looks like Fig. 3-11. Point D on the ruler is between the 4-inch mark and the 5-inch mark. Since the scale of the ruler is in inches, you round off to the nearest inch. And your best answer is that line segment $CD$ is 4 inches long *to the nearest inch* since it is closer to 4 inches than to 5 inches.

Now measure this same segment with a ruler that has subdivisions of $\frac{1}{8}$ inch (Fig. 3-12). On this finer scale, line segment $CD$ measures $4\frac{2}{8}$ inches *to the nearest eighth of an inch*. Thus, the finer the scale, the more precise the measurement. The words "to the nearest eighth of an inch"

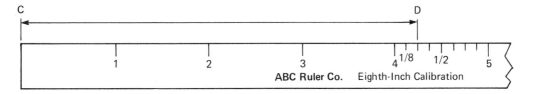

**Figure 3-12**

mean that the measurement is in error by not more than one-half of the $\frac{1}{8}$ inch scale. In this case, the greatest possible error is one-half of $\frac{1}{8}$ which is $\frac{1}{16}$ inch.

When you write measurements you must show this possible error. Do this by using the *plus or minus* sign ($\pm$) between the measurement and the number that shows the possible error. For example, the measurement of line segment $CD$ is written:

$$CD = 4\frac{2}{8} \pm \frac{1}{16} \text{ inch}$$

In general, the *greatest possible error* is **half** of the smallest subdivision shown on the measuring instrument. The *greatest possible error* simply means that the measurement is approximate. As the subdivisions become smaller and smaller, the possible error becomes less and less. If the smallest subdivision is one-fourth, the possible error is one-eighth. If the subdivision is one-eighth, the possible error is one-sixteenth, and so on.

As the subdivision of the basic unit of measurement become smaller, the measurement becomes more precise. *The precision of any measurement depends on the size of the subdivisions of the measuring instrument.*

When a measurement is written in number form, its precision is shown by the place value of the last significant digit on the right.

> **Example 1:** (a) The measurement 34 feet has a precision of 1 foot, because the last significant digit (4) is in the unit's place.
>
> (b) The measurement 450 miles has a precision of 10 miles, since the last significant digit (5) is in the ten's place.
>
> (c) The measurement 3,600 meters has a precision of 100 meters, since the last significant digit (6) is in the hundred's place.

Table 3-1 shows the place value of the last significant digit, the precision, and the greatest possible error.

**TABLE 3-1**

| Measurement | Place value of last significant digit | Precision | Possible error |
|---|---|---|---|
| 5,000 ft | thousand | 1,000 ft | ±500 ft |
| 5,200 ft | hundred | 100 ft | ± 50 ft |
| 5,280 ft | ten | 10 ft | ±  5 ft |

Measurement, precision, and possible error depend on instruments and how they are used. There is always some error. *Only when measurement is obtained by counting is it considered exact.* If you measure the diameter of a piston or the resistance in an electrical circuit, there is some error because instruments are used. But if you *count the number* of pistons in the engine or the *number* of resistors in the circuit, there is no error. The measurement is exact.

## PROBLEMS: Sec. 2   Precision in Measurement

1.  Find the place value of the last significant digit for each measurement. Find the precision of the measurement. Then find the possible error in the measurement. The first row of the table is filled in as an example.

| Given measurement | Place value of last significant digit | Precision | Possible error |
|---|---|---|---|
| 3,400 yards | hundred | 100 | ±50 yd. |
| (a) 45,670 meters | _____ | _____ | _____ |
| (b) 50 centimeters | _____ | _____ | _____ |
| (c) 5,280 feet | _____ | _____ | _____ |
| (d) 5,000,000 cu. mi. | _____ | _____ | _____ |
| (e) 1,100°C | _____ | _____ | _____ |

2.  For (a) through (d) state the precision and the possible error of the measurement.

**Example:** In a measurement of 70 centimeters, what is the the precision and the possible error?

| Measurement | Precision | Error |
|---|---|---|
| 70 cm. | 10 cm. | ±5 cm. |

(a) A surveyor estimates the distance between two markers is 1,500 feet. Measurement: *1,500 feet.*

(b) After careful measurement, the surveyor in problem (a) finds the distance is 1,540 feet. Given measure: *1,540 feet.*

(c) The speed of light is about 186,000 miles per second (mps.). Measurement: *186,000* mps.

(d) The earth is speeding through space at 29,770 meters per second (mps.). Measurement: *29,770 mps.*

3. In (a) through (e) below, a set of measurements is given. Which measures are *exact* and which are *inexact*?

> **Example:** Of the *nine* planets in our solar system, Pluto is the farthest away from the sun. Its distance is *5,900,000,000* kilometers, or about *3,666,000,000* miles.
>
> Exact values: *nine* planets
>
> Inexact values: 5,900,000,000 kilometers and 3,666,000,000 miles

(a) Before going to work, Mr. Emil checked the air pressure in all four tires of his antique car. The tire gauge readings in pounds per square inch (psi) were: *45* psi, *47* psi, *46* psi, and *48* psi.

(b) Mr. James bought a 6-room house at 1067 Ivy Drive. It has 2,750 square feet of floor space. He paid $35,900 for it.

(c) Acme Hardware just received 6 kegs of assorted bolts. Each keg weighs 247 pounds and contains 23,500 bolts.

(d) Mr. Lloyd has two children, John and Ruth. Ruth is 10 years old and weighs 85 pounds. John will be 13 on the second of May. He weighs 126 pounds and is 5 feet 2 inches tall.

(e) Before an overhaul, Mr. Gerlach's 3-cylinder outboard motor had compression readings of 88 psi, 98 psi, and 116 psi. After the overhaul, all cylinders had a reading of 190 psi.

4. For problems (a) through (d) fill in the blanks with the appropriate word, number, etc. Try to answer all questions without referring to the text.

> **Example:** Only measurements made by <u>counting</u> can be considered exact.

(a) If a measurement is obtained by using a ruler, micrometer, meterstick, or vernier caliper, then it is only an _____ of the true value.

(b) The _____ of a measurement depends on the subdivisions of the instrument used.

(c) Generally speaking, the smaller the possible error, the more _____ the measurement.

(d) In a measurement, one-half of the smallest subdivision is called the _____ possible _____.

## SUMMARY AND COMMENTS

The ability to state accurate measurements within the precision of the tools you work with shows that you have mastered the concepts and techniques in this chapter.

The importance of measurement and how it relates to both science and technology was eloquently expressed by Lord Kelvin (1824–1907) of Great

Britain. Lord Kelvin, a leading scientist of that time and one of the founders of the science of thermodynamics (the theory of heat), said:

> I often say that when you can measure what you are speaking about, and express it in numbers, you know something about it; but when you cannot express it in numbers, your knowledge is of a meagre and unsatisfactory kind; it may be the beginning of knowledge, but you have scarcely, in your thoughts, advanced to the stage of Science, whatever the matter may be.

Although Lord Kelvin used the word "Science," the statement applies to all types of measurements, whether they are made in a laboratory with the finest equipment or on the job with a pocket rule.

## Practice Test for  Chapter 3

1. In statements (a) through (g) a unit of measurement is given. Read the statement, then underline the word (or words) that name the units.
   (a) When an airplane is traveling at MACH 1, its speed at sea level is about 750 miles per hour.
   (b) The word *rod* is used by surveyors to measure distance. One rod is equal to $16\frac{1}{2}$ feet.
   (c) Last week Wade got a ticket for parking half an hour in a 10-minute loading zone.
   (d) A machinist must be able to measure a finished part to the nearest thousandth, or ten-thousandth of an inch.
   (e) To convert from pecks to quarts, multiply by eight.
   (f) Last winter the Hongistos burned a cord of oak wood in their fireplace. A cord is a stack measuring $4 \times 4 \times 8$ feet, and is equal to 128 cubic feet.
   (g) One gallon of house paint will cover about 400 square feet of wall.

2. In an Auto Mechanics class the instructor asks you to state the measurable properties of an automobile tire. State at least five properties that can be measured.

   _____  _____  _____  _____  _____

3. A set of various measurements is given in this problem. List the number of significant digits, the place value of the last significant digit, the precision, and the greatest possible error for each value.

| Given value | Number of significant digits | Place value | Precision | Possible error |
|---|---|---|---|---|
| (a) 350 ounces | _____ | ____ | _____ | _____ |
| (b) 2,000 bushels | _____ | ____ | _____ | _____ |
| (c) 64$\bar{0}$ acres | _____ | ____ | _____ | _____ |
| (d) 400 ton | _____ | ____ | _____ | _____ |
| (e) 2,0$\bar{0}$0,000 years | _____ | ____ | _____ | _____ |
| (f) 212°F | _____ | ____ | _____ | _____ |

4. For each of the given amounts, determine the precision.

| *Given amounts* | *Precision* |
|---|---|
| (a) 10̄0 pounds | _____ |
| (b) 750 cubic centimeters | _____ |
| (c) 15 grams | _____ |
| (d) 100 spark plugs | _____ |

5. In each statement fill in the blank (or blanks) with the correct word or number.
   (a) When rounding off a given number to a specified number of *significant digits*, count from _____ to _____. But in rounding off to a specified *place position*, count from _____ to _____.
   (b) Of the two measurements 1,600 feet and 1,648 feet, the second is the more _____ since it has four significant digits.
   (c) The measurements 60 cm., 40 cm., 80 cm., and 90 cm., are all of equal _____.
   (d) If a measurement has a greatest possible error of ±50 feet, the precision is _____ feet.

6. For answers to the parts of this problem, use the *Tables of Useful Equivalents* on page 465.
   (a) EASY SAVE trading stamps measure one inch by one inch and have an area of one square inch. How many stamps are there in one square foot?

**Figure 3-13**

*Answer:* _____

   (b) An inch worm moves forward one inch with each forward thrust of its body. (Fig. 3-14) How many forward motions must it make to travel one mile?

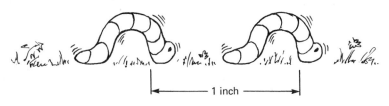

— 1 inch —

**Figure 3-14**

*Answer:* _____

(c) Through shrewd bidding, Captain Campbell bought five tons of surplus military musket balls from the Department of Defense. If each ball weighs one ounce, how many did he buy?

*Answer:*  _____

7. For each measurement in these statements, determine the precision and possible error.
   (a) The radius of the earth is about *3,950 miles*
   Measurement:  *3,950 miles*

   *Answers:*  _____

   _____

   (b) The diameter of the sun is 864,100 miles, and the diameter of the earth is 7,926 miles.
   Measurements:  *864,100 miles*      *7,926 miles*

   *Answers:*  _____  _____

   _____  _____

   (c) On a recent trip, Tom used 39 gallons of gas and traveled 450 miles.
   Measurements: *39 gallons, 450 miles.*

   *Answers:*  _____  _____

   _____  _____

   (d) The engine in a certain car has a displacement of 429 cubic inches and delivers 360 horsepower at 4,600 rpm (rpm = revolutions per minute).
   Measurements: *429 cubic inches, 360 horsepower, 4,600 rpm.*

   *Answers:*  _____  _____

   _____  _____

   _____  _____

8. Statements (a), (b), and (c) give a set of measurements. Determine which are *exact* and which are *inexact*.
   (a) The basic engine specifications for a Mercury Marquis are: *Engine:* V8-429, *Carburetor:* 4-Barrel, *Bore and Stroke:* 4.36 × 3.59 inches, *Piston Displacement:* 429 cubic inches, *Compression Ratio:* 10.5 to 1, *Maximum Brake Horsepower:* 360 hp at 4,600 rpm, *Normal Oil Pressure:* 35 to 60 psi.
   (b) With a full load of passengers and crew totaling 396 people, a Boeing 747 can fly at an altitude of 45,000 feet with an air speed of 650 mph.
   (c) It is estimated that every one and one-half seconds a new automobile rolls off the assembly line. At this rate, 19,200 cars are built in an eight-hour day.

# 4

# Adding and Subtracting Whole Numbers

In Chapter 1 you learned how to write a number in the *decimal system* by using units, tens, hundreds, thousands, millions and so on. Now you will see how these whole numbers are used in everyday calculations. First, consider *addition*. **Addition** *is an operation by which two or more numbers are combined to give a third number*. The numbers that are added together are called *addends*. The result is called the *sum*. The plus sign ($+$) is the instruction to add and form the sum. A table of mathematical symbols used in this text can be found on page xii in front before page 1.

**Example 1:** Find the sum of 6 and 5. This may be written in two forms:

(a) in vertical form:

$$
\begin{array}{r}
6 \longleftarrow \text{addend} \\
+\ 5 \longleftarrow \text{addend} \\
\hline
11 \longleftarrow \text{sum}
\end{array}
$$

(b) In horizontal form:

$$6 \quad + \quad 5 \quad = \quad 11$$
$$\text{(addend)} \quad \text{(addend)} \quad \text{(sum)}$$

In either form the plus sign ($+$) is the instruction to *do the operation called addition*.

The addition of whole numbers can be shown on the number line. Figure 4-1 shows the addition of $6 + 5 = 11$ on the number line.

NOTE: The *addition* of whole numbers involves *counting forward* along the number line. Section 2 shows that subtraction of numbers involves *counting backward* along the number line.

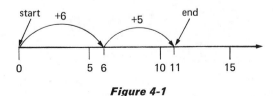

*Figure 4-1*

Sometimes you need to form the sum of more than two numbers.

**Example 2:** If you have four tires, three inner tubes, seven valve cores, and six wheels, how many items do you have?

*Solution:* Form the sum:

| 4 tires |
| 3 tubes |
| 7 cores |
| + 6 wheels |

*Answer:*         Sum = 20 items

NOTE: There are two important ideas in Example 2. First, the sum (20 items) is the same although the *order* in which the addition takes place may be different. The sum stays the same even if the numbers are switched around.

*Changing the order gives the same sum:*

```
    4           4           3
    3           3           4
    7           6           7
  + 6         + 7         + 6
 ────        ────        ────
   20          20          20
```

Second, the sum is the same although the grouping may change. (Use parentheses to group numbers.) When three or more numbers are added together, the sum is the same even if the numbers are regrouped.

$$(4 + 3 + 7) + 6 = 20$$

$$\text{and} \quad (4 + 3) + (7 + 6) = 20$$

$$\text{and} \quad 4 + (3 + 7 + 6) = 20$$

These two ideas are true for all sums. They are stated as rules of addition.

*First Rule of Addition* **The Commutative Law of Addition.** *For any two numbers A and B, the sum $A + B$ or $B + A$ is always the same:*

$$A + B = B + A$$
$$5 + 9 = 9 + 5$$

*Second Rule of Addition* **The Associative Law of Addition.** *For three or more numbers, A, B, and C, the sum:*

$$A + (B + C) = (A + B) + C$$
$$3 + (8 + 7) = (3 + 8) + 7$$

To obtain sums rapidly and without error, you must know certain basic addition facts. Study Table 4-1 to review the necessary facts of addition.

To find the sum of two addends in this table, find one addend in the row on the left, then find the other addend in the column at the top. The sum of these two numbers is where the row and column intersect. In Table 4-1, row 5 and column 8 intersect at the number 13. Note also, that row 8 and column 5 intersect at the number 13.

**TABLE 4-1** BASIC FACTS OF ADDITION

|  | Column number | | | | | | | | | |
|---|---|---|---|---|---|---|---|---|---|---|
| + | 0 | 1 | 2 | 3 | 4 | 5 | 6 | 7 | 8 | 9 |
| row 0 | 0 | 1 | 2 | 3 | 4 | 5 | 6 | 7 | 8 | 9 |
| row 1 | 1 | 2 | 3 | 4 | 5 | 6 | 7 | 8 | 9 | 10 |
| row 2 | 2 | 3 | 4 | 5 | 6 | 7 | 8 | 9 | 10 | 11 |
| row 3 | 3 | 4 | 5 | 6 | 7 | 8 | 9 | 10 | 11 | 12 |
| row 4 | 4 | 5 | 6 | 7 | 8 | 9 | 10 | 11 | 12 | 13 |
| row 5 | 5 | 6 | 7 | 8 | 9 | 10 | 11 | 12 | 13 | 14 |
| row 6 | 6 | 7 | 8 | 9 | 10 | 11 | 12 | 13 | 14 | 15 |
| row 7 | 7 | 8 | 9 | 10 | 11 | 12 | 13 | 14 | 15 | 16 |
| row 8 | 8 | 9 | 10 | 11 | 12 | 13 | 14 | 15 | 16 | 17 |
| row 9 | 9 | 10 | 11 | 12 | 13 | 14 | 15 | 16 | 17 | 18 |

From Table 4-1 you see that:

$$8 + 5 = 13$$

and

$$5 + 8 = 13$$

This is an example of the commutative law of addition.

When you master the addition of several one-digit numbers, you are ready to add numbers that have two, three, four, or more digits. When you add $3 + 7$, it is easy to get the sum 10. But when each addend has several digits, it is better to write them in rows and align their place values vertically in columns. Then add one column at a time.

**Example 3:** The sum of the addends 345, 234, and 654 is easier to find if the addends are written:

$$345$$
$$+ 234$$
$$654$$

*Solution:* To understand the addition of numbers, you must know their *place* value and *face* value. This was explained in Chapter 1. When the numbers are written in expanded form, they are:

$$345 = 3 \text{ hundreds} + 4 \text{ tens} + 5 \text{ units}$$
$$234 = 2 \text{ hundreds} + 3 \text{ tens} + 4 \text{ units}$$
$$654 = 6 \text{ hundreds} + 5 \text{ tens} + 4 \text{ units}$$
$$\text{Sum} = 11 \text{ hundreds} + 12 \text{ tens} + 13 \text{ units}$$

Notice that 13 units $= 1$ ten $+ 3$ units

and          12 tens  $= 1$ hundred $+ 2$ tens

The total is 12 hundreds, 3 tens, and 3 units

*Answer:* When this expanded form is written in the usual way, it is:

$$1,233$$

When adding, do not write out numbers in expanded form. When adding, *carry* any tens from the unit's column to the ten's column, any hundred's from the ten's column to the hundred's column, any thousands from the hundred's column to the thousand's column, and so on until all columns are added.

**Example 4:** 1—thousand carried from hundred's column
1—hundred carried from ten's column
1—ten carried from unit's column

$$
\begin{array}{cccc}
 & 3 & 4 & 5 \\
 & 2 & 3 & 4 \\
 & 6 & 5 & 4 \\
\hline
1, & 2 & 3 & 3 \\
\end{array}
$$

When the sum of a column is 10 or more, keep the digit in the 1 or unit's place. *Carry* the remaining digits to the next column to the left. In example 4, the sum of the 10's column is 13. Only the digit 3 is kept in the 10's place. The digit 1 is carried to the 100's column.

**Example 5:** Find the sum of these measurements:

$$3 \text{ yd. } 2 \text{ ft. } 5 \text{ in.}$$
$$4 \text{ yd. } 1 \text{ ft. } 7 \text{ in.}$$
$$2 \text{ yd. } 1 \text{ ft. } 3 \text{ in.}$$

*Solution:* First, add the columns separately. This gives a total of

$$9 \text{ yd. } 4 \text{ ft. } 15 \text{ in.}$$

You know that there are 12 inches in 1 foot and 3 feet in 1 yard. Convert the above sum:

$$15 \text{ in.} = 1 \text{ ft. } 3 \text{ in.}$$
$$4 \text{ ft.} = 1 \text{ yd. } 1 \text{ ft.}$$

If the values are carried to the separate columns, the total is:

*Answer:* 10 yd. 2 ft. 3 in.

The process of *carrying* is shown in the following diagram:

1 yard carried from feet column
    1 foot carried from inches column

$$3 \text{ yd. } 2 \text{ ft. } 5 \text{ in.}$$
$$4 \text{ yd. } 1 \text{ ft. } 7 \text{ in.}$$
$$2 \text{ yd. } 1 \text{ ft. } 3 \text{ in.}$$
$$\overline{10 \text{ yd. } 2 \text{ ft. } 3 \text{ in.}}$$

When measurements are involved, carry from one column to the next by converting the units. Do not add feet to inches, yards to miles, or "apples to oranges."

At times you need to add measurements with different precision. In order for the sum to have the correct precision, use this rule:

**Rule for Adding Measurements**
*Before adding measurements of unequal precision, round off each value to a precision equal to the **least** precise value. Then find the sum.*

This rule is demonstrated in the following example.

**Example 6:** An aircraft mechanic needs three pieces of tubing measuring 236 cm., $45\overline{0}$ cm., and 230 cm. (Assume no waste occurs in cutting the tubing.) What is the shortest length from which all three pieces can be cut and still allow for error in measurement?

*Solution:* Of these three values, the measurement 230 cm., is the least precise. The 230 has a precision of 10. The other values have a precision of one. Before adding, apply the Rule for Adding Measurements. Round off each value to a precision of 10. Round off 236 to 240. Round off $45\overline{0}$ to

450. Do not change the 230 since it is the least precise value. Add these rounded-off values. The sum is:

$$240 + 450 + 230 = 920 \text{ cm. } \pm 5 \text{ cm.}$$

The value 920 is between the low of 915 and the high of 925 cm.

*Answer:* The shortest length from which all pieces can be cut, and allow for possible error in measurement, is 915 cm.

## PROBLEMS: Sec. 1 Adding Whole Numbers

1. Do these additions in your head. Try to be accurate. Work as rapidly as possible.

   (a)  23    34    53    76    32    19    89
         3     4     5     3     7     2     5

   (b)  32    76    98    43    76    91    15
         2    12     7    18    32     3    51

   (c)  45    19    17    46    97    34    81
         7    31    13    24     4    43    27

   (d)   4    13     6    39    19     9     7
        27    87    35    91    17    23    54

2. Do these additions in your head.

   (a) $23 + 32$        $45 + 9$         $34 + 7$
   (b) $54 + 9$         $34 + 7$         $76 + 9$
   (c) $11 + 31$        $45 + 32$        $67 + 12$
   (d) $3 + 4 + 2$      $17 + 4 + 3$

   **Find the sums in Problems 3 through 14.**

3. (a) 345          (b) 709          (c) 989
       243              19               908
       786              254              990
        34               6               919

4. (a) 1,800        (b) 1,987        (c) 3,608
       675              2,876            4,789
       250              3,567            4,678
       500              4,678            5,897

5. $4,567 + 9,876 + 4,354 + 6,865 + 7,687$

6. $45,897 + 56,788 + 34,654 + 23 + 654 + 1$

7.    165,890        8.    76,900      9.   4,789,543
       23,549               3,867           3,765,897
        3,765             135,972           2,756,908
      230,098                  45           9,870,870
    2,356,097                   4             350,700

| 10. | | 11. | | 12. | |
|---|---|---|---|---|---|
| | 4,800 | | 567,698 | | 3,765,975 |
| | 19,789 | | 345,709 | | 546,654 |
| | 3,766 | | 456,976 | | 46,357 |
| | 53 | | 309,101 | | 4,987 |
| | 198,560 | | 431,006 | | 632 |
| | | | | | 21 |
| | | | | | 5 |

**13.**  23,765,970 + 4,876,546 + 11,965,766

**14.**  119,897,657 + 100,001,090 + 345,800,108

**15.**  Find the missing sums. Add the rows across. Then add the columns.

**Example:**
$$5 + 6 + 8 = \overline{19}$$
$$3 + 5 + 4 = \overline{12}$$
$$4 + 7 + 8 = \overline{19}$$
$$3 + 6 + 7 = \overline{16}$$
$$15 + 24 + 27 = \overline{66} \text{ (Sum of columns = sum of rows)}$$

(a)  $6 + 8 + 8 = $ __?__
    $4 + 5 + 5 = $ __?__
    $7 + 4 + 6 = $ __?__
    $9 + 7 + 3 = $ __?__
    $\underline{?} + \underline{?} + \underline{?} = $ __?__

(b)  $12 + 9 + 6 = $ _____
    $11 + 10 + 8 = $ _____
    $7 + 25 + 16 = $ _____
    $32 + 20 + 25 = $ _____
    $\_ + \_ + \_ = $ _____

(c)  $34 + 78 + 96 = $ _____
    $45 + 98 + 76 = $ _____
    $54 + 67 + 36 = $ _____
    $65 + 19 + 38 = $ _____
    $\_ + \_ + \_ = $ _____

(d)  $156 + 78 + 9 + 34 = $ _____
    $34 + 9 + 17 + 45 = $ _____
    $1,467 + 357 + 76 + 3 = $ _____
    $568 + 4 + 168 + 35 = $ _____
    $\_ + \_ + \_ + \_ = $ _____

**16.**  How many bricks are needed to build a fireplace if it takes 198 for the firebox, 245 for the facing, and 2,795 for the chimney?

**17.**  The weekly gas sales for a service station were: Monday, 3,460 gallons; Tuesday, 3,670 gallons; Wednesday, 3,450 gallons; Thursday, 2,650 gallons; Friday, 4,510 gallons; Saturday, 6,790 gallons; and Sunday, 1,780 gallons. How many gallons were sold that week?

**18.**  Table 4-2 gives the number of passengers on board FBN Airlines for one week.

(a)  Find the number of passengers for each day of the week.

**Example:** Monday: 154 + 134 + 98 + 78 = 464 (sum of 2nd row).

(b)  Find the total number of passengers for each of the four departure times.

(c)  What is the total number of passengers for the week?

(d)  Which is the most popular flight, the 6:45 P.M. or the 9:15 P.M.?

**19.**  Jerry Bilt Construction Company has just completed a luxury office building. Four offices on the second floor have floor space of 540 square feet, 670 square feet, 450 square feet, and 850 square feet. Three offices on the ground floor have floor space of 690 square feet, 680 square feet, and 850

**TABLE 4-2**

| Day of week | *Passengers on board departing at;* | | | |
|---|---|---|---|---|
| | 6: 45 P.M. | 9: 15 P.M. | 10: 00 P.M. | 11: 30 P.M. |
| Sunday | 135 | 156 | 67 | 87 |
| Monday | 154 | 134 | 98 | 78 |
| Tuesday | 76 | 128 | 109 | 56 |
| Wednesday | 101 | 86 | 96 | 75 |
| Thursday | 118 | 108 | 100 | 65 |
| Friday | 187 | 156 | 145 | 123 |
| Saturday | 105 | 118 | 106 | 99 |

square feet. There are three showrooms with 3,450 square feet, 2,650 square feet, and 8,950 square feet.

(a) How many square feet of office space is there on the second floor?

(b) How many square feet of showroom space is there?

(c) What is the *total* square footage of the building, adding all offices and showrooms?

20. Last summer the Jennings family drove across the state of Iowa. The cities they drove through and the mileage between these cities are:

    Council Bluffs to Des Moines   131 miles
    Des Moines to Iowa City   112 miles
    Iowa City to Cedar Rapids   27 miles
    Cedar Rapids to Dubuque   74 miles.

(a) How far did they drive while going across the state?

(b) If the odometer on their car read 45,677 at Council Bluffs, what was the reading at Dubuque?

**Problems 21 through 31.** Add the given values. Be careful when carrying units of measure from one column to the next. Give answers in the simplest form.

21.  5 ft. 6 in.
     4 ft. 9 in.
     3 ft. 7 in.

22.  2 yd. 2 ft. 3 in.
     5 yd. 1 ft. 7 in.
     4 yd. 0 ft. 2 in.

23.  3 yd. 1 ft. 5 in.
     2 yd. 0 ft. 7 in.
     1 yd. 2 ft. 9 in.

24.  2 pt. 4 oz.
     1 pt. 7 oz.
     4 pt. 9 oz.

25.  3 qt. 1 pt.  5 oz.
     2 qt.      10 oz.
     1 qt. 1 pt.  4 oz.

26.  5 gal. 3 qt. 3 pt.
     4 gal. 2 qt. 1 pt.
     1 gal. 5 qt. 2 pt.

27.  14 hr. 45 min.
     13 hr. 23 min.
     10 hr. 10 min.

28.  1 day 18 hr. 35 min.
     2 day 23 hr. 56 min.
     3 day 10 hr. 47 min.

29.  1 yr. 3 mo. 5 day
     2 yr. 6 mo. 6 day
     7 yr. 5 mo. 9 day

30.  2 hr. 34 min. 36 sec.
     3 hr. 56 min. 47 sec.
     5 hr.  6 min.  6 sec.
     7 hr. 23 min. 48 sec.

31.  4 mi. 4,564 ft. 8 in.
     3 mi. 4,539 ft. 7 in.
     7 mi.   489 ft. 9 in.
     6 mi. 3,756 ft. 5 in.

**32.** A Douglas DC-10 left San Francisco Airport at exactly 9:15:23 A.M. (Clock reading was 9 hr., 15 min., 23 sec.). The jet flew to New York in 4 hours, 37 minutes, 56 seconds. What time did the plane land in New York? (Do not adjust for time zones.)

**33.** A time-study analyst wants to know the total time needed to manufacture a certain item. Find the total time to manufacture it if the working times needed for the different stages of manufacture are:

| | |
|---|---|
| Time at shear: | 1 min. 13 sec. |
| Time at punch press: | 1 min. 19 sec. |
| Time at drill press: | 2 min. 35 sec. |
| Time at deburring: | 56 sec. |
| Time at heat treat: | 1 hr. 45 min. 00 sec. |
| Time in cleaner: | 1 hr. 36 min. 00 sec. |
| Time at paint stall: | 5 min. 38 sec. |
| Time in drying oven: | 45 min. 57 sec. |
| Time at final inspection: | 2 min. 45 sec. |

**34.** In one week the New York Stock Exchange recorded the following stock trades:

| | |
|---|---|
| Monday | 14,786,987 shares |
| Tuesday | 13,546,765 shares |
| Wednesday | 11,567,900 shares |
| Thursday | 15,458,324 shares |
| Friday | 13,987,654 shares |

What was the total number of shares traded for the week?

**35.** Mr. Rodriquez owns the Muy-Bueno Garage. Last week his gross sales were:

Monday, $345; Tuesday, $336; Wednesday, $245; Thursday, $187; Friday, $269; Saturday, $465.

What were his total sales for the week?

**36.** An engineering firm has a contract to build a bridge across a river. The bridge will have four spans, each with these measurements:

| | |
|---|---|
| First span: | 450 ft. |
| Second span: | 345 ft. |
| Third span: | 1,600 ft. |
| Fourth span: | 380 ft. |

What is the total length of this bridge? (Use Rule for Adding Measurements.)

**37.** A pipefitter needs five pieces of steam pipe of these lengths:

46 in., 40 in., 56 in., 60 in., and 72 in.

If he orders one long piece of pipe from which to cut all five pieces, what should the length of the pipe be? (Use Rule for Adding Measurements.)

**38.** An electrician has a roll of wire 45 feet long. He needs pieces that are 16 feet, 18 feet, 9 feet, and 2 feet. Add these lengths. Allow for possible error of ±6 inches in each measurement. Does he have enough wire?

**39.** Mr. Berg drives 18 miles to work each day. During the month of April he kept a record of his total driving time for each week.

First week:      3 hr. 46 min. 37 sec.
Second week:    4 hr. 21 min. 46 sec.
Third week:     3 hr. 35 min. 19 sec.
Fourth week:    2 hr. 56 min. 34 sec.

Find the total number of hours, minutes, and seconds Mr. Berg drove.

**40.** In the Spring Quarter, a community college district office recorded these enrollments for the five colleges in the district

College of Alameda      4,568 students
Laney College           7,896 students
Merritt College         5,321 students
Grove Street College     1,789 students
Feather River College     548 students

(a) What was the total number of students in this district?
(b) What is this sum to the nearest hundred?
(c) What is this sum to the nearest thousand?
(d) Round off each of the five numbers to the nearest hundred. Now add the numbers of students. Is the sum larger or smaller than the sum in (b)?

## SEC. 2   SUBTRACTING WHOLE NUMBERS

As addition is the operation by which two or more numbers are combined to form a sum, **subtraction** *is the operation by which the difference between two numbers is determined.* In subtraction the first number is called the *minuend.* The second number is called the *subtrahend.* The amount between the minuend and the subtrahend is called the *difference.* Note that only *two* numbers can be subtracted in one subtraction operation. The minus sign (−) is the instruction to subtract and find the difference.

**Example 7:** The difference between the numbers 12 and 7 is written:

(a) In vertical form     12 ←— minuend

                 − 7 ←— subtrahend

                   5 ←— difference

(b) In horizontal form:

$$12 \quad - \quad 7 \quad = \quad 5$$
(minuend)   (subtrahend)   (difference)

(c) Figure 4-2 shows this subtraction on the number line.

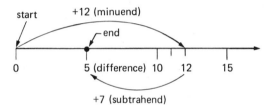

**Figure 4-2**

NOTE: *Subtraction* of 7 from 12 involves counting backward 7 units from the point with coordinate 12.

*Check:* Use this rule to find out if 5 is the correct difference:

*To check a subtraction, add the **difference** to the **subtrahend**. If the sum equals the **minuend**, then the answer is correct.*

$$5 \quad + \quad 7 \quad = \quad 12$$
$$\text{difference} \quad \text{subtrahend} \quad \text{minuend}$$

The subtraction of a number $B$ from another number $A$ gives a third number $C$. This is written as $A - B = C$.

**Example 8:** Subtract 35 from 56.

*Solution:* 35 is the subtrahend. 56 is the minuend. The difference is: $56 - 35 = C$.

*Answer:* $56 - 35 = 21$

*Check:* To check subtraction, add the difference to the subtrahend. The sum of the two should be the same as the minuend. $56 = 35 + 21$.

*Do not reverse the order of the minuend and subtrahend. The result is not the same.* That is, $56 - 35$ is not the same as $35 - 56$. This is discussed more fully in Chapter 6, Signed Numbers.

When subtracting numbers with two or more digits, it may be necessary to *borrow* from the tens, hundreds, thousands, or higher place positions in order to find the difference. Borrowing in subtraction is the opposite of carrying in addition. Again, it is not necessary to write the numbers in expanded form. The usual practice is to write the minuend above the subtrahend then borrow from the columns at the left.

**Example 9:** Find the difference: $1,764 - 483$.

*Solution:* To see what happens, write both numbers in expanded form:

$$1,764 = 1,000 + 700 + 60 + 4$$
$$- \quad 483 = \quad 0 + 400 + 80 + 3$$

In the unit's column the difference of 3 from 4 is 1. In the ten's column the difference $(60 - 80)$ cannot be found until you borrow 100 from the hundred's column and add it to 60.

$$
\begin{array}{r}
600 \\
1,000 + \cancel{700} + 160 + 4 \\
- \quad 0 + 400 + \ 80 + 3 \\
\hline
1,000 + 200 + \ 80 + 1
\end{array}
$$

*Answer:* The minuend 1,764 minus the subtrahend 483 equals the difference 1,281.

When you subtract two measurements, the borrowed value is not always a multiple of ten.

**Example 10:** The NASA tracking station in Guaymas, Mexico, records the orbit times of two satellites. (Fig. 4-3) Satellite A orbits the earth every 1 hour, 13 minutes, 48 seconds. Satellite B orbits every 1 hour, 14 minutes, 37 seconds. Find the difference in orbit times.

*Solution:* You find the difference by subtracting the time for A

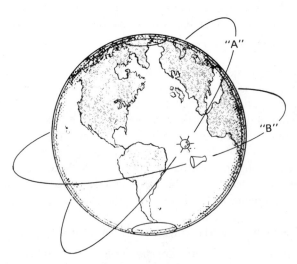

*Figure 4-3*

(1 hour, 13 minutes, 48 seconds) from the time for B (1 hour, 14 minutes, 37 seconds).

$$
\begin{array}{llll}
\text{B} & 1\text{ hr.} & 14\text{ min.} & 37\text{ sec.} \\
\text{A} & -1\text{ hr.} & 13\text{ min.} & 48\text{ sec.} \\
\hline
\end{array}
$$

Since 37 seconds is less than 48 seconds, you must borrow 1 minute (60 sec.) from the minute column. The problem now looks like this.

$$
\begin{array}{llll}
 & & 13\text{ min.} & \\
\text{B} & 1\text{ hr.} & \cancel{14}\text{ min.} & 97\text{ sec.} \\
\text{A} & -1\text{ hr.} & 13\text{ min.} & 48\text{ sec.} \\
\hline
\text{Difference} & 0\text{ hr.} & 0\text{ min.} & 49\text{ sec.}
\end{array}
$$

*Answer:* Satellite B takes 49 seconds longer to orbit the earth than satellite A.

When working measurement problems, it may be helpful to refer to the Tables of Conversion Factors on page 465. If you are in doubt about a conversion value, *look it up!* You will not be paid to make mistakes.

**Example 11:** A carpet layer has a roll of carpet that measures 7 yards, 1 foot, 4 inches long. He carpets a hallway that measures 4 yards, 2 feet, 11 inches long. How much carpet does he have left?

*Solution:* Find the difference by subtraction:

$$
\begin{array}{lll}
7\text{ yds.} & 1\text{ ft.} & 4\text{ in.} \\
-4\text{ yds.} & 2\text{ ft.} & 11\text{ in.} \\
\hline
\end{array}
$$

Here, it is necessary to borrow from both the yard and the feet columns. Borrow 1 yard (3 feet) from the yard column. Then borrow 1 foot (12 inches) from the feet column. The result is:

$$
\begin{array}{lll}
6\text{ yd.} & 3\text{ ft.} & 16\text{ in.} \\
\cancel{7}\text{ yd.} & \cancel{4}\text{ ft.} & \cancel{4}\text{ in.} \\
-4\text{ yd.} & 2\text{ ft.} & 11\text{ in.} \\
\hline
\end{array}
$$

*Answer:*  
$$
\begin{array}{lll}
2\text{ yd.} & 1\text{ ft.} & 5\text{ in.}
\end{array}
$$

To subtract measurements of unequal precision, use this rule:

**Rule for Subtracting Measurements**
*Before subtracting measurements, round off each value to a precision equal to the **least** precise value. Then find the difference.*

**Example 12:** A logging truck has a tare (empty) weight of 23,400 pounds to the nearest 100 pounds. After loading it has a gross (total) weight of 67,000 pounds, to the nearest 1,000 pounds. What is the weight of the load?

*Solution:* The two given values 23,400 and 67,000 do not have the same precision. The least precise value is 67,000. Use the rule for subtracting measurements. Round off the tare weight to 23,000 (nearest 1,000). Find the difference:

$$\begin{array}{r} 67,000 \\ -23,000 \\ \hline 44,000 \end{array}$$

*Answer:* The load weighs 44,000 pounds, to the nearest 1,000 pounds.

## PROBLEMS: Sec. 2 Subtracting Whole Numbers

1. Do these subtractions in your head. Strive for speed and accuracy.

   (a)
   | 14 | 18 | 16 | 12 | 17 | 25 | 26 |
   |----|----|----|----|----|----|----|
   | −3 | −5 | −5 | −3 | −7 | −9 | −8 |

   (b)
   | 34 | 89 | 76 | 65 | 45 | 36 | 65 |
   |----|----|----|----|----|----|----|
   | −12 | −76 | −45 | −63 | −43 | −23 | −56 |

   (c)
   | 145 | 198 | 156 | 465 | 109 | 439 | 198 |
   |-----|-----|-----|-----|-----|-----|-----|
   | −23 | −57 | −47 | −78 | −8 | −254 | −189 |

   (d)
   | 1,567 | 23,567 | 345,789 | 345,897 |
   |-------|--------|---------|---------|
   | −1,235 | −4,675 | −256,897 | −267,569 |

2. Do as many of these subtractions as you can in your head.
   (a) 34 − 3,     45 − 3,     89 − 6,     56 − 5,     33 − 2.
   (b) 23 − 4,     89 − 7,     45 − 8,     78 − 9,     67 − 14.
   (c) 145 − 34,     189 − 56,     456 − 235,     378 − 289.
   (d) 473 − 198,     134 − 125,     1,567 − 1,456,     78,908 − 34,765.

3. Do these subtractions on paper.

   (a)
   | 9,573 | 4,675 | 90,178 | 45,786 |
   |-------|-------|--------|--------|
   | −4,689 | −3,477 | −81,349 | −43,895 |

   (b)
   | 34,765 | 98,876 | 78,945 | 56,498 |
   |--------|--------|--------|--------|
   | −3,786 | −976 | −9,165 | −49 |

   (c) 189,167 − 87,976,     76,945 − 56,800,     43,785 − 34,896.
   (d) 1,896,854 − 987,564,     3,456,821 − 2,876,901

4. Subtract 56,785,987 from 87,987,540.

5. What is the difference between 8,604 and 2,788?

6. Take 5,679 from 10,795.

7. From 56,897 take 42,764.

8. What is the result when 4,657 is decreased by 2,348?

9. Last month the electric meter on Mr. Greene's house read 5,789 kw. This month it reads 6,989 kw. How many kilowatt hours of electricity were used?

10. The gas meter on Mr. Greene's house read 6,784 hundred cubic feet last month. It reads 7,945 hundred cubic feet this month. How many hundred cubic feet of gas was used?

11. Mr. Lee is an engineer. Last year his gross income was $34,678. He paid $5,670 income taxes. What was his income after taxes?

12. The Marianas Trench is the deepest place in the Pacific Ocean. It has a depth of 35,640 feet below sea level. The deepest place in the Atlantic ocean is 30,264 feet below sea level. How much deeper is the Pacific Ocean than the Atlantic Ocean (Fig. 4-4)?

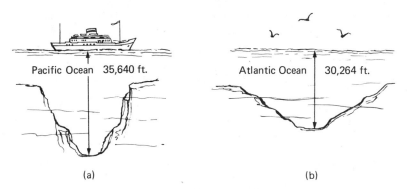

Pacific Ocean   35,640 ft.          Atlantic Ocean   30,264 ft.

(a)                                    (b)

**Figure 4-4**   (a) Pacific Ocean, 35,640 ft; (b) Atlantic Ocean, 30,264 ft

13. Recent measurements show that the moon is 238,857 miles from earth. The distance from the earth to the sun is 92,897,414 miles. Find the shortest distance a satellite would travel on a flight from the moon to the sun.

14. The planet Venus is almost the same size as the earth. The diameter of the earth is 12,756 km, and the diameter of Venus is 12,620 km. By how many kilometers does the diameter of earth exceed the diameter of Venus?

15. The earth takes 365 days to orbit the sun. Pluto, the planet farthest from the sun, takes 90,885 days to orbit the sun. How many more days does it take Pluto to orbit than it takes earth?

16. A nautical mile measures 6,080 feet. A statute mile measures 5,280 feet. How much longer is a nautical mile than a statute mile?

17. Find the differences in problems (a) through (h).

**Example:**   3 yd.  1 ft.  5 in.     (Borrow one yard from the
          −1 yd.  2 ft.  7 in.     yards column. Then borrow one
                                   foot from the feet column)

*Solution:* After borrowing, the problem looks like this:
             2 yd.  3 ft.  17 in.
             3̶ yd.  4̶ ft.  5̶ in.
          −1 yd.  2 ft.  7 in.

*Answer:*      1 yd.  1 ft.  10 in.

|     |                              |     |                                   |
|-----|------------------------------|-----|-----------------------------------|
| (a) | 3 yd.  1 ft.   6 in.<br>−2 yd.  2 ft.   9 in. | (b) | 3 gal.  2 qt.  0 pt.<br>−1 gal.  2 qt.  1 pt. |
| (c) | 5 m.  56 cm.  7 mm.<br>−3 m.  68 cm.  5 mm. | (d) | 1 mi.  3,567 ft.  11 in.<br>−0 mi.  4,567 ft.  10 in. |
| (e) | 4 lb.  13 oz.<br>−2 lb.  15 oz. | (f) | 45°  35 min.  28 sec.<br>−35°  45 min.  52 sec. |
| (g) | 8 hr.  36 min.  47 sec.<br>−6 hr.  47 min.  52 sec. | (h) | 11 mi.  3 furlong  56 yd.<br>−6 mi.  7 furlong  83 yd. |

**18.** Mr. Kelley drives racing cars. In a road race he had a winning time of 1 hr. 36 min. 47 sec. The second place finisher had a time of 1 hr. 37 min. 42 sec. By how many seconds did Mr. Kelley beat the second place driver?

**19.** The road race in which Mr. Kelley drove ended at 9:27:53 A.M., clock time. Subtract Kelley's driving time from this clock time to find what time the race started.

**20.** A carpenter estimates a plank is 22 ft. ± 6 in. He cuts off 10 ft. ± 1 in. What are the smallest and largest possible lengths of the plank that remain?

## SUMMARY AND COMMENTS

This chapter reviews whole number addition and subtraction. Be sure that you understand carrying in addition problems and borrowing in subtraction problems. When carrying or borrowing measurements, use correct conversion values. Memorize the conversion values for:

> yards-feet-inches
> hours-minutes-seconds
> gallons-quarts-pints
> pounds-ounces.

Speed and accuracy in addition and subtraction problems will help make your work in the next chapters much easier. Work all of the word problems. The ability to correctly solve word problems show that you understand the material in this chapter.

## PROBLEMS: Secs. 1 and 2
## Adding and Subtracting Whole Numbers

**1.** Find the sum of 23,567; 34,654; 45,896; and 25,789. Find the sum of 34,678; 56,897; and 26,652. What is the difference between the two sums?

**2.** A carpenter cut these three lengths from a plank 22 feet long:

> 5 ft. 3 in.    4 ft. 7 in.    and    4 ft. 9 in.

After the cuts how much of is left?

3. The Quincy Cannonball left Oakland on track 6 en route to Reno, Nevada, at 6:35 A.M. There was a layover of 1 hour, 34 minutes in Oroville. The train arrived in Reno at 2:35 P.M. the same day. How many hours and minutes did it take to make the trip? (*Omit* the layover.)

4. A Phantom jet flew from San Francisco to New York in 3 hours, 42 minutes, 35 seconds. On the return trip it took 3 hours, 34 minutes, 47 seconds.
   (a) How long did it take to make the trip to New York and return?
   (b) How much longer did it take to fly to New York than to return?

5. Mr. Weston owns a service station. Table 4-3 gives his gross income and total expenses for each day of one business week.

### TABLE 4-3

| Day of week | Expenses | Gross income |
|---|---|---|
| Monday | $165 | $277 |
| Tuesday | $186 | $195 |
| Wednesday | $175 | $246 |
| Thursday | $195 | $204 |
| Friday | $177 | $196 |
| Saturday | $25 | $109 |
| Sunday | $30 | $97 |

   (a) For each day of the week find the net income. (Income − Expenses = Net Income)
   (b) Find the total expenses for the week.
   (c) Find the total gross income for the week.
   (d) Find the net income for the week.

6. A 747-C cargo jet left Travis Air Force Base in California for Hickam Field in Hawaii—a flight of 6 hours. When the plane left, it had enough fuel for 7 hours 30 minutes of flying. How many hours-minutes-seconds of fuel are in the plane after it has been in the air for 3 hr. 47 min. 53 sec.?

7. PDQ Airlines flight 10 is en route from Los Angeles to Seattle. The plane left Los Angeles at 8:45:17 A.M. After a series of hops of 1 hr. 23 min. 45 sec., then 1 hr. 23 min. 00 sec., then 1 hr. 48 min. 35 sec., and a total ground time of 1 hr. 16 min. 46 sec., the plane has *just taken off* from its last stopover and will touch down at its destination.
   (a) How long has the plane been in the air?
   (b) How many hours-minutes-seconds has it been since take off? (Include ground time.)
   (c) At what clock time will the plane arrive in Seattle if the last leg takes 1 hr. 33 min. 54 sec.?

8. A TLC moving van can hold 35,000 lb ±500 lb. In order to have a full load, the van picked up loads of 1,400 lb., 15,500 lb., and 17,000 lb. If all loads are within ±100 pounds, how much more can be loaded and still be within the limit of 35,000 ±500 lb.?

9. The *net worth* of a business is equal to *the assets minus the liabilities*:

   Net Worth = (Sum of Assets) − (Sum of Liabilities)

What is the net worth of the M&M Sheet Metal Co.?

| | Assets | | Liabilities | |
|---|---|---|---|---|
| Cash | $4,678 | Accounts | | |
| Accounts | | payable | $2,657 | |
| receivable | $3,796 | Note (Loan) | $5,000 | |
| Merchandise | $6,545 | Other | | |
| Other assets | $9,785 | liabilities | $7,867 | |

10. The "J" Anthony Toy Company records the daily production of its new, exciting game, "How to Play Tiddlewinks With Manhole Covers." Table 4-4 gives the production record for one week.

**TABLE 4-4**  PRODUCTION RECORD

| | Number of Games Accepted | | Number of Games Rejected |
|---|---|---|---|
| Monday | 1,678 | | 234 |
| Tuesday | 1,756 | | 198 |
| Wednesday | 1,567 | | 97 |
| Thursday | 678 | (Ran out of manhole covers) | 23 |
| Friday | 3,478 | (Worked overtime) | 279 |

(a) What was the total accepted production for the week?

(b) How many games were rejected by the inspectors?

(c) What was the sum of all production for the week? (accepted and rejected)

11. When the Jerry Bilt Construction Company began to build a new home, they had no dimensions for the garage. Can you supply the missing dimensions for the sketch in Fig. 4-5?

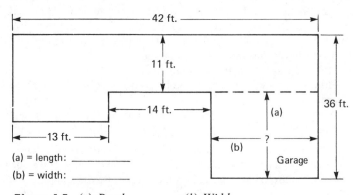

**Figure 4-5**  (a) Depth _____; (b) Width_____

12. The ABC Electric Company is installing lights for a parking lot. Figure 4-6 shows the lot and the lengths of cable needed. The electrician allows an

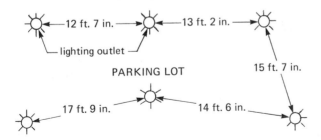

**Figure 4-6**

*additional* 1 foot 6 inches for *each* hookup to a light. How many feet of Romex cable does he need for the job?

13. The J & R Construction Company are home repair specialists. They have contracted to restore an old house. They got the job with a low bid of $8,565. Their estimated costs of labor, materials, and other expenses are:

Labor $4,800
Materials $2,575
Miscellaneous cost $ 450

What is the estimated profit to the company?

## Practice Test for Chapter 4

1. Find the sums.
   (a)  34,789
        45,897
         5,765
       455,752                                  *Answer:* (a) _____
   (b) 37,876 + 1,987,109 + 456,885 = ?         *Answer:* (b) _____

2. Find the differences.
   (a)  9,167
       −4,875                                    *Answer:* (a) _____
   (b) 1,009,010
       −754,887                                  *Answer:* (b) _____
   (c) 98,756 − 46,845 = ?                       *Answer:* (c) _____

3. Subtract 6,848 from 36,653.

4. Find the missing sums. Add horizontally and vertically. (Sum of rows must equal the sum of the columns.)   *Answer:* _____

$$35 + 89 + 6 + 65 + 12 = \underline{\hspace{2cm}}$$
$$57 + 34 + 91 + 8 + 10 = \underline{\hspace{2cm}}$$
$$23 + 8 + 64 + 39 + 3 = \underline{\hspace{2cm}}$$
$$\underline{\ } + \underline{\ } + \underline{\ } + \underline{\ } + \underline{\ } = \underline{\hspace{2cm}}$$

5.  Find the sums.
    (a) 21 yd.  2 ft.  11 in.
        3 yd.  1 ft.  9 in.
        5 yd.  1 ft.  5 in.      *Answer:* (a) _____

    (b) 1 gal.  3 qt.  1 pt.
        3 gal.  2 qt.  0 pt.
        5 gal.  2 qt.  1 pt.      *Answer:* (b) _____

6.  In 1950 the U.S. population was 152,347,978. In 1960 the population was 176,479,845.
    (a) Round off each number to the nearest thousand. Give your *estimate* for the population increase.      *Answer:* (a) _____
    (b) What is the *actual* increase in the 10-year period?
                                                  *Answer:* (b) _____

    **Problems 7 through 10.** Table 4-5 shows the production, production cost, sales, and income for a three-month period of the E&R Tool and Die Company.

**TABLE 4-5**

| Month | Production (Number of Tools) | Production Cost | Sales (Number of Tools) | Income |
|-------|------------------------------|-----------------|-------------------------|--------|
| April | 45,875 | $95,675 | 39,865 | $109,875 |
| May   | 42,985 | $89,762 | 41,347 | $112,755 |
| June  | 39,752 | $82,575 | 38,579 | $104,570 |

7.  (a) Find the total production for the three-month period.
                                     *Answer:* (a) _____
    (b) Find the total production cost for the three-month period.
                                     *Answer:* (b) _____

8.  (a) Find the total number of tools sold for the three-month period.
                                     *Answer:* (a) _____
    (b) Find the total income for the three-month period.
                                     *Answer:* (b) _____

9.  (a) How many tools does the company have left at the end of June? (Use answers 7 (a) and 8 (a).)      *Answer:* (a) _____
    (b) What is the net income to the company for the three-month period? (Use answers 7 (b) and 8 (b).)      *Answer:* (b) _____

10. During which month does the company have the largest *net* income?
                                     *Answer:* _____

# 5

# Multiplying and Dividing Whole Numbers

## SEC. 1  MULTIPLYING WHOLE NUMBERS

Many problems require that the same number be added to itself several times.

> **Example 1:** Most six-cylinder automobile engines use three rings on each piston. How many rings are needed for a complete ring job?
>
> *Solution:* Three rings are needed for each piston, and there are six pistons. Therefore:
>
> $$3 + 3 + 3 + 3 + 3 + 3 = 18 \text{ rings are needed.}$$
> (six 3's)
>
> *Answer:* 18 rings
>
> NOTE: (3 rings per piston) $\times$ (6 pistons) equals 18 rings.

The result is more easily obtained by multiplying than it is by adding.

*Multiplication* *of whole numbers is the process of carrying out repeated additions of the same addend.* When multiplying two numbers together, such as the 3 and 6, the answer (18) is called the *product.* The numbers 3 and 6 are called *factors.* The process is called *multiplication.*

Multiplication may be written in vertical or horizontal form:

(a) vertical form:      6 ←— factor

$\times$ 3 ←— factor

—————

18 ←— product

(b) horizontal form:      3   $\times$   6   =   18
                    (factor)   (factor)   (product)

In contrast to addition, which uses only the plus (+) sign, and subtraction, which uses only the minus sign (−), multiplication uses three signs. The most common sign is the cross ($\times$), or the dot ($\cdot$). Sometimes pairs of parentheses ( ) ( ) are used with no sign between them. [See page xii for a table of mathematical symbols used in this text.]

**Example 2:** Write the multiplication of the two factors 8 and 6.

*Solution:* This may be expressed in any of three forms:

(a) $8 \times 6$   (b) $8 \cdot 6$   (c) (8) (6)

NOTE: When NO sign of operation is between two sets of parentheses, it is an instruction to multiply.

The ability to multiply rapidly and without error depends on knowledge of certain multiplication facts. Table 5-1 shows some of these important facts. Study this table and refresh your skills in multiplication.

Table 5-1 may be used to find the product of any two numbers up to $12 \times 12 = 144$. To determine the product, find the factors (the numbers to be multiplied) in the row and column. You find the product where the row and column intersect. In the table the product of 9 and 7 is 63; $9 \times 7 = 63$ and $7 \times 9 = 63$.

**TABLE 5-1**   TABLE OF MULTIPLICATION FACTS

*Factor column*

| $\times$ | 0 | 1 | 2 | 3 | 4 | 5 | 6 | 7 | 8 | 9 | 10 | 11 | 12 |
|---|---|---|---|---|---|---|---|---|---|---|---|---|---|
| 0 | 0 | 0 | 0 | 0 | 0 | 0 | 0 | 0 | 0 | 0 | 0 | 0 | 0 |
| 1 | 0 | 1 | 2 | 3 | 4 | 5 | 6 | 7 | 8 | 9 | 10 | 11 | 12 |
| 2 | 0 | 2 | 4 | 6 | 8 | 10 | 12 | 14 | 16 | 18 | 20 | 22 | 24 |
| 3 | 0 | 3 | 6 | 9 | 12 | 15 | 18 | 21 | 24 | 27 | 30 | 33 | 36 |
| 4 | 0 | 4 | 8 | 12 | 16 | 20 | 24 | 28 | 32 | 36 | 40 | 44 | 48 |
| 5 | 0 | 5 | 10 | 15 | 20 | 25 | 30 | 35 | 40 | 45 | 50 | 55 | 60 |
| 6 | 0 | 6 | 12 | 18 | 24 | 30 | 36 | 42 | 48 | 54 | 60 | 66 | 72 |
| 7 | 0 | 7 | 14 | 21 | 28 | 35 | 42 | 49 | 56 | 63 | 70 | 77 | 84 |
| 8 | 0 | 8 | 16 | 24 | 32 | 40 | 48 | 56 | 64 | 72 | 80 | 88 | 96 |
| 9 | 0 | 9 | 18 | 27 | 36 | 45 | 54 | 63 | 72 | 81 | 90 | 99 | 108 |
| 10 | 0 | 10 | 20 | 30 | 40 | 50 | 60 | 70 | 80 | 90 | 100 | 110 | 120 |
| 11 | 0 | 11 | 22 | 33 | 44 | 55 | 66 | 77 | 88 | 99 | 110 | 121 | 132 |
| 12 | 0 | 12 | 24 | 36 | 48 | 60 | 72 | 84 | 96 | 108 | 120 | 132 | 144 |

*Factor row*

Learn *three* important rules from this Table of Multiplication Facts.

1. *Multiplying by **zero** always results in a product of **zero***. Look at the top row and first column. When one factor is zero (0), the product is zero. This is true for any multiplication by zero. If the letter $N$ is used to represent any given number, the *zero rule* may be written:

$$N \times 0 = 0 \times N = 0$$

2. **The Commutative Law of Multiplication** *shows that for any two numbers A and B it is always true that:*

$$A \times B = B \times A$$

*When the factors in multiplication problems are reversed, the product is the same.* Reversing the order of the factors is similar to results shown in the Table of Addition Facts in Chapter 4. For example, the product of 4 times 5 is the same as 5 times 4: $4 \times 5 = 5 \times 4 = 20$

3. **The Associative Law of Multiplication** *shows that for any three numbers A, B, and C it is always true that:*

$$A \times (B \times C) = (A \times B) \times C$$
$$3 \times (2 \times 5) = (3 \times 2) \times 5$$

You may verify this law of *grouping* three or more factors by using Table 5-1.

$$2 \times (3 \times 4) = (2 \times 3) \times 4 = 24$$

Memorize this Multiplication Table and work the practice problems. It will increase your accuracy and efficiency in multiplication.

When you use the Table of Multiplication Facts, you can find the product of the factors in the body of the table. You can also reverse the process. If you know one factor and the product, you can find the other factor.

**Example 3:** If the *product* of two numbers is 36 and one factor is 9, find the other factor.

*Solution:* To determine the other factor, look across the top row until you find 9. Go down the 9-column until you find 36. Look to the *factor* row and find the factor opposite 36.

*Answer:* It is 4.

*Check:* If $\underline{\quad ? \quad} \times 9 = 36$ use the table to find that

$$\underline{\quad 4 \quad} \times 9 = 36$$

Practice this method on these five short problems. Find the missing factor to replace the question mark.

1. $\underline{\quad ? \quad} \times 6 = 54 \qquad\qquad ? = \underline{\quad\quad}$
2. $8 \times \underline{\quad ? \quad} = 72 \qquad\qquad ? = \underline{\quad\quad}$

3. $12 \times$ ____?____ $= 132$          ? = ____

4. $7 \times$ ____?____ $= 56$            ? = ____

5. ____?____ $\times 7 = 49$            ? = ____

The ability to find one or more unknown factors in a multiplication problem is important. Use this procedure when you work problems in percentage, ratios, distance-time, etc.

Some problems require both multiplication and addition. To see what is involved in problems of this type, look at these examples.

**Example 4:** Find the result: $(\underline{3} \times 4) + (\underline{3} \times 5) = ?$

*Solution:* Do the multiplication first.

$(3 \times 4) = 12$

$(3 \times 5) = 15$

Then do the addition

$12 + 15 = 27$

*Answer:* 27

NOTE: $(\underline{3} \times 4) + (\underline{3} \times 5) = \underline{3} \times (4 + 5) = 3 \times 9 = 27$

**Example 5:** Find the result: $(\underline{5} \times 7) + (\underline{5} \times 6) = ?$

*Solution:* Do the multiplication first.

$(5 \times 7) = 35$

$(5 \times 6) = 30.$

Then do the addition.

$35 + 30 = 65$

*Answer:* 65

NOTE: $(\underline{5} \times 7) + (\underline{5} \times 6)$

$= \underline{5} \times (7 + 6)$

$= \underline{5} \times 13 = 65$

**Example 6:** Find the result: $(\underline{9} \times 3) + (\underline{9} \times 4) = ?$

*Solution:* Do the multiplication first

$(9 \times 3) = 27$

$(9 \times 4) = 36$

Then do the addition.

$27 + 36 = 63$

*Answer:* 63

NOTE: $(\underline{9} \times 3) + (\underline{9} \times 4)$

$$= \underline{9} \times (3 + 4)$$
$$= \underline{9} \times 7 = 63$$

The results of these three examples show that:

$$(\underline{3} \times 4) + (\underline{3} \times 5) = \underline{3} \times (4 + 5)$$
$$(\underline{5} \times 7) + (\underline{5} \times 6) = \underline{5} \times (7 + 6)$$
$$(\underline{9} \times 3) + (\underline{9} \times 4) = \underline{9} \times (3 + 4)$$

These examples illustrate a law of arithmetic called:

### The Distributive Law

*For any three numbers A, B, and C*
*it is always true that*

$$(A \times B) + (A \times C) = A \times (B + C)$$

In other words, **the distributive law** states: "*When two addends have a common factor (A in the above), then the sum may be written as a single product with the common factor A as one of the factors.*"

**Example 7:** Find the result: $(\underline{5} \times 6) + (\underline{5} \times 9) = ?$

*Solution:* The 5 is the common factor of both addends and, according to the Distributive Law, the result can be written:

$$\underline{5} \times (6 + 9) = \underline{5} \times (15)$$

*Answer:* 75

The Distributive Law makes it possible to find products of numbers such as $7 \times 54$, $9 \times 45$, and $32 \times 67$. To see how this rule is used to compute products such as these, look at the problem: $9 \times 356$. Write this in vertical form which gives

$$\begin{array}{r} 3\ 5\ 6 \\ \times \qquad 9 \\ \hline \end{array}$$

The first *partial product* is $9 \times 6$, which is 54.
The next *partial product* is $9 \times 50$, which is 450.
The last *partial product* is $9 \times 300$, which is 2,700.

When these *partial products* are added, you get:

$$\begin{array}{r} 3\ 5\ 6 \\ \times \qquad 9 \\ \hline 5\ 4 \\ 4\ 5\ 0 \\ 2\ 7\ 0\ 0 \\ \hline 3,2\ 0\ 4 \end{array}$$

$\begin{aligned} &= 9 \times 6 \quad (partial\ product) \\ &= 9 \times 50 \quad (partial\ product) \\ &= 9 \times 300 \ (partial\ product) \\ &= (\text{sum of } partial\ products) \end{aligned}$

When products of larger numbers are involved, take into account the *place* value as well as the *face* value of the numbers you are multiplying.

**Example 8:** Find the product: 421 × 234

*Solution:* Write this as:     421

$\times 234 = 200 + 30 + 4$

Multiply by the 4.   $4 \times 421 = 1,684.$
Multiply by the 30.   $30 \times 421 = 12,630.$
Multiply by 200.   $200 \times 421 = 84,200.$

```
    4 2 1
   ×2 3 4
   ─────────
   1 6 8 4 =    4 × 421 (partial product)
 1 2 6 3 0 =   30 × 421 (partial product)
 8 4 2 0 0 = 200 × 421 (partial product)
 ─────────
 9 8,5 1 4          (sum of partial products)
```

You may simplify these steps. Omit the zero in the units place in the partial product 12,630. Omit the two zeros in the unit's and tens place in the partial product 84,100.

```
      4 2 1
     ×2 3 4
     ───────
     1 6 8 4
     1 2 6 3        0 omitted
     8 4 2         00 omitted
     ───────
     9 8,5 1 4
```

*Answer:*

NOTE: If you use this shortcut method of omitting the zeros, indent the partial products one place to the left for each multiplier:

1. Indent one place for a factor in the 10's place.
2. Indent two places for a factor in the 100's place.
3. Indent three places for a factor in the 1,000's place, and so on.

**Example 9:** Find the product: 479 × 1,653.

*Solution:*
```
      1,6 5 3
    ×     4 7 9
    ───────────
      1 4 8 7 7
    1 1 5 7 1      indent one place
    6 6 1 2        indent two places
    ───────────
```
*Answer:* 7 9 1,7 8 7

Accurate multiplication is essential. Study the following examples before working the problems of this section.

**Example 10:** A plumber makes $986 per month. He saves $92 per
month.
(a) What is his annual salary?
(b) How much does he save per year?

*Solution* (a):  His annual salary is the product of:
(amount per month) × (number of months in 1 year)

which is:

$$
\begin{array}{r}
\$ 9\ 8\ 6 \\
1\ 2 \\
\hline
1\ 9\ 7\ 2 \\
9\ 8\ 6 \\
\end{array}
$$

*Answer* (a):  $1 1,8 3 2   (annual salary)

*Solution* (b):  The amount saved in one year is the product of:
(amount saved per month) × (number of months in
1 year)

*Answer* (b):  ($92) × (12) = $1,104 saved per year.

**Example 11:** On a road map 1 inch represents a distance of 75
miles. How many miles apart are the cities of
Dubuque and Le Mars, Ohio? (See Fig. 5-1.)

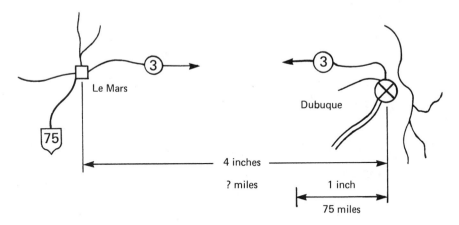

**Figure 5-1**

*Solution:*  The number of miles between the two cities is the product
of:

(number of miles per inch) × (number of inches)

(75 miles per inch) × (4 inches) = 300 miles

*Answer:* 300 miles

# PROBLEMS: *Sec. 1  Multiplying Whole Numbers*

1. Do the multiplications in your head. Strive for accuracy. Work as rapidly as possible.

   (a)
   | 9 | 7 | 8 | 5 | 6 | 4 | 7 |
   |---|---|---|---|---|---|---|
   | ×4 | ×5 | ×6 | ×7 | ×8 | ×9 | ×8 |

   (b)
   | 11 | 12 | 9 | 10 | 9 | 6 | 4 |
   |---|---|---|---|---|---|---|
   | ×11 | × 7 | ×12 | × 9 | × 8 | ×11 | ×12 |

   (c)
   | 21 | 16 | 15 | 32 | 60 | 25 | 18 |
   |---|---|---|---|---|---|---|
   | × 2 | × 3 | × 4 | × 2 | × 5 | × 3 | × 3 |

   (d) $3 \times 4$      $7 \times 8$      $9 \times 12$      $7 \times 12$

   (e) $2 \times (3 \times 4)$      $(4)(9)$      $(2 \times 4)(4 \times 3)$

2. Find the products.

   (a)
   | 345 | 264 | 840 | 229 | 638 |
   |---|---|---|---|---|
   | × 3 | × 2 | × 5 | × 4 | × 7 |

   (b)
   | 45 | 32 | 54 | 28 | 91 |
   |---|---|---|---|---|
   | ×21 | ×14 | ×36 | ×22 | ×73 |

   (c)
   | 245 | 897 | 765 | 439 | 849 |
   |---|---|---|---|---|
   | × 34 | × 54 | × 98 | × 67 | × 32 |

   (d)
   | 1,975 | 3,657 | 8,974 | 3,458 | 5,864 |
   |---|---|---|---|---|
   | × 675 | × 398 | × 254 | × 908 | × 456 |

3. Use the Distributive Law to find the products.

   **Example:** $(5 \times 6) + (5 \times 7) = 5 \times (6 + 7)$   (Distributive Law)
   $$= 5 \times 13 = 65$$

   (a) $(4 \times 6) + (4 \times 8)$
   (b) $(9 \times 5) + (9 \times 7)$
   (c) $(7 \times 5) + (7 \times 8)$
   (d) $(4 \times 5) + (4 \times 6)$

4. Use the Distributive Law to find the products.

   **Example:** $7 \times 23 = 7 \times (20 + 3)$
   $$= (7 \times 20) + (7 \times 3) \quad \text{(Distributive Law)}$$
   $$= 140 + 21$$
   $$= 161$$

   (a) $6 \times 42 = 6 \times (40 + 2) = (6 \times ?) + (6 \times ?)$
   (b) $8 \times 21 = 8 \times (20 + 1)$
   (c) $9 \times 36 =$
   (d) $8 \times 47 =$

5. Find the missing factors:

   **Example:** $\underline{\ \ ?\ \ } \times 5 = 45$. The table of multiplication facts shows that $9 \times 5 = 45$. Therefore, the missing factor is 9.

(a) $\underline{\quad ? \quad} \times 6 = 72$
(b) $7 \times \underline{\quad ? \quad} = 42$
(c) $9 \times \underline{\quad ? \quad} = 54$
(d) $56 \times \underline{\quad ? \quad} = 0$

6. Verify the Associative Law of Multiplication (p. 71) by computing both of the expressions. Do the multiplication in parentheses first.

   **Example:** $3 \times (4 \times 5) = 3 \times (20) = 60$

   $(3 \times 4) \times 5 = (12) \times 5 = 60$

   Since both products are the same, the Associative Law is verified.

   (a) $3 \times (6 \times 7) = (3 \times 6) \times 7$
   (b) $8 \times (9 \times 5) = (8 \times 9) \times 5$
   (c) $11 \times (5 \times 6) = (11 \times 5) \times 6$

7. Find the products:

   (a)
   | 234 | 367 | 579 | 985 | 397 | 496 | 834 |
   |---|---|---|---|---|---|---|
   | 7 | 5 | 4 | 2 | 7 | 6 | 8 |

   (b)
   | 1,567 | 3,867 | 4,762 | 5,982 | 9,740 | 4,500 |
   |---|---|---|---|---|---|
   | 3 | 6 | 8 | 8 | 5 | 6 |

   (c)
   | 23,574 | 16,098 | 18,905 | 23,700 | 14,080 |
   |---|---|---|---|---|
   | 4 | 6 | 7 | 3 | 8 |

   (d) $2 \times 1,896 \qquad 5 \times 10,785 \qquad 19,604 \times 8$

8. Find the products:

   (a)
   | 315 ft. | 69 in. | 458 yd. | 48 gal. | 2,306 cm. |
   |---|---|---|---|---|
   | 12 | 11 | 13 | 24 | 24 |

   (b)
   | 35 days | 6,789 miles | 3,406 lb. | 8,075 m. | 784 cm. |
   |---|---|---|---|---|
   | 16 | 25 | 50 | 32 | 16 |

   (c)
   | 12 doz. | 13 acres | 45 oz. | 82 mm. | 908 sec. |
   |---|---|---|---|---|
   | 179 | 346 | 309 | 576 | 403 |

   (d)
   | 4,789 gm. | $3,080 | 1,075 qt. | 18,608 in. |
   |---|---|---|---|
   | 10 | 45 | 39 | 56 |

9. Find the products:

   (a) $2 \times 4 \times 6 \qquad 8 \times 5 \times 9 \qquad 7 \times 9 \times 6 \qquad 3 \times 5 \times 4$
   (b) $14 \times 5 \times 8 \qquad 3 \times 18 \times 5 \qquad 2 \times 6 \times 23 \qquad 5 \times 15 \times 24$
   (c) $2 \times 34 \times 20 \qquad 7 \times 18 \times 16 \qquad 21 \times 4 \times 56 \qquad 15 \times 7 \times 89$
   (d) $23 \times 189 \times 4 \qquad 25 \times 156 \times 78 \qquad 32 \times 678 \times 54 \qquad 178 \times 890 \times 108$

10. Find the products:

    (a)
    | 2,457 | 23,578 | 12,509 | 18,900 | 35,808 |
    |---|---|---|---|---|
    | 158 | 265 | 462 | 235 | 675 |

    (b)
    | 7,005 | 45,090 | 45,206 | 20,106 | 30,605 |
    |---|---|---|---|---|
    | 204 | 204 | 992 | 205 | 779 |

    (c)
    | 3,009 | 39,703 | 23,705 | 50,800 | 35,705 |
    |---|---|---|---|---|
    | 1,678 | 1,652 | 14,076 | 20,700 | 67,068 |

(d) 1,895,756     2,507,964     23,789,909     34,060,952
      238         1,852           405,065         17,755

**11.** Find the products. Multiply each quantity separately. Express the answer in the simplest form.

**Example:** Find the product.

    4 gal.  3 qt.  1 pt.
        ×      7

    28 gal.  21 qt.  7 pt.
    28 gal.  24 qt.  1 pt.        (7 pt. = 3 qt. 1 pt.)
*Answer:*  34 gal.   0 qt.  1 pt.        (24 qt. = 6 gal.)

(a) 2 week  3 day  4 hr.    (b) 3 week  5 day  19 hr.
        × 7               × 12

(c) 4 ft.  5 in.   (d) 5 yd. 2 ft.  8 in.   (e) 7 yd. 1 ft.  10 in.
    × 4         × 14           × 45

(f) 2 yr.  4 mo.  15 days    (g) 4 yr.  11 mo.  18 days
       × 5              × 16
(use 30 days = 1 mo.)

(h) 3 m. 35 cm.   (i) 35 m. 89 cm.   (j) 16 m. 47 cm.
    × 15         × 62        × 19

**12.** If one set of spark plugs costs $8, how much do twelve sets cost?

**13.** A Cessna 170 can fly at 145 miles per hour. How many miles can it fly in 6 hours? Note: The distance traveled is the product of

(speed of plane) × (hours in air).

**14.** There are 12 inches in 1 foot. How many inches are there in 26 feet?

**15.** There are 27 cubic feet in 1 cubic yard. How many cubic feet are there in 5 cubic yards?

**16.** Bernard's car averages 14 miles per gallon of gas. How many miles can it travel if the gas tank holds 22 gallons?

**17.** A gross is equal to 12 dozen. How many dozen are there in 9 gross?

**18.** Carol can take shorthand at 95 words per minute. How many words can she take in 12 minutes?

**19.** The average person sleeps 8 hours per day. How many hours does he or she sleep in 1 year? (1 year = 365 days)

**20.** Each person in the U.S. uses about 15 gallons of water a day for washing, drinking, cooking, etc. The city of San Francisco has a population of 720,675 people. How many gallons of water are used each day by the people in San Francisco?

**21.** FBN Airlines advertises a flight from Milpitas to Honeydew, California—

a distance of 275 miles. If the single air fare is $54, how much does the air-line charge for a flight of 94 people?

22. Jerry Bilt Construction Company put in a bid to build a new office building. The building is to have 24,785 square feet of floor space. Jerry Bilt estimates that it will cost $26 per square foot to build the building. What is the total estimated cost?

23. Mr. Perrin, a journeyman bricklayer, can lay 145 bricks an hour. (a) How many can he lay in an eight-hour day? (b) In a forty-hour week?

24. One gallon of TUFKOT exterior house paint will cover 468 square feet when used as a primer coat. How many square feet will five gallons cover?

25. A careful time-study shows that it takes a machinist 1 hour, 46 minutes, 18 seconds to complete a very complicated part for a rocket engine. How long does it take the machinist to complete 25 parts?

26. Bungling Buford works for the Jerry Bilt Construction Company. He is paid $975 a month. Buford says his annual salary is $11,690. Is he correct?

27. A JPL jet can fly at an air speed of 189 meters per second. (a) How many meters can it fly in 1 minute? (b) In 1 hour?

28. One gallon of water weighs about 8 pounds. How many pounds of water are there in a swimming pool filled with 26,580 gallons?

29. A certain type of steel reinforcing rod weighs 2 pounds a foot. How much does a piece 26 feet long weigh?

30. Light travels at a speed of 186,000 miles per second. Find, to the nearest million, the number of miles light will travel in 1 hour. (1 hour = 3,600 seconds.)

31. The cost per foot for a modern eight-lane freeway is about $150. What is the cost per mile?

32. One automatic screw machine can produce 2,560 finished brass bushings in 1 hour. How many bushings can 5 machines produce in 8 hours?

33. A high-speed copy machine can produce a single copy of a document in 2 seconds. How many copies can it produce in 4 hours of continuous operation?

34. A salesman for the Mid-Nite Auto Supply earns these commissions.

    $2 for each battery sold
    $1 for each tune-up kit sold
    $3 for each auto tire sold
    $4 for each truck tire sold
    $8 for each set of "Mag" wheels sold

During the first week of May he made the following sales: 14 batteries; 9 tune-up kits; 4 auto tires; 2 truck tires; 5 sets of "Mag" wheels. What was his commission for the week?

35. A journeyman electrician is paid $7 per hour, which includes all fringe benefits. What is the total cost to the J & R Electric Co. for 5 electricians each one working a 36-hour week?

## SEC. 2   DIVIDING WHOLE NUMBERS

In Chapter 4 you learned that addition and subtraction are closely related. Now you will see that multiplication and division are also closely related. Section 5-1 deals briefly with the problem of finding the missing factor when one factor and the product is known. (See page 71.)

> **Example 12:** If the product of two numbers is 56, and one of the factors is 8, find the other factor.

*Solution:* A simple equation for this is:

$$\underline{\quad ? \quad} \times 8 = 56$$

Refer to the Multiplication Table (Table 5-1, page 70) where you find that 7 is the value of the missing factor.

*Answer:* **7** $\times 8 = 56$.

The process of finding the missing factor is called *division*. In Example 12, the known factor 8 is called the *divisor*. The missing factor 7 is called the *quotient*. The result 56 is called the *dividend*. The usual symbol for the instruction to divide is $\div$. The instruction to find the missing factor (to divide) is $56 \div 8 = \underline{\quad ? \quad}$. When the symbol $\div$ is between two numbers, it is an instruction to divide the *first* number by the *second*. When the question mark (?) is replaced by the number 7 (quotient), the equation looks like this:

$$56 \quad \div \quad 8 \quad = \quad 7$$
$$\textit{dividend} \quad \textit{divisor} \quad \textit{quotient}$$

Here $56 \div 8 = 7$ because $7 \times 8 = 56$.

From Example 12 you can see that *division* is related to *multiplication* in the following way:

Use A, B, and C to represent numbers.
A *divided* by B equals C because A equals the *product* of B and C.

$$A \div B = C \text{ because } A = B \times C.$$

The result may be stated:

"To *divide* A by B means to find a number C so that $A = B \times C$."

For instance,   $81 \div 9 = 9$ because $81 = 9 \times 9$
$72 \div 8 = 9$ because $72 = 8 \times 9$
$42 \div 6 = 7$ because $42 = 6 \times 7$

Use the Table of Multiplication Facts on page 70 as a Table of Division Facts. Leave out the row and column of zeros.
You *CANNOT DIVIDE BY ZERO.*

> **Example 13:** Try to find the quotient: $8 \div 0 = \underline{\quad ? \quad}$

*Solution:* According to the definition of division, to find the quotient means to find a third number C so that $8 = 0 \times C$. But, from the Table of Multiplication Facts, you know that *if any number is multiplied by zero, the result is always zero.* Therefore, since $0 \times C$ will always result in a product of zero, no value for C is possible.

NOTE: Remember: ***NEVER DIVIDE BY ZERO.***

*Answer:* The problem $A \div 0 = $ ? does **NOT** have an answer.

### Practice Problems

Use your knowledge of multiplication facts to find the *quotients* of these division problems. Check the answers by multiplication.

1. $32 \div 8 = $ _____ 　　　　 2. $77 \div 11 = $ _____
3. $45 \div 9 = $ _____ 　　　　 4. $64 \div 8 = $ _____
5. $60 \div 12 = $ _____ 　　　　 6. $36 \div 4 = $ _____
7. $132 \div 11 = $ _____ 　　　 8. $56 \div 7 = $ _____
9. $100 \div 10 = $ _____ 　　　 10. $144 \div 12 = $ _____
11. $36 \div 6 = $ _____ 　　　 12. $81 \div 9 = $ _____
13. $121 \div 11 = $ _____ 　　 14. $25 \div 5 = $ _____
15. $49 \div 7 = $ _____

As you probably know by now, the process of dividing one number (the *dividend*) by another (the *divisor*) is the same as finding the (whole) number of times the divisor is contained in the dividend. In the division problem $63 \div 7 = 9$, the number 7 (*divisor*) is contained 9 times in the number 63 (*dividend*). In this case, 7 is contained an *exact* number of times in the number 63 (*dividend*).

Now look at the problem of dividing 17 by 3. A quick look at the Multiplication Table shows that there is no whole number C which gives $17 = 3 \times C$. But you do see that $3 \times 5 = 15$ and $3 \times 6 = 18$. The *nearest whole multiple* of 3 that is less than 17 is 15. From this you know that there are five 3s and a 2 left over in the number 17. That is:

$$17 = 15 + 2 = (3 \times 5) + 2$$

When there is a quantity left over which is smaller than the divisor, it is called the *remainder*. Thus,

$$17 \quad = \quad (3 \quad \times \quad 5) \quad + \quad 2$$
*dividend　divisor　quotient　remainder*

NOTE: The whole number 5 is still called the quotient.

Division of two quantities uses systematic steps to find the whole number of times the divisor is contained in the dividend. The amount (if any) that is left over is the remainder.

Study Examples 14, 15, and 16 to see how to divide larger numbers.

**Example 14:** Divide 257 by 14.

*Solution:* When problems involve large numbers, use the symbol $\overline{)\quad}$. You probably remember this symbol from grade school. Write the problem with the divisor outside the symbol and the dividend inside it:

$$(divisor) \longrightarrow 14\overline{)257} \quad (dividend)$$

Step 1. Start at the 2 in the *dividend* and read LEFT to RIGHT. STOP when you have a number that 14 (*divisor*) will divide into. Here you stop at 25. Divide 14 into 25. 14 into 25 divides once. Put the 1 over the 5 in the dividend. Multiply 14 by 1. Put the product under the 25 in the dividend:

$$\begin{array}{r} 1 \\ 14\overline{)257} \\ \underline{14} = 1 \times 14 \end{array}$$

Subtract 14 from 25 to get a difference of 11. Bring down the 7 to form the number 117:

$$\begin{array}{r} 1\phantom{00} \\ 14\overline{)\ 257} \\ \underline{-14\downarrow} \\ 117 \end{array}$$

Step 2. Divide 14 into 117. Since $9 \times 14 = 126$, which is larger than 117, and $8 \times 14 = 112$ which is smaller than 117, use 8 as the next digit in the *quotient*. Put the 8 next to the 1 in the quotient. Multiply 14 by 8 and write the *product* under the 117:

$$\begin{array}{r} 18 \quad (quotient) \\ 14\overline{)\ 257} \\ \underline{-14\downarrow} \\ 117 \\ \underline{112} = 8 \times 14 \end{array}$$

Subtract 112 from 117 to get a difference of 5. Since there are no more digits in the dividend to bring down, the number 5 is the *remainder*. After you finish the problem it looks like this:

$$\begin{array}{r} 18 = \text{quotient} \\ 14\overline{)\ 257} \\ -14\downarrow \quad \text{(subtract } 1 \times 14) \\ \overline{117} \quad \text{(bring down 7)} \\ -112 \quad \text{(subtract } 8 \times 14) \\ \overline{5} = \text{remainder} \end{array}$$

*Answer:* 257 divided by 14 gives a *quotient* of 18 with a *remainder* of 5.

*Check:* $(14 \times 18) + 5 = 257$

**Example 15:** Divide 2,356 by 125.

*Solution:* Write the problem:

$$125\overline{)2,356}$$

Step 1. Start at the 2 in the dividend and read LEFT to RIGHT until you have a number that 125 will divide into. Stop at 235. Divide 125 into 235. It goes once. Put the 1 over the 5 in the dividend. Multiply 125 by 1. Put the product under the 235:

$$\begin{array}{r} 1\phantom{25} \\ 125\overline{)2,356} \\ \underline{1\,25} \quad = 1 \times 125 \end{array}$$

Subtract 125 from 235 to get a difference of 110. Bring down the 6 to form the number 1,106:

$$\begin{array}{r} 1\phantom{\,106} \\ 125\overline{)\ 2,356} \\ \underline{-1\,25{\downarrow}} \quad \text{(Subtract } 1 \times 125) \\ 1\,106 \end{array}$$

Step 2. Divide 125 into 1,106. Since $9 \times 125 = 1,125$ and $8 \times 125 = 1,000$, use 8 as the next digit in the quotient. Put the 8 next to the 1 in the quotient. Multiply 125 by 8. Subtract the product from 1,106:

$$\begin{array}{r} 18 \\ 125\overline{)\ 2,356} \\ \underline{-1\,25{\downarrow}} \quad \text{(subtract } 1 \times 125) \\ 1\,106 \quad \text{(bring down 6 from dividend)} \\ \underline{-1\,000} \quad \text{(subtract } 8 \times 125) \\ 106 \end{array}$$

Since there are no more digits in the dividend, you have finished the division. You have a quotient of 18 and remainder of 106.

*Answer:* 2,356 divided by 125 gives a quotient of 18 with a remainder of 106.

**Example 16:** Divide 34,514 by 17.

*Solution:* Write the problem:

$$17\overline{)34,514}$$

Step 1. Start at the 3 in the dividend and read from LEFT to RIGHT until you have a number that 17 will divide into. Stop at 34. Divide 17 into 34. It divides twice. Put the 2 above the 4 in 34. Multiply 17 by 2. Put the product under 34. Subtract 34 from 34 to get a difference of 0. Bring down the 5:

$$
\begin{array}{r}
2\phantom{0000} \\
17)\overline{\phantom{0}34,514} \\
-34\downarrow\phantom{00} \\
\hline
0\,5\phantom{00}
\end{array}
$$
(subtract 2 × 17)
(bring down 5)

Step 2. Since 17 will NOT divide into 5, put a zero next to the 2 in the quotient. Then bring down the digit 1 in the dividend to form the number 51. Divide 17 into 51. It will divide three times. Put a 3 next to the zero in the quotient. Multiply 17 by 3 and subtract the product from 51:

$$
\begin{array}{r}
2\,03\phantom{00} \\
17)\overline{\phantom{0}34,514} \\
-34\downarrow| \phantom{0} \\
\hline
0\,5\downarrow \\
51 \\
-\phantom{0}51 \\
\hline
0
\end{array}
$$
(subtract 2 × 17)
(bring down 5)
(bring down 1)
(subtract 3 × 17)

Step 3. Bring down the 4 from the dividend and put it next to the zero at the bottom. Since 17 will NOT divide into 4 and there are no more digits to bring down, put a zero in the quotient over the 4. The quotient is 2,030 and the remainder is 4. After you finish the problem, it looks like this:

$$
\begin{array}{r}
2,030 = \text{quotient} \\
17)\overline{\phantom{0}34,514} \\
-34\downarrow|| \\
5\downarrow| \\
51| \\
-\phantom{0}51\downarrow \\
\hline
04 \\
4 = \text{remainder}
\end{array}
$$
(subtract 2 × 34)
(bring down 5)
(bring down 1)
(subtract 3 × 17)
(bring down 4)

*Answer:* 34,514 divided by 17 gives a quotient of 2,030 with a remainder of 4.

NOTE: After the digit 2 in the quotient there is a digit in the quotient above every digit in the dividend. If you leave out either of the zeros in the quotient (2,030), you will get a wrong answer.

Considerable practice is necessary if you expect to have the speed and accuracy required for many shop problems.

### *Practice Problems*

1.  Find the quotient and remainder for each problem. Check your answers by multiplication.

    **Example:** $54 \div \underline{8} = \underline{6}$ and a remainder of **6** because;

    *Check:*         $54 = (8 \times 6) + 6$

    NOTE: To check your answer: *Multiply the quotient by the divisor and add the remainder. The result must equal the dividend.*

    (a)  $56 \div 9 =$ _____ with a remainder of _____
    (b)  $98 \div 12 =$ _____ with a remainder of _____
    (c)  $186 \div 17 =$ _____ with a remainder of _____
    (d)  $897 \div 23 =$ _____ with a remainder of _____

2.  Divide and check.
    (a)  $3\overline{)44}$        $6\overline{)78}$        $8\overline{)97}$        $5\overline{)65}$
    (b)  $6\overline{)126}$        $8\overline{)896}$        $5\overline{)543}$        $7\overline{)896}$
    (c)  $5\overline{)3467}$        $6\overline{)6547}$        $4\overline{)7650}$        $6\overline{)8769}$

3.  Find the quotient and remainder. Check your results.

    $$\text{Example: } 4\overline{)74} \quad \overset{18 \ \ \text{R } 2}{\phantom{)74}} \qquad \text{(R 2 means:   remainder of 2)}$$

    *Check:* $74 = (4 \times 18) + 2 = 72 + 2 = 74$

    (a)  $7\overline{)217}$        $5\overline{)553}$        $6\overline{)436}$        $8\overline{)645}$
    (b)  $16\overline{)30}$        $15\overline{)66}$        $19\overline{)86}$        $14\overline{)76}$
    (c)  $32\overline{)897}$        $45\overline{)654}$        $32\overline{)546}$        $54\overline{)985}$
    (d)  $6\overline{)1,794}$        $5\overline{)8,975}$        $7\overline{)6,752}$        $9\overline{)10,876}$

Accurate results avoid costly errors and delays in time. Your ability to recognize division problems and work them correctly indicates that you have mastered Chapters 4 and 5.

To get an idea of the variety of problems that call for division, study these examples.

**Example 17:** This ad was in the classified section of the evening paper:

### QUALITY CONTROL TECHNICIAN

High school graduate with technical courses at the Community College level preferred. 1 to 2 yrs visual inspection experience. To be responsible for performance of necessary laboratory testing including material color, weight, finish, and dimension. Salary to $9,000 per year. Call for appointment. 865-5349.

If the salary is $9,000 a year, how much is it a month?

*Solution:* Since there are 12 months in one year, find the monthly salary by dividing:

$$\text{monthly salary} = (\text{annual salary}) \div (12)$$
$$= \$9,000 \div 12$$
$$= \$750$$

*Answer:* $\qquad\qquad = \$750$ a month

**Example 18:** The driver of a delivery truck started at 8:00 A.M. with a full tank of gas. At the end of the day he filled the tank and it took 17 gallons. If he drove 238 miles, how many miles per gallon did the truck get?

*Solution:* To determine the gas mileage divide:

$$\text{gas mileage} = (\text{total miles driven}) \div (\text{gas used})$$
$$= (238 \text{ miles}) \div (17 \text{ gallons})$$

*Answer:* $\qquad\qquad$ 14 miles per gallon.

**Example 19:** The fuel consumption of J. Poore's private plane is 7 gallons per hour. If the tanks on the plane have a capacity of 92 gallons, how many hours, to the nearest whole hour, can the plane stay in the air?

*Solution:* Use division to find the number of hours the plane can remain in the air.

$$\text{Hours in air} = \text{capacity of tanks} \div \text{fuel consumption}$$
$$= 92 \text{ gallons} \div 7 \text{ gallons per hour}$$

*Answer:* $\qquad\qquad = 13$ with a remainder of 1

NOTE: To the nearest whole hour, the plane can remain in the air for 13 hours. The remainder of 1 means there is one-seventh of an hour of flight time (about 9 minutes) remaining after 13 hours in the air.

**Example 20:** The tanker *S. S. Malaga* has a capacity of 98,000 barrels of crude oil. How many hours does it take to load the ship at a rate of 7,000 barrels per hour?

*Solution:* Use division to find the time necessary to load the ship.

$$\text{Hours to load} = \text{total capacity} \div \text{load rate}$$
$$= 98,000 \text{ barrels} \div 7,000 \text{ barrels per hour}$$

*Answer:* $\qquad\qquad = 14$ hours

**Example 21:** Mr. Westbrock, an excellent mechanic, did two complete tune-ups in 5 hours 40 minutes. How long did it take for each tune-up?

*Solution:*  Use division to find the time for one tune-up.

Time for one tune-up:

total time ÷ number of tune-ups done.

In this problem show the division in the form:

$$2)\overline{5 \text{ hr. } 40 \text{ min.}}$$

Divide 2 into 5 for a quotient of 2 hours and a remainder of 1 hour. Convert the 1 hour to 60 minutes and add to the 40 minutes in the minutes column before you divide the minutes by 2.

$$
\begin{array}{r}
2 \text{ hr.} \quad\;\; 50 \text{ min.} \\
\hline
2)\quad 5 \text{ hr.} \quad\;\; 40 \text{ min.} \\
-4 \text{ hr.} \\
\hline
1 \text{ hr.} = 60 \text{ min.} \\
2)\overline{100 \text{ min.}} \\
-100 \text{ min.} \\
\hline
0 \text{ (remainder)}
\end{array}
$$

*Answer:*  It took Mr. Westbrock 2 hours and 50 minutes for each tune-up.

NOTE:  In problems of this type, where several units are involved, a great deal of care must be taken when converting remainders to other units, such as feet to inches, gallons to quarts, or days to hours.

In many problems you will use some or all of the operations of arithmetic to get the final answer. This is especially true when computations involve working with formulas. In Chapter 14 this type of problem is explained. But for now, just learn some of the rules that combine addition, subtraction, multiplication, and division.

### Rules for the Order of Operations in Adding, Subtracting, Multiplying, and Dividing

1. *Always do what is indicated in the parentheses first.*

**Example 22:** $(2 + 5) \times 6 = (7) \times 6 = 42$

2. *Do the multiplications and the divisions in the order that they appear as you read from left to right.*

**Example 23:** $24 \div 6 \times 3 = 4 \times 3 = 12$

3. *Do additions and subtractions in the order that they appear as you read from left to right.*

**Example 24:** $4 + 5 - 3 + 6 = 9 - 3 + 6 = 6 + 6 = 12$

4. *Do the multiplying and the dividing before you do the adding and subtracting, unless parentheses show otherwise.*

**Example 25:** (a) $3 \times 5 + 8 \div 4 = 15 + 2 = 17$

**Example 25:** (b) $4 \times (8 - 5) \div 6 = 4 \times (3) \div 6 = 12 \div 6 = 2$

When you change a measurement to other units, sometimes you must use a combination of multiplication and division.

**Example 26:** A car is traveling at 60 miles per hour. How fast is it traveling in feet per second?

*Solution:* To convert miles to feet, multiply by 5,280. (There are 5,280 feet in 1 mile.) Divide the result by 3,600, which converts hours to seconds. The result is:

60 mph $= (60$ mph $\times 5,280$ ft. per mile$) \div 3,600$ sec. per hr.

$= 316,800$ ft. per hr. $\div 3,600$ sec.

*Answer:* 60 mph $= 88$ ft. per sec.

So far you have not had to worry about the precision of the values when you were multiplying or dividing. But since most technical problems deal with measurements, you must know certain rules to be sure that your results do not have a precision that is not true.

The rules for multiplying and dividing measurements are similar to the rules of adding and subtracting measurements given in Chapter 4. They are useful for almost all on-the-job problems. Sometimes you may get a slight error from a certain step in a problem, but the rules are adequate for most problems. If there is an exact number in a problem, it keeps its precision throughout all stages of the computation.

### Rules for Multiplying and Dividing Measurements

1. *Determine the least precise measurement in the set of given values. (Find the measurement with the fewest number of significant digits.)*
2. *Round off all other measurements to a precision ONE DEGREE GREATER than that of the least precise measurement in Step 1. (All other measurements may have ONE more significant digit than the value obtained in Step 1.)*
3. *Do the indicated operation (multiplication or division).*
4. *Round off the result to the SAME precision as the least precise value in Step 1. (The answer must have only as many significant digits as the value found in Step 1.)*

**Example 27:** To find the area of a rectangle, *multiply* the length by the width. Find the area of the rectangle in Fig. 5-2.

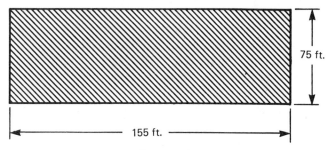

75 ft.

155 ft.

*Figure 5-2*

*Solution:* Find the area by multiplying

$$\text{Area} = (\text{length}) \times (\text{width})$$

Of these two measurements, the width (75 ft.) has the fewest number (2) of significant digits, and is the value to use for Step 1. The other value is now rounded off to have no more than three significant digits, which is ONE MORE than the number of significant digits in 75 (Step 2). In this case the other value (155) already satisfies Step 2 of the rule. Now, form the product:

$$(75) \times (155) = 11{,}625 \qquad (\text{Step 3})$$

Round off this result (11,625) to as many significant digits as the value in Step 1. Thus you round off 11,625 to 12,000 (Step 4).

*Answer:* Area = 12,000 sq. ft. (two significant digits)

**Example 28:** A freight train runs from New Orleans, Louisiana, to Los Angeles, California—a distance of 2,936 miles. If the train travels at an average speed of 35 mph, how many hours is it en route?

*Solution:* The measurement for distance (2,936) has four significant digits, and the measurement for the speed (35) has two significant digits. Apply the rules. Round off the value for distance to three significant digits to 2,940 (Steps 1 and 2). Find the time by dividing:

$$\text{Travel time} = (\text{total distance}) \div (\text{average speed})$$
$$= (2{,}940) \div (35)$$

*Answer:*                          = 84 hours.

NOTE: The answer (84 hours) has two significant digits and does not have to be rounded off.

When you use these rules you get results that are only estimates to possibly more precise answers. If you are *sure* of the precision of your given values, then you may get results that are more precise than those you get by using these rules. The rules are intended for general use. In some cases you may get an error. Analyze each problem separately. Apply critical judgment to obtain the result. This is something you must learn to do before working problems. There are exceptions to these rules. Use good, sound judgment when working all problems. If these rules do not apply to a particular problem, you may change them to meet your needs.

## PROBLEMS: Sec. 2
### Dividing Whole Numbers

Do these divisions in your head. Strive for accuracy.

1. (a) $5\overline{)65}$    (b) $3\overline{)36}$    (c) $7\overline{)42}$    (d) $8\overline{)96}$    (e) $12\overline{)24}$

2. (a) $7\overline{)77}$    (b) $4\overline{)64}$    (c) $9\overline{)108}$    (d) $7\overline{)63}$    (e) $11\overline{)121}$

3. (a) $5\overline{)42}$    (b) $6\overline{)37}$    (c) $8\overline{)81}$    (d) $3\overline{)32}$    (e) $10\overline{)110}$

4. (a) $36\overline{)72}$    (b) $25\overline{)100}$    (c) $20\overline{)80}$    (d) $32\overline{)64}$    (e) $50\overline{)150}$

**Problems 5 through 16. Do the divisions and check your answers.**

**Example:**

$$
\begin{array}{r}
17 \quad r = 10 \\
27\overline{)\ 469} \\
-27\downarrow \quad \text{(subtract } 1 \times 27) \\
\hline
199 \quad \text{(bring down 9)} \\
-189 \quad \text{(subtract } 7 \times 27) \\
\hline
10 \ \text{remainder}
\end{array}
$$

*Check:* $(27 \times 17) + 10 = 469$

5. (a) $23\overline{)345}$    (b) $32\overline{)658}$    (c) $13\overline{)246}$    (d) $21\overline{)875}$

6. (a) $41\overline{)568}$    (b) $33\overline{)765}$    (c) $76\overline{)875}$    (d) $90\overline{)456}$

7. (a) $37\overline{)1,986}$    (b) $45\overline{)12,785}$    (c) $35\overline{)10,874}$

8. (a) $198\overline{)10,865}$    (b) $3,468\overline{)567,975}$    (c) $9,987\overline{)349,972}$

9. (a) $60\overline{)89}$    (b) $45\overline{)90}$    (c) $56\overline{)78}$    (d) $19\overline{)76}$

10. (a) $45\overline{)356}$    (b) $28\overline{)579}$    (c) $83\overline{)332}$    (d) $24\overline{)984}$

11. (a) $32\overline{)986}$    (b) $29\overline{)766}$    (c) $14\overline{)298}$    (d) $13\overline{)356}$

12. (a) $45\overline{)3,457}$    (b) $25\overline{)4,568}$    (c) $17\overline{)3,589}$    (d) $15\overline{)3,579}$

13. (a) $25\overline{)7,090}$    (b) $34\overline{)5,897}$    (c) $46\overline{)4,009}$

14. (a) $32\overline{)45,789}$    (b) $21\overline{)34,606}$    (c) $45\overline{)358,800}$

15. (a) $46\overline{)574,302}$    (b) $64\overline{)34,608}$    (c) $82\overline{)54,789}$

16. (a) $15\overline{)33,000}$    (b) $12\overline{)24,024}$    (c) $17\overline{)18,900}$

**Find the quotients and remainders for problems 17 through 28.**

17. (a) $15\overline{)3,015}$      (b) $24\overline{)48,001}$      (c) $24\overline{)58,890}$

18. (a) $213\overline{)4,270}$      (b) $128\overline{)896}$      (c) $427\overline{)3,569}$

19. (a) $273\overline{)89,876}$      (b) $456\overline{)457,000}$      (c) $285\overline{)9,698}$

20. (a) $229\overline{)10,109}$      (b) $765\overline{)987,679}$      (c) $764\overline{)56,908}$

21. (a) $10\overline{)15,789}$      (b) $100\overline{)23,658}$      (c) $102\overline{)206,900}$

22. (a) $200\overline{)85,200}$      (b) $200\overline{)90,080}$      (c) $300\overline{)390,100}$

23. (a) $500\overline{)680,900}$      (b) $700\overline{)456,987}$      (c) $400\overline{)606,600}$

24. (a) $1,000\overline{)345,900}$      (b) $2,000\overline{)480,800}$      (c) $3,000\overline{)568,895}$

25. (a) $5,280\overline{)45,890}$      (b) $2,789\overline{)45,896}$      (c) $9,547\overline{)457,987}$

26. (a) $8,367\overline{)780,679}$      (b) $1,728\overline{)1,567,483}$      (c) $1,987\overline{)34,789,730}$

27. (a) $2,245\overline{)123,456,789}$      (b) $8,975\overline{)35,050,909}$      (c) $3,400\overline{)54,700,805}$

28. (a) $20,002\overline{)400,006}$      (b) $4,050\overline{)109,007}$      (c) $11,056\overline{)345,000,709}$

**Do the divisions in problems 29 through 32.** Be careful when converting units.

**Example:**

```
      2 yd.      1 ft.      10 in.
3)    7 yd.      2 ft.       6 in.
     -6 yd.     +3 ft.     +24 in.
      1 yd.   3) 5 ft.   3) 30 in.
                 -3 ft.     -30 in.
                  2 ft.       0 remainder
```

*Answer:* 2 yd. 1 ft. 10 in.

29. (a) $5\overline{)15 \text{ ft.} \quad 10 \text{ in.}}$      (b) $4\overline{)5 \text{ yd.} \quad 1 \text{ ft.} \quad 8 \text{ in.}}$

30. (a) $3\overline{)10 \text{ gal.} \quad 2 \text{ qt.}}$      (b) $3\overline{)7 \text{ gal.} \quad 3 \text{ qt.} \quad 1 \text{ pt.}}$

31. (a) $6\overline{)13 \text{ yr.} \quad 2 \text{ mo.} \quad 4 \text{ wk.}}$      (b) $9\overline{)10 \text{ min.} \quad 3 \text{ sec.}}$

32. (a) $3\overline{)9 \text{ hr.} \quad 27 \text{ min.} \quad 36 \text{ sec.}}$      (b) $7\overline{)15 \text{ miles} \quad 600 \text{ ft.} \quad 14 \text{ in.}}$

33. Divide 56,890 by 53.

34. Find the quotient and remainder when 6,784 is divided by 78.

35. If the dividend is 345 and the divisor is 23, what is the quotient? The remainder?

36. Divide 56 into 9,874.

37. How many sevens are there in 602?

38. In a certain division problem the divisor is 22, the quotient is 31, and the remainder is 13. What is the dividend?

39. There are 1,760 yards in 1 mile. There are 220 yards in 1 furlong. How many furlongs are there in 1 mile?

40. There are 43,560 square feet in 1 acre. There are 9 square feet in 1 square yard. How many square yards are there in 1 acre?

**41.** Mr. Hindley harvested 76,050 bushels of wheat from 975 acres. What was the yield in bushels per acre?

**42.** Is it possible to distribute 3,117 marbles among 19 boys in such a way that each boy gets the same number and there are none left over?

**43.** When loading a cargo ship with small, foreign cars, each car takes about 225 cubic feet of space. How many cars can be loaded in the ship if its capacity is 84,500 cubic feet?

**44.** The safe-load capacity of a railroad flat car is 145,000 pounds. It is to be loaded with coils of sheet steel that weigh 3 tons each. How many coils can be safely loaded on the flat car? (1 ton = 2,000 pounds.)

**45.** In one year the budget for the Peralta Community College District was $24,375,000. The total enrollment was 19,500 students. What was the cost to the district to educate one student?

**46.** Jerry Bilt Construction Company instructed Bungling Buford to find the total number of truck loads of earth needed for a certain land fill. B. B. knows that a large dump truck will carry 6 cubic yards per load. The land fill requires 685,000 cubic yards. He told his boss that it would take 114,150 truck loads. Do you agree with him?

**47.** Do the operations to get a single number. (See Rules for Order of Operations, page 87)

    **Example:** $(2 \times 3) + (3 \times 5) - (12 \div 3)$

    *Solution:* Do the multiplications and the division FIRST.

    *Answer:* $6 + 15 - 4 = 17$

    (a) $4 \times (2 \times 3) + 18(3 \times 3) = ?$
    (b) $2 \times (3 + 4) + 3 \times (5 - 2) = ?$ (In this problem, do the additions and subtractions inside the parentheses FIRST.)
    (c) $(2 + 3) \times (5 + 15 \div (3 \times 5)) = ?$
    (d) $(15 \div 3) \times 5 + 15 \div (3 \times 5) = ?$

**48.** A *hogshead* is an old unit of liquid measure (1 hogshead equals 63 gallons). How many hogsheads (hhd.) are there in 756 gallons?

**49.** *When is a pound not a pound?* When measuring the weight of an object, there are two systems that use the *pound* as a unit of measure. Commerce uses the **avoirdupois** pound which weighs 16 ounces. Jewelers and other people dealing in precious metals use the **troy** pound which weighs only 12 ounces. Thus, a pound troy is *not* a pound avoirdupois because the pound troy is 4 ounces less.

1 pound troy = 12 ounces

1 pound avoir. = 16 ounces

An ingot of gold weighs 36 pounds (avoir.). What is its weight in pounds troy?

**50.** Do you agree with the statement: "A ton of feathers weighs more than a ton of gold"? State your reasons.

**Problems 51 through 54 involve measurements.** Before you do them, review the rules for multiplying and dividing measurements on page 88.

**51.** Determine the electric current (*I*) in this series circuit (Fig. 5-3).

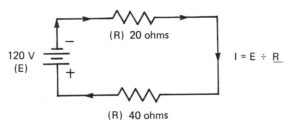

(R) 20 ohms

120 V
(E)

I = E ÷ R

(R) 40 ohms

**Figure 5-3**

NOTE: The total resistance (*R*) is the sum of all resistances.

**52.** A swimming pool that holds 232,500 gallons of water is to be filled at the rate of 150 gallons per minute. How many minutes will it take to fill the pool?

**53.** Mr. Bradley drove 304 miles in 8 hours. What was his speed? (speed = distance ÷ time)

**54.** A bin containing 7,200 pounds of cement is to be emptied into bags which hold 96 pounds each. How many bags are needed?

**55.** Two cities that are 475 miles apart are to be plotted on a chart with a scale: 1 inch = 25 miles. How many inches apart should they be plotted?

**56.** Ferbert bought a freezer costing $367. He paid $25 down, and is to pay the balance in 9 equal payments. What is the amount of each payment?

**57.** A space capsule can travel from the earth to the moon in two and one-half days (60 hours), traveling a distance of 256,800 miles. Find the speed of the capsule in miles per hour.

**58.** The sun is about 93,000,000 miles from the earth. Light travels at a speed of 186,000 miles per second.
(a) How many seconds does it take the light from the sun to reach the earth?
(b) How many minutes does it take the sunlight to reach the earth?

**59.** If no waste is involved, how many 2-ounce fittings can be made from a piece of bar stock weighing 11 pounds? (1 pound = 16 ounces)

**60.** The tanker *S.S. Laingor* has a capacity of 2,436,000 gallons of crude oil. A barrel of oil holds 42 gallons. What is the capacity of the tanker, measured in barrels?

**61.** A submarine is at a depth of 1,482 feet. What is the depth of the submarine, in fathoms? (A *fathom* is a measure of distance, and is equal to 6 feet.)

## PROBLEMS: Secs. 1 and 2
### Multiplying and Dividing Whole Numbers

Practice as much as possible. It will help you solve many problems accurately and quickly.

1. Find the products.

| | | | | |
|---|---|---|---|---|
| (a) $4 \times 6$ | $6 \times 8$ | $5 \times 9$ | $9 \times 7$ | $5 \times 4$ |
| $7 \times 8$ | $9 \times 9$ | $8 \times 7$ | $7 \times 6$ | $4 \times 11$ |
| (b) $7 \times 5$ | $8 \times 9$ | $4 \times 12$ | $5 \times 11$ | $12 \times 8$ |
| $11 \times 11$ | $9 \times 7$ | $8 \times 8$ | $7 \times 7$ | |
| (c) $9 \times 11$ | $8 \times 3$ | $3 \times 12$ | $12 \times 11$ | $4 \times 4$ |
| $7 \times 12$ | $9 \times 6$ | $3 \times 7$ | $6 \times 6$ | $5 \times 9$ |
| (d) $2 \times 3 \times 8$ | $3 \times 4 \times 7$ | $2 \times 4 \times 9$ | $2 \times 6 \times 11$ | |
| $9 \times 4 \times 6$ | $8 \times 5 \times 7$ | $8 \times 9 \times 7$ | $12 \times 12 \times 12$ | |

2. Find each of the products. To check your answers, reverse the order of the factors and multiply again.

**Example:** $6 \times 15 = 90$

Reverse the order of the factors and multiply: $15 \times 6 = 90$
Notice that for multiplication, reversing the factors gives the same answer (Commutative Law of Multiplication).

| | | | |
|---|---|---|---|
| (a) $13 \times 22$ | $12 \times 14$ | $13 \times 15$ | $17 \times 21$ |
| $25 \times 14$ | $21 \times 16$ | $14 \times 35$ | |
| (b) $16 \times 42$ | $27 \times 54$ | $38 \times 27$ | $15 \times 16$ |
| $45 \times 26$ | $65 \times 38$ | $98 \times 64$ | |
| (c) $123 \times 135$ | $168 \times 245$ | $456 \times 187$ | $227 \times 376$ |
| $675 \times 975$ | $465 \times 778$ | | |
| (d) $1,564 \times 2,476$ | $3,655 \times 7,632$ | $1,668 \times 7,784$ | |
| $9,085 \times 1,908$ | $4,506 \times 8,703$ | | |

3. For each problem, find the value of $n$ that makes the equation true.

**Example:** Find the value of $n$ so that $5n = 45$.

*Solution:* Refer to Multiplication Table 5–1 on page 70, which shows that $5 \times \mathbf{9} = 45$.

*Answer:* $n = 9$

| | | | | |
|---|---|---|---|---|
| (a) $4n = 36$ | $3n = 12$ | $6n = 30$ | $7n = 63$ | $9n = 27$ |
| (b) $12n = 108$ | $10n = 50$ | $11n = 132$ | $7n = 42$ | $8n = 64$ |
| (c) $9n = 54$ | $8n = 72$ | $7n = 49$ | $2n = 24$ | $5n = 40$ |

Use the *Associative Law of Multiplication* to check the answers for problems 4 through 7.

**Example:** $2 \times (3 \times 6) = (2 \times 3) \times 6 = 36$

Note the regrouping of the factors.

4. (a) $2 \times (3 \times 4)$        (b) $3 \times (4 \times 7)$        (c) $6 \times (3 \times 9)$
    (d) $8 \times (9 \times 6)$        (e) $7 \times (5 \times 3)$        (f) $6 \times (6 \times 8)$

**5.** (a) $(3 \times 7) \times 8$      (b) $(8 \times 6) \times 9$      (c) $(4 \times 6) \times 3$
    (d) $(8 \times 6) \times 5$      (e) $(5 \times 3) \times 7$      (f) $(12 \times 6) \times 4$

**6.** (a) $12 \times (13 \times 15)$      (b) $15 \times (23 \times 45)$      (c) $(16 \times 32) \times 21$
    (d) $(86 \times 75) \times 42$      (e) $(23 \times 14) \times 98$

**7.** (a) $154 \times (234 \times 142)$      (b) $156 \times (98 \times 457)$
    (c) $(368 \times 550) \times 387$      (d) $(124 \times 975) \times 1,653$

**Find the value in problems 8 through 11.** First, do the operation in the parentheses, and then do any other multiplications, additions, or subtractions.

    **Example:** $(23 + 14) \times 11 + 16 = (37) \times 11 + 16 = 407 + 16 = 423$

**8.** (a) $(15 + 19) \times 6 + 12$          (b) $(23 + 11) \times 17 + 34$

**9.** (a) $(14 \times 17) \times 3 + 25$          (b) $(13 \times 45) + (25 \times 16)$

**10.** (a) $(12 + 5) \times 9 + (18 + 5) \times 6$    (b) $(34 - 15) \times 4 + (45 - 18) \times 2$

**11.** (a) $(12 + 3) \times 4 - (5 + 7) \times 2$      (b) $(15 - 4) \times 5 - (14 - 7) \times 3$

**12.** Do these divisions in your head.

                   $9$
    **Example:** $6\overline{)54}$

     *Check:* $6 \times 9 = 54$

    (a) $4\overline{)12}$     $5\overline{)30}$     $6\overline{)42}$     $7\overline{)56}$     $8\overline{)32}$

    (b) $8\overline{)48}$     $9\overline{)63}$     $4\overline{)36}$     $8\overline{)96}$     $5\overline{)45}$

    (c) $9\overline{)36}$     $8\overline{)64}$     $7\overline{)49}$     $11\overline{)22}$     $9\overline{)108}$

    (d) $12\overline{)60}$     $6\overline{)72}$     $8\overline{)96}$     $11\overline{)121}$     $3\overline{)27}$

    (e) $10\overline{)60}$     $9\overline{)72}$     $7\overline{)0}$     $12\overline{)120}$     $1\overline{)8}$

    (f) $5\overline{)65}$     $2\overline{)42}$     $3\overline{)39}$     $4\overline{)60}$     $7\overline{)140}$

    (g) $13\overline{)39}$     $15\overline{)300}$     $3\overline{)66}$     $7\overline{)490}$     $9\overline{)810}$

**Divide and check the answers in problems 13 through 20.**

    **Example:**     $87$
                $4\overline{)348}$
                $\underline{32}$
                $28$
                $\underline{28}$
                $0$

    *Check:* $4 \times 87 = 348$

**13.** (a) $3\overline{)693}$    (b) $4\overline{)484}$    (c) $6\overline{)366}$    (d) $5\overline{)105}$    (e) $3\overline{)366}$

**14.** (a) $2\overline{)442}$    (b) $4\overline{)348}$    (c) $8\overline{)968}$    (d) $7\overline{)882}$    (e) $6\overline{)468}$

**15.** (a) $2\overline{)2,468}$    (b) $4\overline{)8,484}$    (c) $5\overline{)7,840}$    (d) $3\overline{)9,738}$    (e) $6\overline{)4,584}$

**16.** (a) $3\overline{)69,876}$    (b) $4\overline{)95,860}$    (c) $6\overline{)85,794}$    (d) $5\overline{)37,590}$

**17.** (a) $24\overline{)96}$    (b) $27\overline{)567}$    (c) $26\overline{)8,164}$    (d) $85\overline{)71,995}$

**18.** (a) $48\overline{)61,584}$    (b) $72\overline{)45,018}$    (c) $63\overline{)27,044}$

19.  (a) $144\overline{)864}$     (b) $174\overline{)7,482}$   (c) $298\overline{)11,026}$

20.  (a) $36\overline{)230,400}$   (b) $1,728\overline{)101,658}$    (c) $5,280\overline{)25,789,456}$

21.  There are 12 spark plugs packed in a box. There are 16 boxes in a shipping carton. How many spark plugs are in 8 cartons?

22.  Kevin is a shipping clerk and he can type 42 words per minute. How many words can he type in 28 minutes?

23.  Jon can pedal his racing bike at 27 miles per hour. How many miles will he travel in a bike race that lasts 4 hours?

24.  The Escondido High School marching band has 13 rows with 9 students in each row. How many students are there in the band?

25.  A truck is loaded with 78 cartons that weigh 216 pounds each. If the truck weighs 11,800 pounds empty, what is the weight of the loaded truck?

26.  A clerk in a hardware store is doing an inventory of nuts, bolts, screws, etc. Here are some of the items counted:

     4 boxes of lag screws with 48 to a box
     6 boxes of carriage bolts with 24 to a box
     10 boxes of stove bolts with 48 to a box
     7 boxes of eye bolts with 12 to a box
     18 boxes of assorted machine screws with 120 to a box
     28 boxes of assorted wood screws with 50 to a box

     Find the total number of bolts, screws, etc.

27.  With each revolution of a car's wheels, the car moves forward 6 feet. How far will the car travel if the wheels make 1,094 revolutions?

28.  A certain dictionary defines 47 different words on each page. If the dictionary has 1,179 pages, how many words does it define?

29.  Jerry Bilt Construction Company has just finished a new office building. The building has 54 floors for offices and there is a restaurant on the top floor. If the offices are 16 feet high and the restaurant is 22 feet high, what is the total height of the building?

30.  It took a construction company 42 working days to complete 9,492 feet of freeway. How many feet were completed each day?

31.  Jerry Bilt Construction Company has 9 dump trucks. Each truck can carry 4 cubic yards of earth.
     (a) How many loads will it take to move 26,208 cubic yards of earth?
     (b) If all of the trucks are used, how many loads will each truck haul?

32.  The timecard for a machinist shows that he worked 42 hours and 15 minutes during a five-day week. How many hours and minutes did he work each day?

33.  In one year American Motors produced 49,536 Gremlins. How many were produced per month?

34.  The total passenger car production for one year by GM, Ford, Chrysler–Dodge, and American Motors was 6,550,180. If there were 260 working days, how many cars were produced per day?

**35.** Ruth has a term paper that contains 12,865 words. If she can type 45 words per minute, how long will it take her to type the paper? Give answer to the nearest minute.

**36.** Jon wants to set a long-distance walking record. He plans to walk from San Francisco to New Orleans—a distance of 2,278 miles. Jon can walk 4 miles per hour, and he plans to walk 7 hours each day.
  (a) How many days will it take him to reach New Orleans?
  (b) If he leaves San Francisco on May 2, what day will he arrive in New Orleans?

## Practice Test for Chapter 5

**1.** Do these multiplications.
  (a)    457            (b)    1,067         *Answer:* (a) _____
      $\times 302$              $\times\ \ 753$              (b) _____

**2.** Do these divisions. Check your answer by multiplication.
  (a) $47\overline{)675}$    Check: _____ $\times$ _____ = _____
  (b) $24\overline{)408}$    Check: _____ $\times$ _____ = _____

**3.** Find the quotient and remainder when 4,677 is divided by 36.

    Quotient = _____    Remainder = _____

**4.** Find the value of:

    $(16 + 24) \times 11 + 42 =$ _____

**5.** A fourth-year sheetmetal apprentice makes about $5 an hour.
  (a) How much does he make in an 8-hour day? _____
  (b) How much does he make in a regular 40-hour week? _____
  (c) How much does he make in 23 eight-hour days? _____

**6.** A single brass fitting weighs 3 ounces. A full carton of fittings weighs 10 pounds 13 ounces. The empty carton weighs 14 ounces. How many fittings are in a full carton? _____

**7.** When an automobile is shifted into reverse gear, it takes the engine 117 revolutions to turn the wheels 10 times. How many revolutions will the engine turn, if the wheels turn 90 times? _____

**8.** A truck is traveling at 45 miles per hour. How fast is it traveling in feet per second? _____

**9.** Find this product:

    5 hr.   23 min.   17 sec.
    _____ $\times$ 12 _____        *Answer:* _____

**10.** Do this division:

    $7\overline{)10\ \text{hr.}\quad 42\ \text{min.}\quad 50\ \text{sec.}}$        *Answer:* _____

**11.** A new sports car is advertised to sell for $890 down and the balance in 36 "easy" payments of $187 each. What is the total price of the car?
                *Answer:* _____

# 6

# Signed
# Numbers

## SEC. 1   NEGATIVE NUMBERS

Today's advanced and rapidly changing technology requires that all
first-class tradesmen know how to use *negative numbers*. More and more
we use both positive and negative numbers in our daily lives. Almost
everyone has either made or heard statements such as:

The temperature is *minus* five degrees.
The caster angle should be *minus* two degrees.
The tolerance is plus or *minus* five ten-thousandths.
The angle of depression is *minus* three degrees.
T-*minus*-ten and counting.

Certainly you recognize the last statement, which became familiar with the
televised rocket launches and the space age. These statements show only
a few ways that negative numbers are used in measurements. In this chapter
you will learn about negative numbers, and how to add, subtract, multiply,
and divide with them.

How do you show that a number is negative? The answer is simple.
If you want to show that you are using a negative number, put a
minus sign (−) *in front* of that number. Thus, −1, −2, −3, −4, −5, −6,
−7, . . . represent *negative one, negative two, negative three, negative four,
negative five, negative six, negative seven*, and so on. The minus sign *in*

*front* of the number is part of the symbol. It shows the NEGATIVITY of the number. In a similar manner, a plus sign (+) may be placed *in front* of a positive number to show its POSITIVITY. Thus, +1, +2, +3, +4, ... represent *positive one, positive two, positive three, positive four,* and so on. The plus sign may be omitted from the positive numbers, because it is understood that both 5 and +5 are *positive.* But the minus sign is **never** omitted from negative numbers.

Positive and negative numbers are called *signed numbers.* Here are some examples of signed numbers:

+4°C    −7°F    +13 inches    +9 cm    −67°F    −1,094 feet

Both positive and negative numbers can be associated with points on a *number line.* The positive numbers and zero have already been located on the number line in Fig. 6-1. Positive numbers are associated with points

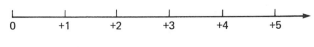

**Figure 6-1**

on the right side of (above) zero. Negative numbers are associated with points on the left side of (below) zero. In Fig. 6-2 you see that positive numbers are above zero and negative numbers are below zero. For each positive number associated with a point to the right of zero, there is a negative number associated with a point to the left of zero, in the same relative position.

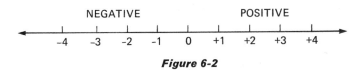

**Figure 6-2**

Negative numbers and the number line may be used to show quantities such as:

    Temperature below zero on a thermometer
    Elevation below sea level
    Deposit and withdrawal of money

It is important for you to remember that:

POSITIVITY is a *quality* of a number, and is represented by

        $+a$    $+b$    $+c$    and so on.

NEGATIVITY is a *quality* of a number, and is represented by

$$-a \qquad -b \qquad -c \qquad \text{and so on.}$$

Figures 6-3 and 6-4 show two ways that positive and negative values are used.

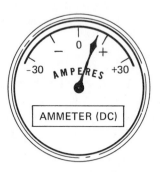

**Figure 6-3**  *Automobile ammeter*

**Figure 6-4**  *Aircraft air temperature gage*

An *ammeter* is an instrument for measuring amperes of electrical current. If the flow is toward the battery, a *positive* value is shown (as in Fig. 6-3). If the flow of current is away from the battery, then a *negative* value is shown. Thus, charge and discharge are indicated by positive and negative numbers.

An aircraft *air temperature gauge* (Fig. 6-4) is used to show the temperature of the air entering the carburetor of an aircraft engine. Its purpose is to detect icing conditions. Note that 0°C is the freezing point as the calibration of the instrument is in degrees Celsius. The instrument records a temperature of minus 12°C, $(-12°C)$.

## PROBLEMS: Sec. 1  Negative Numbers

**1.** What are the coordinates of the points marked *a*, *b*, *c*, and *d* in Fig. 6-5?

**Figure 6-5**

**2.** Use the number line in Fig. 6-6 and give the coordinate of the point:
(a) 4 units to the right of *a*.
(b) 6 units to the left of *b*.

***Figure 6-6***

(c) 5 units to the right of *b*.

(d) 7 units to the left of *a*.

3. Use a number line and give the coordinate of the point:
   (a) four units to the left of $-6$.
   (b) three less than $+5$.
   (c) four greater than $-9$.
   (d) six less than $-7$.

4. Use appropriate signed numbers to show:
   (a) 18° below freezing on Celsius scale.
   (b) 18° below zero on Fahrenheit scale.
   (c) 18° below freezing on Fahrenheit scale.
   (d) 36° below freezing on Fahrenheit scale.

5. Find the coordinate of the final position in problems (a) through (d).
   (a) Start at zero, go three units right, seven units left, two units right, five units left.
   (b) Start at negative four, go two units left, five units right, six units left, ten units right.
   (c) Start at positive seven, go five units left, six units right, four units right, eleven units left.
   (d) Start at positive one, go left five units, right four units, left six units, right seven units.

6. In a football game, how do you show:
   (a) a loss of 5 yards?        (b) a gain of 16 yards?
   (c) a loss of 3 yards?        (d) no gain?

7. Use signed numbers to show:
   (a) a bank deposit of $56.      (b) a withdrawal of $23.
   (c) a debit of $45.           (d) an overdrawn account of $10.

8. Use signed numbers to show:
   (a) an elevation of 182 feet below sea level.
   (b) a minus tide of 2 feet.
   (c) a loss of $185 on the stock market.
   (d) eight seconds before lift-off for a rocket.

## SEC. 2    ADDING SIGNED NUMBERS

In Section 1 of this chapter, you saw that positive and negative numbers are associated with points on a number line. Did you notice that:

- For each positive number $(+a)$, there is a *mirror image* number $(-a)$ on the other side of zero?
  
  and that

- For each negative number $(-a)$, there is a
  *mirror image* number $(+a)$ on the other side of zero?

Thus:      +1 and −1 are mirror images of each other.
           +2 and −2 are mirror images of each other.
           −10 and +10 are mirror images of each other.
           −20 and +20 are mirror images of each other,
           and so on.

Use this idea of *mirror images* to develop an operation in arithmetic called:

### Finding the Negative of a Number N.

Make a clear distinction between *a negative number* and *the negative of a number.*

> *The negative of a number N is the mirror image of
> N on the number line* (relative to zero).

For a number $N$, label *the negative of N* by $-N$.

Thus: The negative of +4, label $-(+4)$ is −4 (see Fig. 6-7).
      The negative of −4, label $-(-4)$ is +4 (see Fig. 6-7).

Here are examples of finding *the negative of a number.*

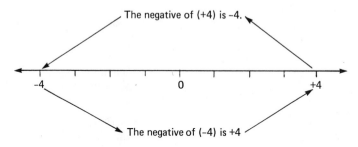

The negative of (+4) is –4.

The negative of (–4) is +4

**Figure 6-7**

**Example 1:** (a) The negative of +6 is −6

because                    $-(+6) = -6$

(b) The negative of −7 is +7

because                    $-(-7) = +7.$

(c) The negative of the negative of +4 is +4

because   $-(-(+4)) = -(-4) = +4$ (Fig. 6-8).

(d) The negative of the negative of −8 is −8

because   $-(-(-8)) = -(+8) = -8$ (Fig. 6-9).

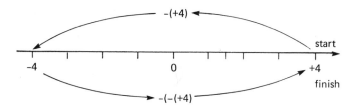

**Figure 6-8**

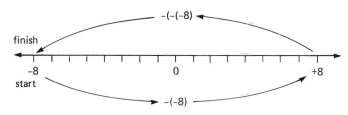

**Figure 6-9**

From these four examples you see that:

(a) The *negative of a positive number* is a negative number.
(b) The *negative of a negative number* is a positive number.
(c) The *negative of the negative* of a positive number is the same positive number (Fig. 6-8).
(d) The *negative of the negative* of a negative number is the same negative number (Fig. 6-9).

The property in (c) and (d) is called *the double negative property. The negative of the negative* of a number is like a double reflection. It gives back the original number.

By now you know that the minus sign (−) is used in three different ways:

1. To show the *operation of subtraction:*

$$10 - 5$$

2. To show the *negative quality of a number:*

$$-15 \text{ (negative 15)}$$

3. To show the operation of finding the *negative of a number:*

$$-(+4) = -4$$
$$-(-4) = +4$$

You must know these three different uses of the minus sign.

In this section and section 3 you will learn how to add and subtract signed numbers. Here are some typical problems:

| | |
|---|---|
| Add positive five to positive four. | $(+5) + (+4)$ |
| Add positive five to negative four. | $(+5) + (-4)$ |
| Add negative five to negative four. | $(-5) + (-4)$ |
| Subtract positive four from positive five. | $(+5) - (+4)$ |
| Subtract negative four from positive five. | $(+5) - (-4)$ |
| Subtract negative four from negative five. | $(-5) - (-4)$ |

As you will see, you can do these problems by paying attention to the QUALITY of the numbers and the OPERATION indicated.

## Adding Numbers with Like Signs

**Example 2:** Find the sum of the two positive numbers:

$$(+5) \text{ and } (+4).$$

*Solution:* Find $(+5) + (+4)$. To do this, use the number line (see Fig. 6-10). Think of adding on the number line as moving the point to the right or left on the number line—For *positive* numbers move to the *right*. For *negative* numbers move to the *left*. Therefore, to add positive five $(+5)$ to positive four $(+4)$ start at zero and go right to $(+5)$, and then go four more $(+4)$ to the right. The result is $(+9)$.

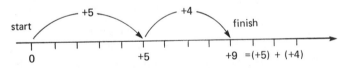

**Figure 6-10**

*Answer:* $(+5) + (+4) = +9$

**Example 3:** Find the sum of $(-4)$ and $(-5)$.

*Solution:* Use the number line in Fig. 6-11. Start at zero and go

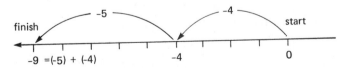

**Figure 6-11**

four units to the left to (−4). Then go five more units to the left.

*Answer:* $(-4) + (-5) = -9$.

Here are some more problems and answers.

$$(+6) + (+5) = +11$$
$$(+6) + (+9) + (+8) = +23$$
$$(-4) + (-6) = -10$$
$$(-5) + (-7) + (-13) = -25$$

From the results, it is possible to state the rule for adding two or more numbers with like signs:

### To Add Numbers with Like Signs

*Ignore the common sign and add the numerals together.*
*Attach the common sign to the result.*

Thus:
$$(+a) + (+b) = +(a + b)$$
$$(-a) + (-b) = -(a + b)$$

Check the answers to **Practice Problems** 1 to 10. Don't skip them even though they may look easy. You must understand the addition of signed numbers with LIKE (the same) signs before you study the addition of numbers with unlike signs.

1.  $(+4) + (+3) = +7$
2.  $(+5) + (+7) = +12$
3.  $(-6) + (-8) = -14$
4.  $(-6) + (-2) + (-3) = -11$
5.  $(+3) + (+6) + (+2) = +11$
6.  $(-78) + (-54) = -132$
7.  $(+1,786) + (+2,459) = +4,245$
8.  $(-789) + (-235) + (-567) = -1,591$
9.  $(-67) + (-45) + (-34) + (-56) = -202$
10. $(+56) + (+145) + (+354) + (+189) = +744$

## Adding Numbers with Unlike Signs

**Example 4:** Find the sum of $(+5)$ and $(-3)$ (see Fig. 6-12).

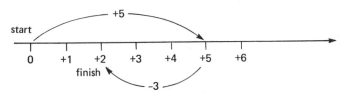

***Figure 6-12***

*Solution:* Use the number line in Fig. 6-12 to find $(+5) + (-3)$. Start at zero and go five units to the right to $(+5)$. Now go to the left three units. The result is the point with coordinate $(+2)$.

*Answer:* $(+5) + (-3) = +2$

**Example 5:** Find the sum: $(-6) + (+4)$.

*Solution:* Use the number line in Fig. 6-13. Start at zero and go six units to the left. Now go four units to the right. The result is the point with coordinate $(-2)$.

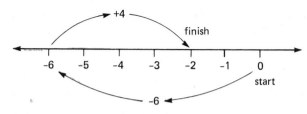

**Figure 6-13**

*Answer:* $(-6) + (+4) = -2$

**Example 6:** Find the sum: $(+5) + (-8)$.

*Solution:* Use the number line in Fig. 6-14. Start at zero and go five units to the right. Now go eight units to the left. The result is the point with coordinate $(-3)$.

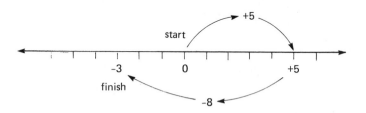

**Figure 6-14**

*Answer;* $(+5) + (-8) = (-3)$

Here are some more problems and answers:

$$(+8) + (-9) = -1 \qquad (-9) + (+9) = 0$$
$$(+15) + (-23) = -8 \qquad (-6) + (+7) = +1$$

In each case, note that:

> The result is the **difference** between the two numbers. The sign of the
> the answer is the sign of the larger number.

From the results of these examples, it is possible to state the rule for
adding numbers with unlike signs:

### To Add Numbers with Unlike Signs

*Ignore the signs and subtract the smaller numeral from the larger.
Affix the sign of the larger of the two numerals to the difference.*

Thus:

$$(-8) + (+6) = -2 \qquad (+12) + (-5) = +7$$
$$(-76) + (+85) = +9 \qquad (+12) + (-134) = -22$$

Use a number line for **Practice Problems** 1 through 5. Use the rule for
problems 6 through 10.

1.  $(+7) + (-8) = -1$
2.  $(-9) + (+3) = -6$
3.  $(-16) + (+14) = -2$
4.  $(+13) + (-16) = -3$
5.  $(-7) + (+7) = 0$
6.  $(-87) + (+98) = +11$
7.  $(+100) + (-45) = +55$
8.  $(-189) + (+234) = +45$
9.  $(-765) + (+675) = -90$
10. $(-3,890) + (+2,895) = -995$

When more than two signed numbers are added, use both rules.

**Example 7:** Find the sum: $(+7) + (-4) + (+3) + (-9)$.

*Solution:* It is easier to add all numbers of *like sign* (First Rule),
and then apply the Second Rule to find the sum of the
*unlike* signed numbers.

Thus: $(+7) + (-4) + (+3) + (-9) = (+10) + (-13)$

[First Rule]

$= -3$ [Second Rule]

*Answer:* $(+7) + (-4) + (+3) + (-9) = -3$

The result can also be seen on a number line (Fig. 6-15).

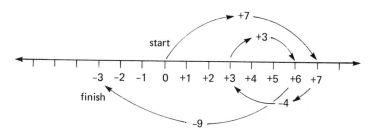

**Figure 6-15**

Here are some examples of how addition of signed numbers occur in practical problems.

**Example 8:** This table shows the gains and losses of the price of one share of stock in International Stump. The price on Monday was $54. Find the price of one share on Friday.

| Monday | Tuesday | Wednesday | Thursday | Friday |
|--------|---------|-----------|----------|--------|
| +$3    | −$5     | −$2       | +$6      | −$8    |

*Solution:* The price on Friday is $54 plus the increase or decrease for each day of the week. This result is the sum of the values given in the table:

$$
\begin{aligned}
\text{Net change} &= \underbrace{(+3) + (-5)}_{(-2)} + \underbrace{(-2) + (+6)}_{(+4)} + (-8) \\
&= \underbrace{(-2) \qquad + \qquad (+4)}_{(+2)} \qquad + (-8) \\
&= (-6)
\end{aligned}
$$

*Answer:*  $54 + (-$6) = $48

**Example 9:** Kirchhoff's Law of Voltage states:

> *The sum of the emf's (positive voltage) and the (negative) voltage drops around a series loop circuit must add up to zero.*

An emf (like a battery) is positive and is given a plus (+) value. A resistor ($R$) is negative and is given a minus (−) value. Start at point A on the diagram of Fig. 6-16 and verify Kirchhoff's law.

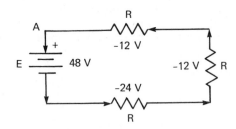

*Figure 6-16*

*Solution:* Start at the point marked A and go around the loop following the arrows. The sum of the emf (+) and (R) drops (−) is:

$$\text{Sum} = (+48) + (-24) + (-12) + (-12)$$
$$= (+24) \quad + \quad (-24) = 0$$

*Answer:* Sum $= 0$

Therefore, Kirchhoff's law is verified.

# PROBLEMS: Sec. 2
## Adding Signed Numbers

**1.** Add these signed numbers on a number line (Fig. 6-17).

**Example:** Add: $(+4) + (+2)$.

*Solution:* First locate $(+4)$ on the line. Add the signed number $(+2)$ by going in the positive direction two units. (Fig. 6-17)

*Answer:* $(+4) + (+2) = +6$

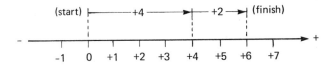

**Figure 6-17**

(a) $(+5) + (+3)$ $+8$      (b) $(+3) + (+6)$
(c) $(+10) + (+4)$ $14$      (d) $(-5) + (-3)$
(e) $(-8) + (-5)$      (f) $(+4) + (-2)$
(g) $(-14) + (+12)$ $-2$      (h) $(+4) + (-2)$
(i) $(+3) + (+5) + (-4)$ $4$      (j) $(-7) + (+6) + (-9)$

**2.** Find the sum of these signed numbers.

**Example:** Add: $(+13) + (+25) + (-18) + (-5) = \underline{\ ?\ }$.

*Solution:* Adding the two positive values gives:

$$(+13) + (+25) = +38 \quad \text{(partial sum)}.$$

Adding the two negative values gives:

$$(-18) + (-5) = -23 \quad \text{(partial sum)}.$$

Adding these two partial sums gives:

*Answer:* $\quad (+38) + (-23) = +15$

(a) $(-17) + (-14) + (+34) + (+21)$
(b) $(+15) + (-17) + (-19) + (+45)$
(c) $(-167) + (-345) + (+678) + (-745)$
(d) $(-190) + (-345) + (+345) + (-190)$
(e) $(+1,768) + (-4,574) + (-3,578) + (+1,832)$
(f) $(-90) + (+1,678) + (-23,675) + (+56,874)$

3. Find the sum of the signed numbers. Note that the plus sign for the positive numbers is not given.
   (a) $(-2) + 5 + 12 + (-15)$
   (b) $26 + 67 + (-23) + (-86) + 15$
   (c) $(-1) + 4 + (-2) + (-1)$
   (d) $(-13) + (-45) + 78 + 92 + (-14)$
   (e) $(-12) + (-12) + 36 + (13) + 1$
   (f) $1,867 + 908 + (-856) + (-7,546)$
   (g) $2,785,965 + (-3,765,982) + 100,089$
   (h) $(-9,567) + 9,543 + (-325) + 348$

4. Figure 6-18 is for a loop circuit with three batteries (emf) and four resistors. Kirchhoff's law states that the sum of the emf's (positive) and resistance drops (negative) in voltage must be zero. Are the proper values on each component in the circuit?

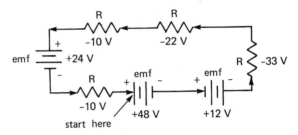

**Figure 6-18**

5. Find the positive value for the missing emf marked $X$ in Fig. 6-19. [Remember, according to Kirchhoff's law, the sum of emf's (battery) and voltage drops (resistors) must be zero.]

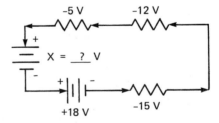

**Figure 6-19**

6. Find the negative value for the missing voltage drop at the resistor marked X in Fig. 6-20.

7. The emf at battery B is twice the emf at battery A in Fig. 6-21. Use Kirchhoff's law to find the values at A and B.

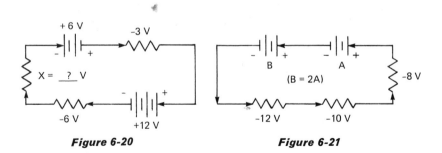

**Figure 6-20**                    **Figure 6-21**

8. A Brinell hardness test gauge is used to measure the hardness of metals. Most metals have a Brinell hardness number. This number shows the ability of the metal to withstand wear. (Brinell hardness numbers are in handbooks on metals.) A manufacturer is producing castings from Monel metal which has Brinell hardness numbers between 110 and 130. For quality control, the manufacturer sets a Brinell hardness number of 120 as a standard for his castings. Inspectors of the castings give signed numbers for gauge readings above (+) or below (−) the standard of 120. For instance, if a casting has a reading of −5, it means the hardness test shows the value to be 120 + (−5) = 115. If a casting shows a hardness number above or below the limits of 110 to 130, it must be rejected. Determine the hardness numbers of the castings listed in Table 6-1. Which castings are to be rejected?

**TABLE 6-1**

| Casting number | Gauge reading | Brinell number | Accepted? Rejected? |
|---|---|---|---|
| * 1 | +6 | 120 + 6 = 126 | Accepted |
| 2 | −9 | 120 + | _____ |
| 3 | +15 | 120 + | _____ |
| 4 | −16 | 120 + | _____ |
| 5 | +1 | 120 + | _____ |
| 6 | −8 | 120 + | _____ |
| 7 | −11 | 120 + | _____ |
| 8 | +9 | 120 + | _____ |
| 9 | −12 | 120 + | _____ |
| 10 | −6 | 120 + | _____ |

*The first row is done as an example.

9. The *tensile strength* of a metal is its resistance to a force which tends to tear the metal apart. A *tensiometer* is an instrument used for measuring the tensile strength of a material. Copper wire (hard drawn) has limits of tensile strength between 49 and 67. A tensiometer with a set standard of 55 is used to test a shipment of copper wire. If the reading is above 55,

a plus value is assigned. If the reading is below 55, a minus value is assigned. Table 6-2 gives values for ten shipments of copper wire. Find the tensile strength of each shipment, and determine which shipment is above the limit of 67 or below the limit of 49 and must be rejected.

**TABLE 6-2**

| Shipment number | Tensiometer reading | Tensile strength | Accept Reject |
|---|---|---|---|
| *1 | +20 | 55+(+20)=75 | Reject |
| 2 | +14 | _____ | _____ |
| 3 | −11 | _____ | _____ |
| 4 | −14 | _____ | _____ |
| 5 | −9 | _____ | _____ |
| 6 | +12 | _____ | _____ |
| 7 | −16 | _____ | _____ |
| 8 | +13 | _____ | _____ |
| 9 | −21 | _____ | _____ |
| 10 | −19 | _____ | _____ |

*The first row is done as an example.

10. A weather balloon is sent aloft to determine the temperature at various altitudes. At 2,500 feet the temperature is +3°C, and at 4,800 feet the temperature drops 19°C. What is the temperature at the higher altitude?

11. A pilot flying at 14,000 feet observes that the outside temperature is −16°F. After climbing to an altitude of 17,500 feet the temperature drops another 12°F. What is the reading on the thermometer?

12. A certain compound of silicone melts at −118°C and boils at +53°C. How many degrees of temperature are between the boiling and melting points?

13. Carbon dioxide ($CO_2$), which is dry ice, melts (changes from a solid to a gas) at −76°F. How many degrees colder is a piece of dry ice than a piece of water ice which has a melting point of +32°F?

## SEC. 3   SUBTRACTING SIGNED NUMBERS

In Chapter 4 you subtracted a smaller (positive) whole number from a larger (positive) whole number. You will now learn to subtract signed numbers such as:

$$(+5) - (+8) \qquad (+9) - (-5) \qquad (-8) - (-5)$$

To carry out subtractions such as these, you need a clear understanding of *exactly* what it means to subtract one number from another. From Chapter 4, (p. 58) recall that to check a subtraction result, you *add* the *difference* (D) to the *subtrahend* (S) to see if the *sum is equal to the minuend* (M).

Thus:     $12 - 8 = 4$     because     $12 = 8 + 4$

When letters are used to represent numbers you have

$$M - S = D \qquad \text{because} \qquad M = S + D$$

Or, written in vertical form,

| | | | | | |
|---|---|---|---|---|---|
| $M$ | minuend | | | $S$ | subtrahend |
| $-S$ | subtrahend | | Because | $+D$ | difference |
| $D$ | difference | | | $M$ | minuend |

From this you see that when you subtract one number from another, you are asking, "What number must be **added** to the subtrahend to get the minuend?"

To subtract 9 from 12 is to ask, "What do I **add** to 9 to get 12?" This is written:

$$12 - 9 = \quad ? \quad \text{or} \quad \begin{array}{r} 12 \\ -\ 9 \\ \hline ? \end{array}$$

Note that when the *difference* is added to the *subtrahend*, the sum is the *minuend*. Using this fact, how do you answer the subtraction question:

$$(+7) - (+9) = \ ?$$

Here, signed numbers are used to emphasize the *Quality* of the numbers involved. The difference ($D$) is the signed number that you **add** to $(+9)$ to get $(+7)$. To find the value of $D$, use the number line in Fig. 6-22. Start

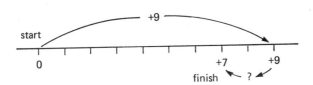

**Figure 6-22**

at zero and go to the point with coordinate $(+9)$. Now look at the point with coordinate $(+7)$. What do you **ADD** to $(+9)$ to get to $(+7)$? Move two units to the left to get to $(+7)$ from $(+9)$. From your knowledge of adding signed numbers, you know that this must be $(-2)$.

Therefore: $(+7) - (+9) = -2$ because $(+9) + (-2) = +7$.

Now look at this subtraction problem:

Subtract negative 6 from positive 4

$$(+4) - (-6) = \ ?$$

What do you add to negative six to get positive four? Use the number line in Fig. 6-23.

$$(-6) + (\text{difference}) = +4$$

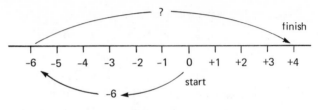

**Figure 6-23**

Start at zero and go to the left six units as shown by the negative quality of (−6). Look at the point with coordinate (+4). How far is it to the right? Start at (−6) and go (+10) units to the right to get to (+4).

Therefore:                    $(-6) + (+10) = (+4)$

and                           $(+4) - (-6\ ) = +10.$

Before stating a rule for the subtraction of signed numbers, do one more problem.

Find the difference:     $(-6) - (-10) = \underline{\quad ? \quad}$ Use the number line in Fig. 6-24 to solve the problem. What do you add to negative 10 to get negative 6?

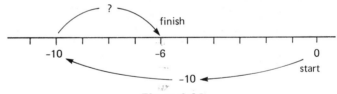

**Figure 6-24**

Start at zero and go 10 units to the left as indicated by the number (−10). What do you add to get to the point with coordinate (−6)? Add (+4) to get to (−6). Therefore:

$$(-6) - (-10) = +4 \text{ because } (-10) + (+4) = (-6).$$

Look at the three problems you have solved:

1. $(+7) - (+9) = -2$
2. $(+4) - (-6) = +10$
3. $(-6) - (-10) = +4$

Notice that *to subtract* S (subtrahend) *from* M (minuend) you added the **negative** *of* S, (−S), *to* M.

Thus:                    $M - S = M + (-S) = D$

This operation is the

### Rule for Subtracting Signed Numbers

*To **subtract** two signed numbers, add the **negative** of the subtrahend to the minuend. This sum is the answer.*

Here are some additional problems with answers.

1. Subtract $(+9) - (+5)$. The negative of $(+5)$ is $- (+5) = -5$. Therefore: $(+9) - (+5) = (+9) + (-5) = +4$. The answer $(+4)$ is found by applying the rules for adding signed numbers.
2. Subtract $(+7) - (-11)$. The negative of $(-11)$ is $- (-11) = +11$. Therefore: $(+7) - (-11) = (+7) + (+11) = +18$.
3. Subtract $(-8) - (-15)$. The negative of $(-15)$ is $- (-15) = +15$. Therefore: $(-8) - (-15) = (-8) + (+15) = +7$.
4. Subtract $(-7) - (+12)$. The negative of $(+12)$ is $- (+12) = -12$. Therefore: $(-7) - (+12) = (-7) + (-12) = -19$.

### Practice Problems

Use the Rule for Subtracting Signed Numbers and verify these problems. Refer to a number line to check your results.

1. $(-9) - (+15) = -24$
2. $(+14) - (-11) = +25$
3. $(-13) - (-15) = +2$
4. $(-15) - (-12) = -3$

Notice that in every case:

$M$ (minuend) $- S$ (subtrahend) $= M$ (*minuend*) $+ (-S)$ (*Negative of subtrahend*) $= D$ (difference).

**Example 10:** A jet flying at 45,000 feet recorded an air temperature of minus fifty-six degrees Fahrenheit $(-56°F)$. Upon landing in San Francisco, the air temperature was sixty-four degrees Fahrenheit $(+64°F)$. What is the difference in temperature?

*Solution:* The temperature difference is:

(temperature at San Francisco) $-$ (temperature 45,000 ft.)

$$(+64°F) - (-56°F) = +120°F$$

*Answer:* There is a 120-degree difference in temperature.

**Example 11:** With all of the electrical equipment (lights, radio, map light, cigar lighter, etc.) turned on and the engine not running, an automobile ammeter shows a discharge of minus 24 amps. When the engine is started, the ammeter shows a charge of 12 amps. What is the amperage output of the alternator?

*Solution:* You find the output of the alternator by taking the difference between the two ammeter readings.

Alternator output = (engine running) − (engine not running)

$$= (+12 \text{ amps}) - (-24 \text{ amps})$$

$$= (+12) + (+24); \text{ (add the negative of } -24)$$

*Answer:*     Output = 36 amps

**Example 12:** Cryogenics is the science of low-temperature events. A technician, working in a cryogenic laboratory, observes that oxygen is a gas at temperatures above −183°C. It is a liquid at temperatures between −183° and −218°C. At temperatures below −183°C it is a solid. The technician wants to know the number of degrees change between −183°C and −218°C. How many degrees must he lower the temperature to change oxygen from a liquid to a solid form?

*Solution:* Find the number of degrees change by taking the difference between the *final* desired value (solid state) and the *initial* value (liquid state):

Drop in temp = $(-218°C) - (-183°C)$

$$= (-218°C) + (+183°C); \text{ (add the negative of}$$

$$= -35°C \qquad\qquad -183)$$

*Answer:* He must lower the temperature 35 degrees Celsius.

Restudy the examples and rules of this section before you do the problems.

## PROBLEMS: Sec. 3
### Subtracting Signed Numbers

1. Find the differences of the signed numbers. To subtract signed numbers, *add the negative of the subtrahend to the minuend:*

$$M - S = M + (-S) = D$$

**Examples:**     $5 - (-3) = 5 + (+3) = 8$

$$7 - (+5) = 7 + (-5) = 2$$

$$-6 - (-4) = -6 + (+4) = -2$$

(a) $8 - (+6)$                  (b) $9 - (+8)$

(c) $-8 - (+5)$             (d) $-10 - (+4)$

(e) $-19 - (+10)$         (f) $109 - (+14)$

2. Find these differences:

(a) $7 - (-5)$        (b) $16 - (-8)$        (c) $54 - (-45)$

(d) $-36 - (-14)$        (e) $-21 - (-46)$        (f) $35 - (-89)$

(g) $-23 - (-17) - (-15) - (+75)$

(h) $45 - (+16) - (-15) - (+32) - (-45)$

3. Table 6-3 shows the increase (+) or decrease (−) in temperature from an initial reading at 5:00 A.M. at the Aniak, Alaska, weather station.

**TABLE 6-3**  TEMPERATURE CHANGE AT 2-HOUR INTERVALS (°F)

| *Initial temperature* 5 A.M. | 7 A.M. | 9 A.M. | 11 A.M. | 1 P.M. | 3 P.M. | 5 P.M. |
|---|---|---|---|---|---|---|
| $-18°F$ | $+6°F$ | $+8°F$ | $+10°F$ | $-4°F$ | $-9°F$ | $-13°F$ |

(a) What is the sum of all the increases and decreases?

(b) What was the temperature at 11 A.M.?

(c) What was the temperature at 3 P.M.?

(d) What was the temperature at 5 P.M.?

4. A weather balloon is sent aloft to measure changes in temperature. At ground level the temperature is 29°C. At 5,000 meters the temperature is −14°C. What is the temperature difference between ground level and 5,000 meters?

5. Table 6-4, taken from a surveyor's notebook, shows the increase (+) or decrease (−) in elevation from an initial reading of 1,568 feet at point *A*.

**TABLE 6-4**  CHANGE IN ELEVATION (IN FEET) AT VARIOUS POINTS

| *Elevation at point A* | point B | point C | point D | point E | point F | point G |
|---|---|---|---|---|---|---|
| 1,568 ft. | $+176$ | $+45$ | $-68$ | $-85$ | $+79$ | $-177$ |

(a) What is the sum of the positive and negative changes in elevation?

(b) How much higher is point *B* than point *F*?

(c) What is the elevation of point *G*?

## SEC. 4   MULTIPLYING SIGNED NUMBERS

In the section on multiplying whole numbers you learned that multiplication is a shortcut for finding the sum of repeated additions of the same number. For instance, the repeated addition:

$$4 + 4 + 4 + 4 + 4 + 4 + 4 = 28$$

is written as a multiplication:

$$7 \times 4 = 28$$

In the case where the addend is negative, for instance $(-4)$, repeated addition gives:

$$(-4) + (-4) + (-4) + (-4) + (-4) + (-4) + (-4) = -28.$$

When written as a multiplication, this is:

$$7 \times (-4) = -28.$$

Since the number seven could be written $(+7)$, this last result shows that:

$$(+7) \times (-4) = -28$$

This answer is the basis for

### Rule for Multiplying Signed Numbers—Unlike Signs

*When one of the factors in a multiplication of two numbers is negative and the other factor is positive, the product is negative.*

*or*

*The product of a positive and a negative is a negative.*

For instance    $(+5) \times (-7) = -35$    $(-6) \times (+8) = -48$

The order of the factors is not important. In multiplication problems it is possible to change the order and still get the same answer. It is an application of The Commutative Law of Multiplication. Thus:

$$(-7) \times (+5) = (+5) \times (-7) = -35$$

In order to establish a rule for the multiplication of two negative numbers, look at the result of continued subtractions of the same negative number.

**Example 13:** Find the result: $100 - (-5) - (-5) - (-5) - (-5)$

*Solution:* Subtract $-5$ from 100 four times. Use the rule for subtracting negative numbers. The result is:

$$100 + (+5) + (+5) + (+5) + (+5)$$

In each subtraction, the negative of $-5$, which is $-(-5) = +5$, is added. For the time being, ignore the number 100 and notice that:

$$-(-5) - (-5) - (-5) - (-5) = +(+5) + (+5) + (+5) + (+5)$$
$$(-4) \times (-5) = (+4) \times (+5) = +20$$

*Answer:* $100 - (-5) - (-5) - (-5) - (-5) = 100 + (4 \times 5)$
$$= 120$$

NOTE: Since the two sides of the equality represent the same number, then $(-4) \times (-5) = (+4) \times (+5)$

Example 13 is the basis for the

### Rule for Multiplying Two Negative Numbers

*The product of a negative with a negative is a positive.*

For instance,   $(-9) \times (-6) = +54$

$(-13) \times (-15) = +195$

$(-16) \times (-16) = +256$

The rules for multiplication of signed numbers may be expressed by using only signs. They are:

$$(+) \times (+) = (+)$$
$$(-) \times (-) = (+)$$ like signs

$$(+) \times (-) = (-)$$
$$(-) \times (+) = (-)$$ unlike signs

**Example 14:**

$(+6) \times (+8) = +48$   (positive times positive is positive)

$(-6) \times (-8) = +48$   (negative times negative is positive)

$(+6) \times (-8) = -48$   (positive times negative is negative)

$(-6) \times (+8) = -48$   (negative times positive is negative)

An easy way to remember these rules is:

*When the signs are the same, the product is positive. When the signs are different, the product is negative.*

### Practice Problems

Work these problems before going on. Do not skip them. You must know the rules for signed numbers and be able to use them correctly.

1. $(+4) \times (-6)$
2. $(-8) \times (+6)$
3. $(-9) \times (+9)$
4. $(-15) \times (-4)$
5. $(-45) \times (-1)$
6. $(-68) \times (-4)$
7. $(-4) \times (+6) \times (+2)$
8. $(-2) \times (-3) \times (+4)$
9. $(-3) \times (-2) \times (-5)$
10. $(+4) \times (-5) \times (-1)$
11. $(5) \times (6) \times (-5)$
12. $(-15) \times (24) \times (-2)$

Sometimes the multiplication sign ($\times$) is omitted in a multiplication problem. For instance, $(-9)(4)$ is a short way of writing $(-9) \times (4)$.

13. $(-14)(-5)(5)$
14. $(3)(45)(-2)$
15. $(10)(-10)(-10)(10)$
16. $(-1)(-1)(-1)(-1)$

Answers to Practice Problems.

1. $-24$   2. $-48$   3. $-81$   4. $+60$   5. $+45$   6. $+272$
7. $-48$   8. $+24$   9. $-30$   10. $+20$   11. $-150$
12. $+720$   13. $+350$   14. $-270$   15. $+10,000$   16. $+1$

Here are some examples of multiplication of signed numbers.

Certainly all of us remember the fun of riding on a see-saw or teeter-totter. This familiar playground equipment is a long plank, balanced at a point called the *fulcrum*. A person rides on each end of the plank. When one end goes up, the other end goes down. Remember what happened when your partner on the other end suddenly got off the see-saw? The result was probably a jarring of teeth and a bruise here or there. The force with which you hit the ground is called *torque*. *Torque* is a measure of the tendency to produce rotation (or twisting) about a point. Physicists have a formula for measuring torque. It is:

$$\textbf{\textit{Torque}} = (\textbf{\textit{weight of object}}) \times (\textbf{\textit{distance from fulcrum}})$$

If the distance is in feet, and the weight is in pounds, the torque is measured in *pound-feet*.

The distance from the fulcrum can be either positive or negative. The rules for finding the distance from the fulcrum to the weighted object are:

1. *If the distance is to the right of the fulcrum, it is considered positive* (+).
2. *If the distance is to the left of the fulcrum, it is considered negative* (−).

Thus, torque can have either positive or negative values. Figure 6-25 shows these rules.

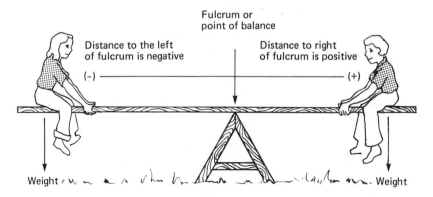

Fulcrum or
point of balance

Distance to the left
of fulcrum is negative

Distance to right
of fulcrum is positive

(−) ——————————————————— (+)

Weight ‧‧‧‧‧‧‧‧‧‧‧‧‧‧‧‧‧‧‧‧‧‧‧‧‧‧‧‧‧‧ Weight

*Figure 6-25*

**Example 15:** If the boy on the right weighs 75 pounds and is 4 feet from the fulcrum, he produces $(+4) \times (+75) = +300$ pound-feet of torque. If the girl on the left weighs 60 pounds and is 5 feet from the fulcrum, she produces $(-5) \times (+60) = -300$ pound-feet of torque. When the *sum of all torques is zero*, then the system is said to be in *equilibrium* (or in balance). In this case the

$$\text{Sum of all torques} = (+300) + (-300)$$
$$= 0$$

Therefore, the system is balanced.

A torque-wrench (Fig. 6-26) is a tool used by auto mechanics, diesel mechanics, and machinists. It has a long handle (the lever arm) and a dial at one end that shows the torque produced (in pound-feet).

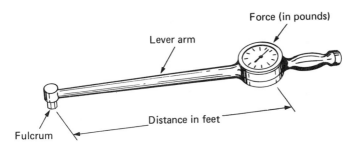

Force (in pounds)

Lever arm

Distance in feet

Fulcrum

**Figure 6-26**

The fulcrum for a torque-wrench is at the point where the socket is attached to the wrench. The force is applied at the other end. The torque (sometimes called the *moment of force*) is:

$$\text{Torque} = (\text{force}) \times (\text{distance})$$

The dial on the tool shows the torque produced. It is not necessary to figure it out.

**Example 16:** What will the dial read for a force of 90 pounds located 2 feet from the socket?

*Solution:*  Torque = (90 lb.) × (2 ft.)

*Answer:*          = 180 pound-feet

Technicians in the aeronautical trades must be able to figure out *moments of force* (torque). In order to know the effect of adding or removing objects from an aircraft, they must find the resulting moment, because it shows whether the aircraft will be nose or tail heavy. In Fig. 6-27 the *datum line* is an imaginary vertical line from which all horizontal measurements are made. The datum line is the fulcrum for the aircraft. The distance from the datum to a weight is called the *moment arm*. Distance is positive if it is to the right (rearward). Distance is negative if it is to the left (forward). If a weight is added it has a positive value. If a weight is removed it has a negative value.

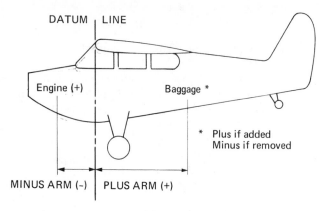

*Figure 6-27*

**Example 17** (a): If 60 pounds of baggage is loaded into the plane at a distance of $+5$ feet from the datum line, what is the moment of force?

*Solution* (a): $(+60 \text{ pounds}) \times (+5 \text{ ft.}) = +300$ pound-feet.

*Answer* (a): Moment $= +300$ pound-feet.

**Example 17:** (b): If 50 pounds of baggage is removed, what is the resulting moment?

*Solution* (b): $(-50 \text{ lb.}) \times (+5 \text{ ft.}) = -250$ pound-feet.

*Answer* (b): Moment $= -250$ pound-feet

Positive moments tend to make the plane tail heavy. Negative moments tend to make the plane nose heavy. The total moment (sum of all moments) is used to find the overall stability of the aircraft.

Learn to use the rules of signs for distance and weight.

1. Items added *forward* of the datum:
   $(+)$ weight $\times$ $(-)$ arm $= (-)$ moment
2. Items added *rearward* of the datum:
   $(+)$ weight $\times$ $(+)$ arm $= (+)$ moment
3. Items removed *rearward* of the datum:
   $(-)$ weight $\times$ $(+)$ arm $= (-)$ moment
4. Items removed *forward* of the datum:
   $(-)$ weight $\times$ $(-)$ arm $= (+)$ moment

The **total** *moment of the aircraft* is the sum of the weight moment of the aircraft and all of the individual moments.

**Example 18:** Table 6-5 shows a calculation used to find the moment of an aircraft. In this example the accepted values are: Gasoline = 6 pounds per U.S. gallon. Oil (lubricating) = 8 pounds per U.S. gallon. Crew and passengers = 170 pounds per person. (See Fig. 6-28.)

**TABLE 6-5**

|  | *Weight* (lb.) | × | *Arm* (in.) | = | *Moment* (lb.-in.) |
|---|---|---|---|---|---|
| Aircraft | +1,170 | | +12 | | +14,040 |
| Oil | +16 | | −48 | | −768 |
| Fuel | +95 | | +24 | | +2,280 |
| Pilot | +170 | | +12 | | +2,040 |
| Passenger | +170 | | +48 | | +8,160 |
| Baggage | +100 | | +84 | | +8,400 |

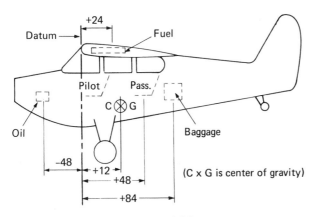

**Figure 6-28**

*Answer:* Total moment of aircraft = +34,152 lb.-in.

# PROBLEMS: *Sec. 4*
## *Multiplying Signed Numbers*

**1.** Use the rules of signs to find the products.

**Example:**

(+24) × (+68) = +1,632;  (positive) × (positive) = positive
(−11) × (−68) = + 748;  (negative) × (negative) = positive
(+45) × (−17) = − 765;  (positive) × (negative) = negative
(−81) × (+32) = −2,592;  (negative) × (positive) = negative

(a) $(+101) \times (+35)$            (b) $(-165) \times (+234)$
(c) $(-86) \times (-3,456)$        (d) $(-8) \times (+9) \times (-6)$
(e) $(-98) \times (+60)$           (f) $(+43) \times (-65)$
(g) $(-59) \times (-34)$           (h) $(+11) \times (+45) \times (-63)$

2.  In some of these problems, the plus sign $(+)$ is omitted; in others, the product sign $(\times)$ is omitted. Find the product in each problem.

    **Example:** $(14)(-12)(5) = -840$.

          Note that when no sign of operation appears between the parentheses it means to multiply.

(a) $(-68)(+56)(-10)$
(b) $(101)(-8)(-56)$
(c) $(25) \times (36)(18)$
(d) $(+56) \times (-17)(-10)(-8)$

3.  Use a number scale to show that
(a) $(-6) \times (+5) = (+6) \times (-5)$
(b) $(+3) \times (+5) = (+5) \times (+3)$
(c) $(+7) \times (-4) = (+4) \times (-7)$

4.  Commercial flight 406 is instructed by the air traffic controller to descend at the rate of 600 feet per minute for 12 minutes. If the airplane is at 32,600 feet before starting descent, what will the altitude be at the end of 12 minutes?

5.  An airplane flying at 45,000 feet has an outside temperature of $-54°$F. The pilot knows that the temperature *increases* 3 degrees for each 1,000 feet *decrease* in altitude. If the pilot decreases 21,000 feet, what will the temperature be?

6.  A see-saw is balanced when the sum of the moments adds up to zero. Which of the following see-saws is balanced? (See Figs. 6-29, 6-30, 6-31, 6-32.)

(a)

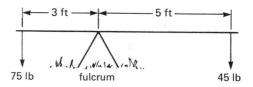

**Figure 6-29**

(b)

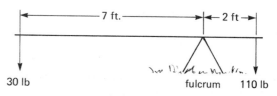

**Figure 6-30**

(c) NOTE: When there are two or more weights on one side, each weight must be multiplied by its distance from the fulcrum. The sum of the moments on one side must equal the sum of the moments on the other for balance. (Fig. 6-31.)

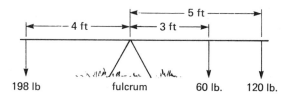

**Figure 6-31**

(d) (See Fig. 6-32.)

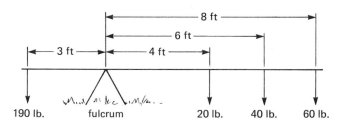

**Figure 6-32**

7. In Fig. 6-33 a workman has just lifted the rock off the ground by using the lever-fulcrum principle. Use the values in the figure and find the weight of the rock.

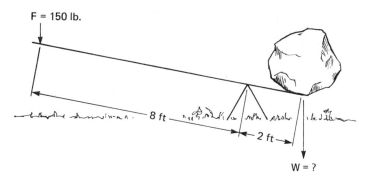

**Figure 6-33**

**8.** Fill in the missing values in Table 6-6. Use Fig. 6-34 to get the distances. Total Moment of Aircraft = _____

**TABLE 6-6**

|  | *Weight* (lb.) | × | *Arm* (in.) | = | *Moment* (lb.-in.) |
|---|---|---|---|---|---|
| (a)  Aircraft | +1,870 | | +24 | | _____ |
| (b)  Oil | +    28 | | | | _____ |
| (c)  Fuel | +  190 | | | | _____ |
| (d)  Pilot | +  170 | | +12 | | _____ |
| (e)  Passenger | +  170 | | | | _____ |
| (f)  Baggage | +  140 | | | | _____ |

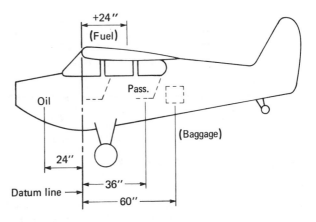

*Figure 6-34*

**9.** An aircraft has just undergone certain equipment changes and relocations. The original moment of the aircraft was 21,420 lb.-in. Use the values in Table 6-7 to find the new moment. (Study Examples 17 and 18 in the text before starting this problem.)

*FORWARD* _____ *REARWARD*

Landing light (new) (3 lb. at 72 in.)

Wheel pants (9 lb. at 12 in.)

Landing light (old) (4 lb. at 72 in.)

Radio (old) (19 lb. at 12 in.)

Fuel tank (30 lb. at 12 in.)

Radio (new) (16 lb. at 12 in.)

Rear seat (14 lb. at 36 in.)

DATUM

LINE

**TABLE 6-7**

| | *Weight* (lb.) $\times$ | *Arm* (in.) $=$ | *Moment* (lb.-in.) |
|---|---|---|---|
| *Added* | | | |
| Landing light (new) | $+\ 3$ | _____ | _____ |
| Fuel tank (new) | $+30$ | _____ | _____ |
| Radio (new) | $+16$ | _____ | _____ |
| Wheel pants | $+\ 9$ | _____ | _____ |
| *Removed* | | | |
| Radio (old) | $-19$ | _____ | _____ |
| Rear seat | $-14$ | _____ | _____ |
| Landing light (old) | $-\ 4$ | _____ | _____ |

Net change = _____    _____

New Moment = _____.

## SEC. 5   DIVIDING SIGNED NUMBERS

Every division problem is equivalent to a multiplication problem. The same rules of signs work for both. For instance,

1. $\dfrac{+100}{+\ 25} = +4$  because $+100 = (+4) \times (+25)$

2. $\dfrac{-75}{+\ 3} = -25$ because $-75 = (+3) \times (-25)$

3. $\dfrac{+180}{-\ 9} = -20$ because $+180 = (-9) \times (-20)$

4. $\dfrac{-240}{-\ 60} = +4$  because $-240 = (-60) \times (+4)$

### Rule for Dividing Signed Numbers

*When dividing two signed numbers, the quotient is* **positive** *if the numbers are of* **like** *sign. The quotient is* **negative** *if the numbers are of* **unlike** *sign.*

If only the signs are used, the rule is:

$$\left.\begin{array}{l} \dfrac{(+)}{(+)} = (+) \\[2mm] \dfrac{(-)}{(-)} = (+) \end{array}\right\} \text{like signs} \qquad \left.\begin{array}{l} \dfrac{(+)}{(-)} = (-) \\[2mm] \dfrac{(-)}{(+)} = (-) \end{array}\right\} \text{unlike signs}$$

Another statement of the rule is:

*When the dividend and divisor have the same sign, the quotient is positive. When the signs are different, the quotient is negative.*

### Practice Problems

Use the rule for dividing signed numbers. Do as many as you can without using paper and pencil.

1. $\dfrac{+180}{+10} = $ _____   $\dfrac{+160}{+20} = $ _____   $\dfrac{+86}{+43} = $ _____

2. $\dfrac{+170}{-17} = $ _____   $\dfrac{+16}{-8} = $ _____   $\dfrac{+42}{-2} = $ _____

3. $\dfrac{-186}{+3} = $ _____   $\dfrac{-2,000}{+2,000} = $ _____   $\dfrac{-288}{+12} = $ _____

4. $\dfrac{-45}{-9} = $ _____   $\dfrac{-81}{-9} = $ _____   $\dfrac{-300}{-60} = $ _____

5. Check answers 1 through 4 by multiplication.

Division of signed numbers is used to find either distances or forces in problems involving torque.

**Example 19:** How much force must be applied to the end of a torque wrench that has a handle 2 feet long to produce 100 pound-feet of torque?

*Solution:* From the formula for torque:

$$\text{(Torque) } T = \text{(Distance) } D \times \text{(Force) } F$$

$$T = D \times F$$

If the torque ($T$) is 100 pound-feet, and the distance ($D$) is 2 feet, find the force ($F$). From the formula you can see that the force is the result of dividing $T$ by $D$:

$$F = \frac{T}{D} \qquad \text{since } F \times D = T$$

Insert the given values:

$$F = \frac{100 \text{ lb.-ft.}}{2 \text{ ft.}}$$

$$F = 50 \text{ lb.}$$

*Answer:* 50 pounds of force must be applied to the handle of the wrench to produce the desired torque.

**Example 20:** When you use a pair of pliers you know that as you apply force to the handles you increase the holding force. In Fig. 6-35, a force of 50 pounds is applied to the handles 5 inches from the fulcrum (the point where the handles are connected). What is the resulting force 2 inches from the fulcrum?

*Solution:* The torque produced by gripping must equal the torque at the point of application.

$$(5 \text{ in.}) \times (50 \text{ lb}) = (2 \text{ in.}) \times (F \text{ lb.})$$

$$250 \text{ lb.-in.} = 2 \text{ in.} \times F \text{ lb.}$$

The value for $F$ is found by dividing 250 by 2. The result is:

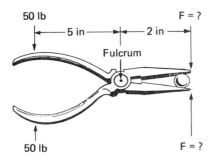

**Figure 6-35**

$$F = \frac{250 \text{ lb.-in.}}{2 \text{ in.}}$$

$$F = 125 \text{ lb.}$$

*Answer:* The gripping force of the jaws is 125 pounds. This is an increase of two and one-half times the applied force.

## COMMENTS

Man has long been aware of the mechanical advantage gained through the application of the lever and the fulcrum. Ever since he found that he could move a large object by using a pole and a rock, he has been improving his tools by including this principle. Figures 6-36, 6-37, 6-38, and 6-39, show some uses of the lever arm-fulcrum principle in today's tools.

**Figure 6-36** *Spanner wrench*

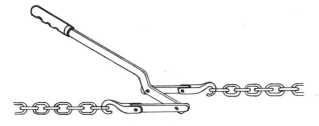

**Figure 6-37** *Chain binder*

**Figure 6-38**   *Shop pliers*

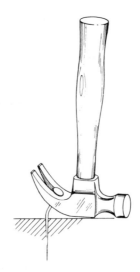

**Figure 6-39**   *Claw hammer*

## PROBLEMS: Sec. 5 Dividing Signed Numbers

**1.** Use the rules of signs to signs to find the quotients.

**Example**   $\dfrac{+45}{+15} = +3$   positive ÷ positive = positive

$\dfrac{-75}{-25} = +3$   negative ÷ negative = positive

$\dfrac{+60}{-20} = -3$   positive ÷ negative = negative

$\dfrac{-81}{+27} = -3$   negative ÷ positive = negative

(a) $\dfrac{+65}{+13}$   (b) $\dfrac{-85}{+17}$   (c) $\dfrac{+56}{-8}$   (d) $\dfrac{-100}{-25}$

(e) $\dfrac{-135}{+5}$   (f) $\dfrac{-132}{-12}$   (g) $\dfrac{+150}{-75}$   (h) $\dfrac{+180}{+60}$

**2.** Find the quotients.

(a) $\dfrac{190}{-5}$   (b) $\dfrac{-1,750}{-25}$   (c) $\dfrac{1,500}{-50}$

(d) $\dfrac{-875}{-25}$   (e) $\dfrac{-750}{+75}$   (f) $\dfrac{10,600}{-20}$

3. Combine the rules for multiplying and dividing signed numbers to find the value of these problems.

   **Example:** $\dfrac{(+24) \times (-56)}{(-7) \times (+12)} = \dfrac{-1,344}{-84} = +16$

   (a) $\dfrac{(-45) \times (-25)}{+75}$

   (b) $\dfrac{(+75) \times (-90)}{-45}$

   (c) $\dfrac{-100}{(-5) \times (-4)}$

   (d) $\dfrac{(-35) \times (-60) \times (+10)}{(-20) \times (-5) \times (-7)}$

4. A force of 50 pounds produces a torque of $-450$ lb-ft. How far is the force from the fulcrum?

5. An object was added to an airplane, producing a *negative* moment. Was the object placed forward or rearward of the datum line?

6. An object was removed from an airplane, producing a *positive* moment. Was the object located forward or rearward of the datum line?

7. A weather balloon recorded a drop in temperature of 56°F in rising 8,000 feet in 7 minutes. What was the rate of decrease in temperature in degrees per minute?

8. During a severe blizzard the temperature fell from 18°F above zero to 22°F below zero in 5 hours. What was the rate of temperature decrease in degrees per hour?

9. A Polaris missile was fired from a submarine 500 feet below the surface of the ocean. The missile rose to an altitude of 1,000 feet above sea level in 6 seconds. What was its rate of climb in feet per second?

10. Table 6-8 gives the engine-tightening specifications for a late-model automobile. Assume that a torque wrench with a handle two feet long will be used. Find the amount of force ($F$) to be applied to give the required torque value. Fill in the blanks in the table with the necessary force values. The first value (12 to 18 lbs.) is shown as an example. Note that it is necessary to *divide* the required torque values by the length of the handle (moment arm).

**TABLE 6-8** ENGINE TIGHTENING SPECIFICATIONS*

| | Spark plugs | Cylinder head bolts | Exhaust manifold | Intake manifold | Flywheel to crankshaft |
|---|---|---|---|---|---|
| | 24–36 lb.-ft. | 64–84 lb.-ft. | 14–20 lb.-ft. | 40–60 lb.-ft. | 50–70 lb.-ft. |
| Force ($F$) = 12–18 lb. | (a) _____ | (b) _____ | (c) _____ | (d) _____ | |

*Torque specifications for clean and lightly oiled threads only. Dry or dirty threads produce increased friction which prevents accurate measurement of tightness.

11. Figure 6-40 shows an engine rocker arm assembly. To open the valve to the proper position, 120 pound-inches of torque is needed. Use the dimensions on the figure to find the force needed on the push rod to open the valve.

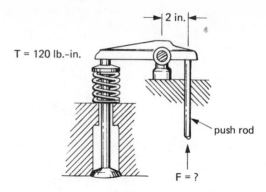

*Figure 6-40*

12.  In working some formulas you may have to do several operations using signed numbers. Find each value in these problems. Pay careful attention to the rules of signs.

**Example:** The value of:

$$\frac{(-4)(-6)}{(+10) + (-2)} = \frac{+24}{+8} = +3$$

Note that the multiplication in the numerator and the addition in the denominator are done before the division.

(a) $\dfrac{(-6) \times (+12)}{(-2) \times (-3) + (-5) \times (-6)}$    (b) $\dfrac{(-4) + (-10)}{(+3) \times (-5) + (+1)}$

(c) $\dfrac{(+12) \times (-6) + 10}{(-6)(-6) - (+5)}$    (d) $\dfrac{(-3)(-3) + (-4)}{(-4) \times (-5) - (5)(5)}$

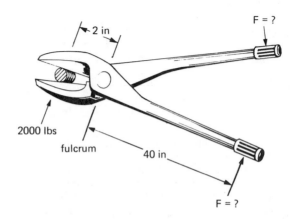

*Figure 6-41*

13. An airplane descended from an altitude of 48,000 feet to an altitude of 16,000 feet in 16 minutes. Assume the rate of descent was constant. Use signed numbers to show the rate of descent.

14. How much force must be applied to the handles of the bolt cutter in Fig. 6-41 to produce a cutting force of 2,000 pounds?

## PROBLEMS: Secs. 1-5 Signed Numbers

1. Do the operations indicated for these problems.
   (a) $(-3) + (-4) - (-5) =$
   (b) $(5)(-2) + (-5)(2) =$
   (c) $(-26) - (-26) + 10 =$
   (d) $(-5)(2 + 3) + (-10) =$
   (e) $\dfrac{(4)(-16) + (-5)(-8)}{(-3)(4) + 4} =$
   (f) $\dfrac{(10)(-1) + (-3)(-6)}{-1 - (+1)} =$

2. Fill in the blanks with the correct word.
   (a) When adding numbers of like sign, find their numerical sum and attach the _____ sign.
   (b) When subtracting signed numbers, add the negative of the _____ to the _____.
   (c) When multiplying two signed numbers, the product is _____ if the signs are the same and _____ if the signs are different.
   (d) The rules for dividing signed numbers are the same as those for _____ of signed numbers.

3. During a football game, the fullback for the Honeydew Hornets scored the following gains and losses:

| Play number | Gain (+) | Loss (−) |
|:-----------:|:--------:|:--------:|
| 1  | 3  |   |
| 2  |    | 1 |
| 3  | 5  |   |
| 4  |    | 7 |
| 5  | 18 |   |
| 6  | 6  |   |
| 7  |    | 4 |
| 8  | 2  |   |
| 9  |    | 1 |
| 10 | 6  |   |

   (a) How many yards did he gain?
   (b) How many yards did he lose?
   (c) What was his net yardage for the ten plays?

4. Apply Kirchhoff's law to the loop circuit in Fig. 6-42. (NOTE: The sum of voltage gains and losses must add up to zero.)

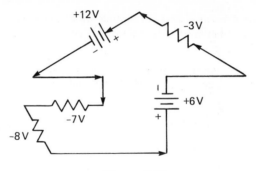

**Figure 6-42**

5.  Apply Kirchhoff's law to the loop circuit to find the value for the unknown emf X in Fig. 6-43.

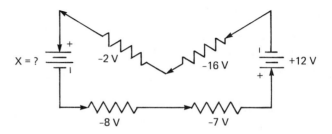

**Figure 6-43**

6.  Table 6-9 gives electric current requirements (in amps) for the electrical equipment on an automobile.

**TABLE 6-9**

| Equipment item | Current required (amps) |
|---|---|
| 1. Turn signal lights | 5 |
| 2. Door lights (courtesy-dome) | 3 |
| 3. Head lights (hi-beam) | 14 |
| 4. Head lights (lo-beam) | 13 |
| 5. Electric window (up or down) | 5 |
| 6. Air conditioner | 10 |
| 7. Heater, low | 8 |
| 8. Heater, high | 10 |
| 9. De-icer | 16 |
| 10. Power seat | 15 |
| 11. Radio | 1 |
| 12. Emergency warning lights | 14 |

In problems (a) through (f) some items are labeled ON. This means they are drawing current. These items are listed by the numbers in Table

6-9. The current produced by the alternator is indicated by (+___amps). In each problem find the net current drain on the battery.

**Example:** Items ON 1, 4, 7, 11.

$$(\text{current drain} = -5 + (-13) + (-8) + (-1)$$
$$= -27 \text{ amps}$$

Alternator output = +21 amps.
*Net* current drain = $-27 + (+21) = -6$ amps.

(a) Items ON 3, 6, 7, 9.
   Alternator output = +28 amps.
(b) Items ON: 2, 6, 11, 12.
   Alternator output = +9 amps.
(c) Items ON: 3, 6, 5, 12.
   Alternator output = +26 amps.
(d) Items ON: 4, 7, 10, 11.
   Alternator output = +16 amps.
(e) Items ON: 2, 4, 11, 12.
   Alternator output = 0 amps (engine off).
(f) Items ON: 5, 6, 11.
   Alternator output = +10 amps.

7. In a *closed traverse* survey the sum of the angles of depression and elevation must add up to zero. An angle measured below the horizontal is called an *angle of depression*, and it has a negative value. An angle measured above the horizontal is called an *angle of elevation*, and it has a positive value. See Fig. 6-44. Bungling Buford did a closed traverse survey. Figure 6-45 is a sketch of his survey. Do you agree with his values?

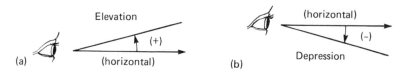

**Figure 6-44**

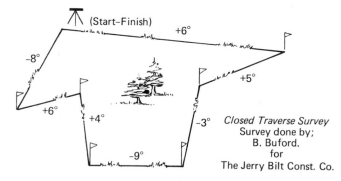

**Figure 6-45**

8. The fulcrum for a railway crane is on a line midway between the rails. The counterweight **W** weighs 28 tons and is located 6 feet from the fulcrum. When the crane is in the position shown in Fig. 6-46, find the maximum load **L** that can be lifted. (Use the principle derived for the see-saw.)

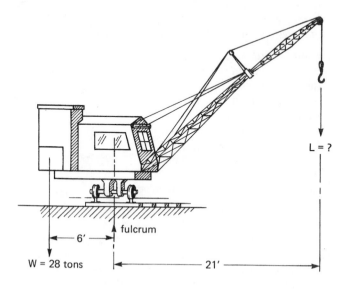

L = ?

fulcrum

← 6' →

W = 28 tons      |←———— 21' ————→|

**Figure 6-46**

9. The torque of an engine (gas or diesel) is the force exerted *one foot* from the center of the crankshaft. If an engine develops 100 pound-feet of torque, it exerts a force of 100 pounds one foot from the center of the crankshaft. What is the force (in pounds) one foot from the crankshaft of a diesel engine producing 750 pound-feet of torque?

10. Automotive engineers use a formula for finding the torque developed by an engine. The formula uses the rpm (revolutions-per-minute) and the brake horsepower (**BHP**) (the actual power available at the flywheel). The formula is:

$$\text{Torque } (T) = \frac{5,252 \times \text{BHP}}{\text{rpm}} \quad \text{(pound-feet)}$$

**Example:** Find the torque produced by an engine which develops 200 BHP at 1,300 rpm.

*Solution:* Substituting into the formula gives:

$$T = \frac{5,252 \times 200}{1,300} = \frac{1,050,400}{1,300} = 808 \text{ lb.-ft.}$$

*Answer:* 808 pound-feet.

(a) Find the torque produced by an engine that develops 300 BHP at 2,600 rpm.

(b) Find the torque produced by an engine that develops 200 BHP at 800 rpm.

(c) Find the BHP for an engine that develops 520 pound-feet of torque at 1,010 rpm.

11. Disc type brakes are on many cars and trucks. Use the idea of torque to measure the stopping ability of a disc brake. To find the torque created by the pad pressing against the disc find the distance from the axle to the pad. Multiply this distance by the frictional force of the pad on the disc. In Fig. 6-47 the distance from the center of the axle to the pad is 6 inches. The frictional force exerted by the pad on the disc is 1,500 pounds. Find the torque produced.

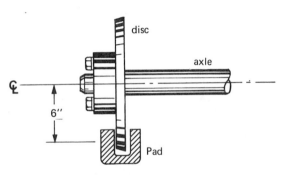

disc

axle

₵

6″

Pad

**Figure 6-47**

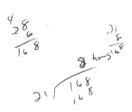

## Practice Test for Chapter 6

1. Add the signed numbers.
   (a) $(+16) + (-5) + (+8) =$ _____
   (b) $(-14) + (+15) + (-23) + (-4) =$ _____

2. Find the value of:
   (a) $(-4)(+7) + (-3)(-6) =$ _____
   (b) $\dfrac{(+4)(-5) - (-8)(-4)}{(+6) - (+10)} =$ _____

3. At altitudes above 10,000 feet the temperature decreases by 2°F for each 500 feet of additional altitude. If the temperature is −18°F at 10,000 feet, what is the temperature at 28,500 feet?

4. A man tries to lift a 350-pound casting by applying the lever-fulcrum principle. If the length of force arm is 7 feet, how much force ($F$) must he apply? See Fig. 6-48.

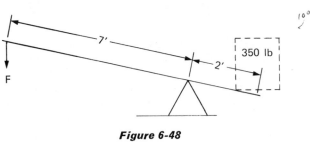

**Figure 6-48**

5. Verify Kirchhoff's law for the loop circuit shown in Fig. 6-49.

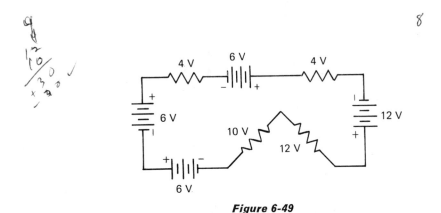

**Figure 6-49**

6. Liquid oxygen boils (that is, changes from a liquid to a gas) at $-329°F$. Ordinary water boils at $+212°F$. How many degrees difference is there between the boiling points of liquid oxygen and water?

7. Find the moment for the light plane. Use Table 6-10.

8. How much torque is exerted on a nut by a wrench 18 inches long with a force of 45 pounds at the end of the wrench?

9. The pad on a disc brake is 6 inches from the axle. How much force must be applied to the pad to produce 7,650 pound-inches of torque?

**TABLE 6-10**

| | *Weight (in pound)* $\times$ *Arm (in inches)* = *Moment (in lb.-in.)* | | |
|---|---|---|---|
| Aircraft | +1,478 | +24 | = _____ |
| Oil | +17 | −36 | = _____ |
| Battery (remove) | −28 | +60 | = _____ |
| Battery (new) | +26 | −24 | = _____ |
| Fuel tank (reserve) | +27 | −30 | = _____ |
| Fuel in reserve tank | +48 | −30 | = _____ |
| Remove rear seat | −64 | −18 | = _____ |
| | Total moment of aircraft = | | _____ |

# 7

# Common Fractions, Ratios, Proportions, and Averages

## SEC. 1   FRACTIONS

In many problems in the shop or on the job you will have to use fractions. In this section you will learn

1. what fractions are,
2. how to recognize when two or more fractions are equal,
3. how to express fractions in simplest form, and
4. how to express the same fraction in different ways.

First, you will study what fractions are and what they represent. Fractions occur when a counting process involves a whole (unit) quantity and equal parts of the whole. To see what is involved, use the number line that you studied earlier.

Figure 7-1 shows a unit of a number line that is divided into four parts of equal length. Select two of the four equal parts. In Fig. 7-2 the

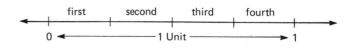

*Figure 7-1*

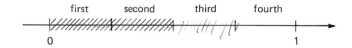

*Figure 7-2*

first and second parts are shown by the slant lines (////////). The fact that two of the four equal parts have been chosen is shown by using the symbol $\frac{2}{4}$. This symbol is called a *fractional number*, or a *fraction*. It is read *"two-fourths."*

A fraction is a number that lets you show that a certain number of equal subdivisions of a whole have been chosen. If you take three of the equal parts of the line segment, show it by the fraction $\frac{3}{4}$ (*three-fourths*).

If you take only one part, show it by the fraction $\frac{1}{4}$ (*one-fourth*), and so on.

In each of the fractions $\frac{1}{4}, \frac{2}{4}, \frac{3}{4}$, the number above the bar is called the *numerator* of the fraction. The number (4) below the bar is called the *denominator* of the fraction. The numerator of a fraction shows the number of equal subdivisions of the whole that have been selected. The denominator shows the *total* number of subdivisions of the whole. For example, the fraction $\frac{7}{8}$ means that 7 of 8 equal parts of the whole are chosen. In similar fashion:

$\frac{2}{3}$ means that 2 of 3 equal parts of the whole are chosen.

$\frac{5}{16}$ means that 5 of 16 equal parts of the whole are chosen.

$\frac{0}{6}$ means that none (zero) of 6 equal parts of the whole are chosen.

In a fraction:

$$\frac{N}{D} \begin{array}{l} \longleftarrow \text{numerator} \\ \longleftarrow \text{denominator} \end{array}$$

The numerator $N$ shows the number of equal subdivisions of the whole that are chosen. The denominator $D$ shows the **total** number of subdivisions of the whole.

Look at the number line again, and this time take 3 segments of *equal* length as shown in Fig. 7-3. Notice that each segment is subdivided into 3 equal parts. From the first segment, take 2 of the 3 equal parts. From the second segment, take all 3 of the equal parts. From the third segment, take 1 of the 3 equal parts. The selections are shown in Fig. 7-4. Each

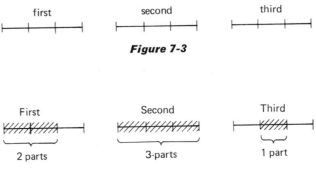

part represents $\frac{1}{3}$ (one-third) of a whole unit, and 6 equal subdivisions are chosen. This is shown by the fraction with numerator 6 and denominator 3, which is written $\frac{6}{3}$.

When the numerator of a fraction is larger than the denominator, the fraction is called an *improper fraction*. For instance, the *improper fraction* $\frac{8}{3}$ means that 8 parts from several equal whole units are chosen, with each unit divided into 3 equal subdivisions.

The *improper fraction* $\frac{9}{5}$ means that 9 parts from several equal whole units are chosen, with each unit divided into 5 equal subdivisions.

When all of the equal parts of a unit are chosen, the numerator and denominator of the fraction are equal. Thus, fractions such as $\frac{2}{2}, \frac{5}{5}, \frac{8}{8}$, and $\frac{10}{10}$ mean that the total unit is chosen. The whole unit, however, can be named by the number one (1).

Therefore, *fractions with numerator and denominator equal are names for the whole number 1:*

$$\frac{N}{N} = 1$$

Here are some ways that fractions are used in technology.
1. A container holds 8 quarts of oil. If 3 quarts are drained off, write the amount removed as a fraction of the total.

   Answer:   Of the 8 equal parts, 3 are chosen. This is written $\frac{3}{8}$.

2. A piece of bar stock is divided into 12 pieces of equal length. A machin-

ist uses 7 of the pieces. What fraction shows the number of parts used?

Answer: Of the 12 equal parts, the machinist uses 7. This is written $\frac{7}{12}$.

Fractions let you express that portion of a whole that is chosen. To see how several fractions can all represent the same amount of the total, look at the number line. Take a segment of the number line. Divide it into 2 equal subdivisions, as in Fig. 7-5. Choose one of the subdivisions.

**Figure 7-5**

Each subdivision represents $\frac{1}{2}$ (one-half) of the total. Now, subdivide the segment into 4 equal subdivisions, as shown in Fig. 7-6. Choose 2 of the subdivisions. You can see from these figures that the fraction $\frac{1}{2}$ and the fraction $\frac{2}{4}$ are the same. If you subdivide the segment into 6, 8, 10, 12,

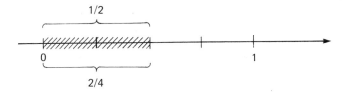

**Figure 7-6**

etc., number of parts and choose the same portion of the segment each time, you will see that the fractions

$$\frac{1}{2} \quad \frac{2}{4} \quad \frac{3}{6} \quad \frac{4}{8} \quad \frac{5}{10} \quad \frac{6}{12} \quad \frac{7}{14} \quad \cdots$$

all represent the same amount.

From this observation the following statement can be made:

*When two or more fractions represent the same portion of the total, the fractions are called EQUAL.*

Therefore:

$$\frac{1}{2} = \frac{2}{4} = \frac{3}{6} = \frac{4}{8} = \frac{5}{10} = \frac{6}{12} = \frac{7}{14} \text{ and so on.}$$

Similar reasoning shows that:

$$\frac{1}{3} = \frac{2}{6} = \frac{3}{9} = \frac{4}{12} = \frac{5}{15} = \frac{6}{18} \text{ and so on}$$

and

$$\frac{1}{4} = \frac{2}{8} = \frac{3}{12} = \frac{4}{16} = \frac{5}{20} = \frac{6}{24} \text{ and so on.}$$

Now, look again at the equal fractions:

$$\frac{1}{2} \quad \frac{2}{4} \quad \frac{3}{6} \quad \frac{4}{8} \quad \frac{5}{10} \quad \frac{6}{12} \quad \cdots$$

Do you see that in each case:

$$\frac{1}{2} = \frac{1 \times 2}{2 \times 2} = \frac{2}{4} ?$$

$$\frac{1}{2} = \frac{1 \times 3}{2 \times 3} = \frac{3}{6} ?$$

$$\frac{1}{2} = \frac{1 \times 4}{2 \times 4} = \frac{4}{8} ?$$

$$\frac{1}{2} = \frac{1 \times 5}{2 \times 5} = \frac{5}{10} ?$$

From this observation, it is possible to state an important rule about equal fractions:

### Rule for Obtaining Equal Fractions by Multiplication

*Multiplying the numerator and denominator of a fraction by the **same** nonzero number results in another equal fraction.*

Thus:

$$\frac{N}{D} = \frac{N \times \text{(any nonzero number)}}{D \times \text{(same nonzero number)}}$$

**Example 1:** Find four fractions equal to the fraction $\frac{2}{5}$.

*Answer:*  $\frac{2}{5} = \frac{2 \times 3}{5 \times 3} = \frac{6}{15}$  (Multiply numerator and denominator by 3.)

$\frac{2}{5} = \frac{2 \times 4}{5 \times 4} = \frac{8}{20}$  (Multiply numerator and denominator by 4.)

$\frac{2}{5} = \frac{2 \times 6}{5 \times 6} = \frac{12}{30}$  (Multiply numerator and denominator by 6.)

$\frac{2}{5} = \frac{2 \times 10}{5 \times 10} = \frac{20}{50}$  (Multiply numerator and denominator by 10.)

Multiply BOTH numerator and denominator by the **same** number. If you do not, the result is an unequal fraction.

Instead of multiplying both numerator and denominator by the same nonzero number, you can *divide* both by the same nonzero number. The result is a fraction equal to the original. To see this look at the equal fractions:

$$\frac{3}{9} \quad \frac{5}{15} \quad \frac{6}{18}$$

All of these fractions are equal to $\frac{1}{3}$ because:

When both numerator and denominator of $\frac{3}{9}$ are divided by 3,

the result is $\frac{1}{3}$.  $\quad \frac{3 \div 3}{9 \div 3} = \frac{1}{3}$

When both numerator and denominator of $\frac{5}{15}$ are divided by 5,

the result is $\frac{1}{3}$.  $\quad \frac{5 \div 5}{15 \div 5} = \frac{1}{3}$

When both numerator and denominator of $\frac{6}{18}$ are divided by 6,

the result is $\frac{1}{3}$.  $\quad \frac{6 \div 6}{18 \div 6} = \frac{1}{3}$

This result is stated as another rule of equal fractions:

### *Rule for Obtaining Equal Fractions by Dividing*

*Dividing the numerator and denominator of a fraction by the **same** nonzero number results in another equal fraction.*

Thus:

$$\frac{N}{D} = \frac{N \div \text{(any nonzero number)}}{D \div \text{(same nonzero number)}}$$

These two rules are often combined in the single statement:

*Multiplying or dividing both numerator and denominator of a fraction by the **same** nonzero number results in an equal fraction.*

**Example 2:** Find four fractions equal to the fraction $\frac{18}{36}$.

*Answer:* 1. *Divide* both numerator and denominator by 2:

$$\frac{18 \div 2}{36 \div 2} = \frac{9}{18}$$

2. *Divide* both numerator and denominator by 6:

$$\frac{18 \div 6}{36 \div 6} = \frac{3}{6}$$

3. Divide both numerator and denominator by 18:

$$\frac{18 \div 18}{36 \div 18} = \frac{1}{2}$$

4. Multiply both numerator and denominator by 5:

$$\frac{18 \times 5}{36 \times 5} = \frac{90}{180}$$

A fraction $\frac{N}{D}$ is in its *simplest form* when the **ONLY** number that will divide BOTH numerator and denominator is 1. Thus, the fractions $\frac{1}{4}$, $\frac{3}{5}$, $\frac{6}{7}$, $\frac{9}{16}$, $\frac{11}{32}$ are in their *simplest* form because the only number that will divide both the numerator and denominator of each fraction is 1.

## PROBLEMS: Sec. 1 Fractions

1. Fill in the blanks in these tables. The first row is filled in as an example.

**TABLE 7-1**

| | Fraction | Numerator | Denominator |
|---|---|---|---|
| | $\frac{2}{3}$ | 2 | 3 |
| (a) | $\frac{5}{8}$ | | |
| (b) | $\frac{7}{16}$ | | |
| (c) | $\frac{15}{32}$ | | |

**TABLE 7-2**

| | Fraction | Numerator | Denominator |
|---|---|---|---|
| | $\frac{17}{32}$ | 17 | 32 |
| (a) | $\frac{15}{64}$ | | |
| (b) | $\frac{?}{16}$ | 7 | 16 |
| (c) | $\frac{?}{?}$ | 11 | 64 |
| (d) | $\frac{25}{?}$ | | 128 |

2. Figures 7-7 through 7-11 show a unit divided into an equal number of parts. Find the *total number of parts for each unit and express the shaded portion as a fraction of the total.*

**Example:**

Total number of parts = 8.
Number of parts shaded = 5.
Fraction of total shaded = $\frac{5}{8}$.

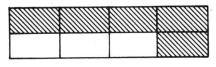

*Figure 7-7*

(a)

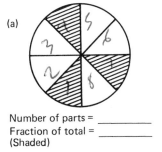

Number of parts = _____
Fraction of total = _____
(Shaded)

*Figure 7-8*

(b)

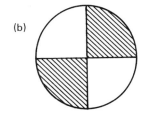

Number of parts = _____
Fraction of total = _____
(Shaded)

*Figure 7-9*

(c)

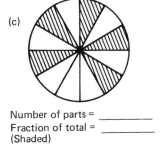

Number of parts = _____
Fraction of total = _____
(Shaded)

*Figure 7-10*

(d)

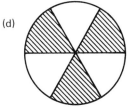

Number of parts = _____
Fraction of total = _____
(Shaded)

*Figure 7-11*

**3.** What fractional part of each figure is shaded?

(a)

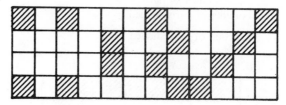

*Figure 7-12*     Fraction = _____

(b)

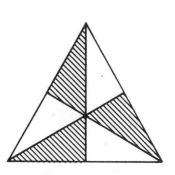

*Figure 7-13*     Fraction = _____

(c)

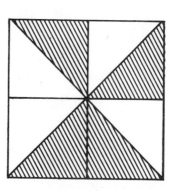

*Figure 7-14*     Fraction = _____

4. Figures 7-15 through 7-17 show a unit divided into an equal number of parts. Find the amount of one part and fill in the missing values on the drawing. Write all fractions in reduced form.

**Example:** Amount of one part is $\frac{1}{6}$ (one-sixth).

The values for *A*, *B*, and *C* are;

$$A = \frac{2}{6} = \frac{1}{3}; \quad B = \frac{3}{6} = \frac{1}{2}; \quad C = \frac{5}{6}$$

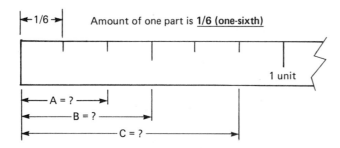

*Figure 7-15*

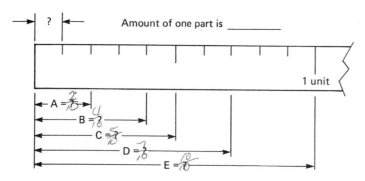

*Figure 7-16*

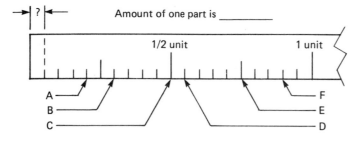

*Figure 7-17*

5.  Find the amount of one part. Determine the fractional amounts of *A*, *B*, and *C*. See Fig. 7-18.

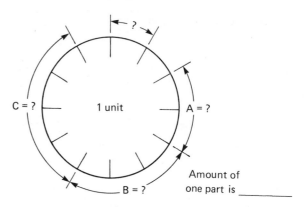

**Figure 7-18**

6.  Test yourself on reading the ruler. Put answers in the left-hand column. Write all fractions in the simplest form (Fig. 7-19).

7.  Each of the fractions in this problem can be reduced to lowest terms by dividing. Find the simplest form for each fraction.

    **Example:** $\dfrac{24 \div 12}{36 \div 12} = \dfrac{2}{3}$ (Divide numerator and denominator by 12.)

    (a) $\dfrac{45}{100}$   (b) $\dfrac{12}{16}$   (c) $\dfrac{14}{32}$   (d) $\dfrac{26}{64}$   (e) $\dfrac{5}{10}$   (f) $\dfrac{36}{64}$   (g) $\dfrac{50}{250}$

    (h) $\dfrac{50}{25}$   (i) $\dfrac{16}{16}$   (j) $\dfrac{125}{1,000}$

8.  Multiply both numerator and denominator by the factors 2, 4, 8, and 16 to get four fractions equal to each of the fractions.

    **Example:** Given $\dfrac{1}{2}$. Equal fractions are $\dfrac{1 \times 2}{2 \times 2} = \dfrac{2}{4}$

    $$\dfrac{1 \times 4}{2 \times 4} = \dfrac{4}{8} \qquad \dfrac{1 \times 8}{2 \times 8} = \dfrac{8}{16} \qquad \dfrac{1 \times 16}{2 \times 16} = \dfrac{16}{32}$$

    (a) $\dfrac{3}{4}$   (b) $\dfrac{5}{8}$   (c) $\dfrac{7}{10}$   (d) $\dfrac{3}{16}$   (e) $\dfrac{7}{2}$

TEST on reading the RULE

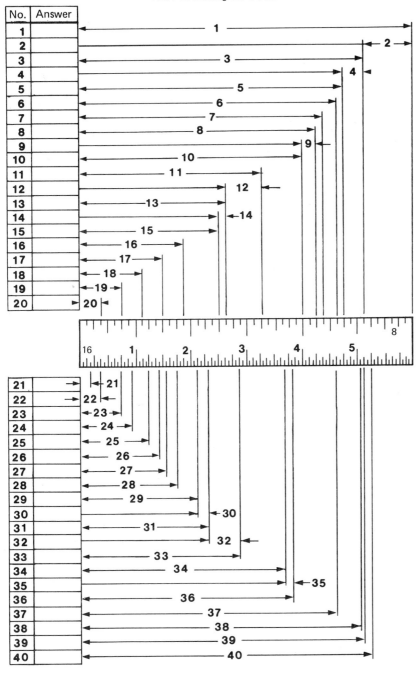

*Figure 7-19*

**9.** Reduce each fraction to its simplest form.

**Example:** $\dfrac{10}{16} = \dfrac{10 \div 2}{16 \div 2} = \dfrac{5}{8}$ (Divide numerator and denominator by 2.)

(a) $\dfrac{2}{4}$  $\dfrac{6}{12}$  $\dfrac{8}{16}$  $\dfrac{7}{14}$  $\dfrac{5}{10}$  $\dfrac{4}{8}$  $\dfrac{9}{18}$  $\dfrac{3}{6}$

(b) $\dfrac{2}{6}$  $\dfrac{6}{18}$  $\dfrac{5}{15}$  $\dfrac{7}{21}$  $\dfrac{8}{24}$  $\dfrac{4}{12}$  $\dfrac{9}{27}$  $\dfrac{3}{9}$

(c) $\dfrac{4}{16}$  $\dfrac{5}{20}$  $\dfrac{6}{24}$  $\dfrac{9}{36}$  $\dfrac{7}{28}$  $\dfrac{8}{32}$  $\dfrac{3}{12}$  $\dfrac{2}{8}$

(d) $\dfrac{4}{32}$  $\dfrac{8}{32}$  $\dfrac{16}{32}$  $\dfrac{18}{32}$  $\dfrac{14}{32}$  $\dfrac{12}{32}$  $\dfrac{10}{32}$

(e) $\dfrac{4}{16}$  $\dfrac{8}{16}$  $\dfrac{14}{16}$  $\dfrac{16}{16}$  $\dfrac{10}{16}$  $\dfrac{2}{16}$  $\dfrac{12}{16}$

(f) $\dfrac{6}{8}$  $\dfrac{8}{8}$  $\dfrac{4}{8}$  $\dfrac{2}{8}$  $\dfrac{10}{8}$  $\dfrac{14}{8}$

(g) $\dfrac{8}{10}$  $\dfrac{6}{15}$  $\dfrac{20}{36}$  $\dfrac{6}{20}$  $\dfrac{10}{25}$  $\dfrac{15}{30}$  $\dfrac{14}{18}$

(h) $\dfrac{9}{12}$  $\dfrac{20}{32}$  $\dfrac{24}{32}$  $\dfrac{16}{64}$  $\dfrac{32}{64}$  $\dfrac{24}{64}$  $\dfrac{48}{64}$

**Problems 10 through 15.** Reduce each fraction to its simplest form.

**10.** (a) $\dfrac{2}{4}$ inch  (b) $\dfrac{14}{16}$ inch  (c) $\dfrac{6}{8}$ inch  (d) $\dfrac{22}{32}$ inch  (e) $\dfrac{40}{64}$ inch

**11.** (a) $\dfrac{4}{8}$ foot  (b) $\dfrac{6}{12}$ foot  (c) $\dfrac{3}{12}$ foot  (d) $\dfrac{18}{36}$ yard  (e) $\dfrac{24}{36}$ yard

**12.** (a) $\dfrac{220}{1,760}$ mile  (b) $\dfrac{1,760}{5,280}$ mile  (c) $\dfrac{3,960}{5,280}$ mile

**13.** (a) $\dfrac{6}{16}$ pound  (b) $\dfrac{8}{16}$ pound  (c) $\dfrac{14}{16}$ pound  (d) $\dfrac{20}{32}$ quart

**14.** (a) $\dfrac{30}{60}$ hour  (b) $\dfrac{36}{60}$ minute  (c) $\dfrac{14}{24}$ day  (d) $\dfrac{9}{24}$ day  (e) $\dfrac{8}{12}$

**15.** (a) $\dfrac{15}{100}$ centimeter  (b) $\dfrac{150}{1,000}$ kilometer  (c) $\dfrac{850}{1,000}$ liter

**Problems 16 through 32 involve the use of fractional parts.** Express each answer as a fraction reduced to its simplest form.

**Example:** Thirty-six minutes (36 min) is what fractional part of an hour (60 min)?

*Solution:* The whole unit is 60 minutes (1 hour). The number of equal parts of the whole is 36. Express this as a fraction:

*Answer:* Fractional part $= \dfrac{36}{60} = \dfrac{36 \div 12}{60 \div 12} = \dfrac{3}{5}$ hr.

**16.** Six inches is what fractional part of a foot? (1 ft. = 12 in.)

**17.** Eight hours is what fractional part of a day? (1 day = 24 hr.)

**18.** Two feet is what fractional part of a yard? (1 yard = 3 ft.)

**19.** Sixteen inches is what fractional part of a yard? (1 yd = 36 in.)

**20.** Eighteen inches is what fractional part of a foot?

21. Three months is what fractional part of a year? (1 year = 12 months.)

22. Thirty centimeters is what fractional part of a meter? (1 meter = 100 cm.)

23. Twelve ounces is what fractional part of a pound? (1 pound = 16 oz.)

24. During a track meet at a local community college the following races were run:
    (a) 220 yard run                (b) 440 yard run
    (c) 880 yard run                (d) 1,760 yard run
    Write each of these distances as fractional parts of a mile. (1 mile = 1,760 yards.)

25. A carpenter cut off 8 feet from a board 14 feet long.
    (a) What fractional part of the board did he cut off?
    (b) What fractional part was left?

26. Mr. Whitney drives a delivery truck. On Monday he filled the gas tank, which has a capacity of 24 gallons. At the end of the week he again filled the tank, and it took 18 gallons of gas. Write the amount of gas used as a fractional part of a full tank.

27. The average person spends 8 hours a day sleeping, 2 hours eating, and 3 hours relaxing. Write each amount of time spent on these activities as fractional parts of a day. (1 day = 24 hours)

28. Last week Mr. Fujii spent $36 for food, $9 for gas, $18 for clothes, $8 for entertainment, and $19 for other expenses. Write each amount spent as a fraction of the total amount spent.

29. Last season the Milpitas Maulers won 18 games and lost 12. What fraction of *total* games played did they win? What fraction of *total* games played did they lose?

30. On the last history test Henni got 34 correct answers out of a possible 42. What fraction of the test did she get wrong?

31. In a class taking Technical Mathematics, 24 students are enrolled in auto mechanics, 6 in diesel mechanics, 9 in machine shop, 3 in metal shop, and 3 in other trades. Show the number of students enrolled as a fraction of the total class.

32. An AC Transit bus left Hayward at 12:30 P.M. It arrived in Berkeley at 2:04 P.M. The total amount of time for all stops was 16 minutes. What fraction of the total travel time was used at stops?

# SEC. 2   MORE ABOUT FRACTIONS

In Section 1 of this chapter you learned what fractions are. You also learned how to get equal fractions from other fractions by multiplying or dividing numerator and denominator by the same nonzero number. In this section you will learn to compare fractions to find which is larger and which is smaller. You will also study methods of combining whole numbers with fractions.

You know that two fractions are equal when they represent the same portion of the total unit. Thus $\frac{4}{10}$ and $\frac{6}{15}$ are equal since they both equal the fraction $\frac{2}{5}$. Instead of reducing both fractions to their simplest form to see the equality, you can do this:

Multiply both numerator and denominator of $\frac{4}{10}$ by 3 to get

$$\frac{4 \times 3}{10 \times 3} = \frac{12}{30}.$$

Multiply both numerator and denominator of the fraction $\frac{6}{15}$ by 2 to get $\frac{6 \times 2}{15 \times 2} = \frac{12}{30}.$

Since both results give the same fraction $\frac{12}{30}$, the original fractions are equal. In this case you found a *common denominator* for the two fractions by multiplication. This method of finding a common denominator of two or more fractions by multiplication is an important procedure. Study it carefully.

**Example 3** (a): Write the fractions $\frac{1}{2}$, $\frac{2}{3}$, and $\frac{3}{4}$ with a common denominator.

*Solution* (a): Multiply both the numerator and denominator by the numbers 6, 4, and 3. The result is a common denominator of 12 for the three fractions.

*Answer* (a):
$$\frac{1}{2} = \frac{1 \times 6}{2 \times 6} = \frac{6}{12}$$
$$\frac{2}{3} = \frac{2 \times 4}{3 \times 4} = \frac{8}{12}$$
$$\frac{3}{4} = \frac{3 \times 3}{4 \times 3} = \frac{9}{12}$$

**Example 3** (b): Write the fractions $\frac{3}{4}$, $\frac{5}{6}$, and $\frac{7}{9}$ with a common denominator.

*Solution* (b): Write each of the fractions as a fraction with denominator 36.

*Answer* (b):
$$\frac{3}{4} = \frac{3 \times 9}{4 \times 9} = \frac{27}{36}$$
$$\frac{5}{6} = \frac{5 \times 6}{6 \times 6} = \frac{30}{36}$$
$$\frac{7}{9} = \frac{7 \times 4}{9 \times 4} = \frac{28}{36}$$

The smallest common denominator is called the *least common denominator* (LCD). The least common denominator (LCD) is not always an easy number to find when there are several fractions involved. Fortunately, for most of your work with fractions you need to find only **any** common denominator. Whether it is the smallest or not will not make much difference.

**Example 4:** Find a common denominator for the fractions $\frac{2}{5}, \frac{3}{8}, \frac{5}{6}$.

*Solution:* Find a number that 5, 6, and 8 will divide into evenly. Such a number is their product: $5 \times 6 \times 8 = 240$. Use this as a common denominator.

*Answer:* $\frac{2}{5} = \frac{2 \times 48}{5 \times 48} = \frac{96}{240}$  (NOTE $240 \div 5 = 48$)

$\frac{5}{6} = \frac{5 \times 40}{6 \times 40} = \frac{200}{240}$  (NOTE $240 \div 6 = 40$)

$\frac{3}{8} = \frac{3 \times 30}{8 \times 30} = \frac{90}{240}$  (NOTE $240 \div 8 = 30$)

NOTE: Actually, the number 120 is the LCD because 5, 6, and 8 all divide it evenly, and no smaller number will work.

Fractions are written with the same denominator in order to compare their size. From the definition of a fraction, you see that *if two fractions have the* **same** *denominator, the one with the larger numerator is the larger fraction.* Thus $\frac{23}{25}$ is larger than $\frac{21}{25}$ because $\frac{23}{25}$ means that 23 parts out of 25 are chosen. $\frac{21}{25}$ means that 21 parts out of 25 are chosen. Since 23 is larger than 21, the fraction $\frac{23}{25}$ represents more than the fraction $\frac{21}{25}$.

When two fractions are compared for size and you know which is the larger, you can show this by using the symbol $>$ in this way:

The symbol $>$ means *is greater than* and is placed between the numbers. The point is toward the smaller of the pair. For instance,

$$\frac{23}{25} > \frac{21}{25}$$

When the symbol is reversed, it means *less than:*

$$\frac{21}{25} < \frac{23}{25}$$

Note that the point is always toward the smaller number.

Here is how you use the ideas just discussed to compare two or more fractions.

**Example 5:** Compare the fractions $\frac{2}{3}$ and $\frac{3}{4}$.

*Solution:* Write both fractions with a common denominator of 12.

$$\frac{2}{3} = \frac{2 \times 4}{3 \times 4} = \frac{8}{12}$$

$$\frac{3}{4} = \frac{3 \times 3}{4 \times 3} = \frac{9}{12}$$

Since $\frac{8}{12}$ is less than $\frac{9}{12}$ then,

*Answer:* $\frac{2}{3} < \frac{3}{4}$

**Example 6:** Arrange the fractions $\frac{3}{4}$, $\frac{5}{6}$, $\frac{7}{9}$ in *increasing* order.

*Solution:* From Example 3 (b) in this section you know that:

$$\frac{3}{4} = \frac{27}{36} \qquad \frac{5}{6} = \frac{30}{36} \qquad \frac{7}{9} = \frac{28}{36}$$

*Answer:*  $\qquad\qquad\qquad \frac{3}{4} < \frac{7}{9} < \frac{5}{6}$

You must know how to combine whole numbers with fractions. When a whole number and a fraction are shown together, it is called a *mixed number*. Numbers such as:

$$3\frac{1}{2} \qquad 4\frac{1}{4} \qquad 5\frac{7}{8} \qquad 17\frac{9}{16}$$

are mixed numbers. A mixed number such as $2\frac{1}{2}$ means that 2 whole units and $\frac{1}{2}$ are taken. The ruler in Fig. 7-20 shows measurements of $3\frac{1}{2}$ inches and $5\frac{1}{4}$ inches.

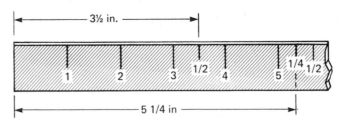

*Figure 7-20*

Mixed numbers which show the combination of whole numbers with fractions may be written as improper fractions. For example, look at the mixed number $3\frac{1}{2}$. To express this in fraction form, note that the whole number 3 $\left(\text{which may be written as } \dfrac{3}{1}\right)$ is equal to the fraction $\dfrac{6}{2}$ because

$$\frac{3}{1} = \frac{3 \times 2}{1 \times 2} = \frac{6}{2}$$

The mixed number $3\frac{1}{2}$ now represents a total of $\dfrac{6}{2} + \dfrac{1}{2}$ (six halves plus one-half) or $\dfrac{7}{2}$ in all. Therefore, $3\frac{1}{2} = \frac{7}{2}$.

A mixed number is changed to an improper fraction by this process.

### Changing Mixed Numbers to Improper Fractions

*Multiply the whole number by the denominator of the fraction. Add this product to the numerator of the fraction. Write this sum over the denominator of the fraction.*

**Example 7:** Change the mixed number $4\frac{1}{3}$ to an improper fraction.

*Solution:* Multiply the whole number 4 by the denominator 3. Add the product (12) to the numerator (1) of the fraction which is 13. This sum is the numerator of the improper fraction. The number 3 is the denominator.

*Answer:* $\qquad\qquad 4\frac{1}{3} = \dfrac{(4 \times 3) + 1}{3} = \dfrac{13}{3}$

When changing a mixed number, such as $5\frac{3}{8}$, to an improper fraction, say to yourself: "Multiply the 5 by the 8 and add the 3. This gives 43 for the numerator. The denominator of the fraction is the 8. Therefore, $5\frac{3}{8}$ is equal to $\dfrac{43}{8}$."

$$5\frac{3}{8} = \frac{43}{8}$$

Use this process to change these mixed numbers to improper fractions.

**Example 8:** (a) $4\dfrac{3}{5} = \dfrac{(5 \times 4) + 3}{5} = \dfrac{20 + 3}{5} = \dfrac{23}{5}$

(b) $10\dfrac{1}{4} = \dfrac{(4 \times 10) + 1}{4} = \dfrac{40 + 1}{4} = \dfrac{41}{4}$

(c) $5\dfrac{3}{16} = \dfrac{(16 \times 5) + 3}{16} = \dfrac{80 + 3}{16} = \dfrac{83}{16}$

### Practice Problems

Are the following answers correct?

**1.** $5\dfrac{2}{3} = \dfrac{17}{3}$     **5.** $8\dfrac{5}{32} = \dfrac{261}{32}$

**2.** $6\dfrac{3}{8} = \dfrac{51}{8}$     **6.** $3\dfrac{6}{10} = \dfrac{36}{10}$

**3.** $11\dfrac{5}{16} = \dfrac{181}{16}$     **7.** $25\dfrac{3}{25} = \dfrac{628}{25}$

**4.** $9\dfrac{11}{64} = \dfrac{587}{64}$     **8.** $35\dfrac{15}{64} = \dfrac{2,255}{64}$

Improper fractions may be changed to mixed numbers. To see how to do this, look at the improper fraction $\dfrac{27}{5}$. This fraction shows a total of twenty-seven fifths. Since there are five-fifths in each whole unit, twenty-five fifths represents five whole units:

$$\frac{25}{5} = \frac{5}{1} = 5$$

Since there are twenty-seven fifths in all, then there are five whole units and two fifths left over. This is written as the mixed number $5\dfrac{2}{5}$.

This process is stated as:

### Rule for Changing Improper Fractions to Mixed Numbers

*To change an improper fraction to a mixed number, divide the numerator by the denominator to find the number of whole units in the fraction. If the division has a remainder, write the remainder over the denominator to show the fraction left over.*

**Example 9:** Change the improper fraction $\dfrac{35}{8}$ to a mixed number.

*Solution:* Divide 35 by 8 to get a quotient of 4 and a remainder of 3. The 4 is the number of whole units in the fraction. The 3 shows that there are three eighths left over.

*Answer:* $\dfrac{35}{8} = 4\dfrac{3}{8}$

Sometimes there is no remainder. In this case, the improper fraction shows a whole number of units. For instance, the improper fraction $\dfrac{42}{6}$ is equal to the whole number 7 because

$$\frac{42}{6} = \frac{42 \div 6}{6 \div 6} = \frac{7}{1} = 7$$

### Practice Problems

In each of these problems, the improper fractions have been changed to mixed numbers. Prove the results.

**1.** (a) $\frac{5}{2} = 2\frac{1}{2}$  (b) $\frac{7}{4} = 1\frac{3}{4}$  (c) $\frac{9}{4} = 2\frac{1}{4}$  (d) $\frac{7}{2} = 3\frac{1}{2}$

**2.** (a) $\frac{11}{4} = 2\frac{3}{4}$  (b) $\frac{17}{8} = 2\frac{1}{8}$  (c) $\frac{19}{16} = 1\frac{3}{16}$  (d) $\frac{23}{16} = 1\frac{7}{16}$

**3.** (a) $\frac{25}{8} = 3\frac{1}{8}$  (b) $\frac{45}{32} = 1\frac{13}{32}$  (c) $\frac{33}{16} = 2\frac{1}{16}$  (d) $\frac{53}{32} = 1\frac{21}{32}$

**4.** (a) $\frac{45}{16} = 2\frac{13}{16}$  (b) $\frac{75}{4} = 18\frac{3}{4}$  (c) $\frac{35}{2} = 17\frac{1}{2}$  (d) $\frac{53}{4} = 13\frac{1}{4}$

**5.** (a) $\frac{145}{8} = 18\frac{1}{8}$  (b) $\frac{97}{64} = 1\frac{33}{64}$  (c) $\frac{32}{8} = 4$  (d) $\frac{64}{4} = 16$

**6.** (a) $\frac{345}{16} = 21\frac{9}{16}$  (b) $\frac{547}{8} = 68\frac{3}{8}$  (c) $\frac{237}{2} = 118\frac{1}{2}$

## PROBLEMS: Sec. 2   More About Fractions

**Problems 1 through 15.** (a) State the least common denominator for the pair of fractions. (b) Use this least common denominator to find fractions equal to the given fractions.

**Example:** Given fractions $\frac{2}{3}$ $\frac{4}{5}$.

(a) Least common denominator is $3 \times 5 = 15$

(b) $\frac{2}{3} = \frac{2 \times 5}{3 \times 5} = \frac{10}{15}$   $\frac{4}{5} = \frac{4 \times 3}{5 \times 3} = \frac{12}{15}$

**1.** $\frac{1}{2}$ $\frac{1}{4}$  (a) _____  (b) _____

**2.** $\frac{1}{3}$ $\frac{1}{4}$  (a) _____  (b) _____

**3.** $\frac{2}{3}$ $\frac{5}{6}$  (a) _____  (b) _____

**4.** $\frac{1}{2}$ $\frac{1}{6}$  (a) _____  (b) _____

**5.** $\frac{5}{6}$ $\frac{1}{9}$  (a) _____  (b) _____

**6.** $\frac{3}{4}$ $\frac{5}{8}$  (a) _____  (b) _____

**7.** $\frac{7}{4}$ $\frac{7}{6}$  (a) _____  (b) _____

**8.** $\frac{1}{10}$ $\frac{3}{5}$  (a) _____  (b) _____

**9.** $\frac{7}{12}$ $\frac{1}{8}$  (a) _____  (b) _____

**10.** $\frac{3}{4}$ $\frac{4}{5}$  (a) _____  (b) _____

**11.** $\dfrac{3}{5}$ $\dfrac{5}{6}$   (a) _____   (b) _____

**12.** $\dfrac{9}{4}$ $\dfrac{7}{6}$   (a) _____   (b) _____

**13.** $\dfrac{2}{3}$ $\dfrac{5}{16}$   (a) _____   (b) _____

**14.** $\dfrac{3}{5}$ $\dfrac{5}{8}$   (a) _____   (b) _____

**15.** $\dfrac{1}{3}$ $\dfrac{2}{7}$   (a) _____   (b) _____

**Problems 16 through 25.** Put these fractions in order from *smallest* to *largest*.

**Example:** Given fractions: $\dfrac{5}{8}$ $\dfrac{5}{6}$ $\dfrac{2}{3}$.

*Solution:* Write these fractions with a common denominator of 24.

$$\frac{5}{8} = \frac{15}{24} \qquad \frac{5}{6} = \frac{20}{24} \qquad \frac{2}{3} = \frac{16}{24}$$

Put in order from smallest to largest.

$$\frac{15}{24} < \frac{16}{24} < \frac{20}{24}$$

*Answer:* In terms of the original fractions:

$$\frac{5}{8} < \frac{2}{3} < \frac{5}{6}$$

**16.** $\dfrac{3}{4}$   $\dfrac{5}{8}$

**17.** $\dfrac{2}{3}$   $\dfrac{5}{6}$

**18.** $\dfrac{5}{4}$   $\dfrac{4}{3}$

**19.** $\dfrac{4}{10}$   $\dfrac{41}{100}$

**20.** $\dfrac{4}{11}$   $\dfrac{1}{3}$

**21.** $\dfrac{1}{3}$   $\dfrac{1}{4}$   $\dfrac{5}{12}$

**22.** $\dfrac{3}{4}$   $\dfrac{4}{5}$   $\dfrac{13}{20}$

**23.** $\dfrac{3}{5}$   $\dfrac{7}{10}$   $\dfrac{12}{25}$

**24.** $\dfrac{5}{8}$   $\dfrac{11}{16}$   $\dfrac{3}{4}$

**25.** $\dfrac{7}{16}$   $\dfrac{1}{2}$   $\dfrac{3}{5}$

**26.** Which fraction shows the larger measurement?

(a) $\frac{9}{16}$ inch or $\frac{20}{32}$ inch?          (b) $\frac{3}{8}$ inch or $\frac{41}{64}$ inch?

(c) $\frac{17}{32}$ inch or $\frac{5}{10}$ inch?          (d) $\frac{13}{16}$ inch or $\frac{25}{32}$ inch?

**Problems 27 through 36.** Write each *improper* fraction as a mixed number.

**27.** (a) $\frac{21}{4} =$ _____   (b) $\frac{15}{4} =$ _____   (c) $\frac{19}{4} =$ _____

**28.** (a) $\frac{13}{2} =$ _____   (b) $\frac{15}{2} =$ _____   (c) $\frac{23}{2} =$ _____

**29.** (a) $\frac{25}{8} =$ _____   (b) $\frac{47}{8} =$ _____   (c) $\frac{17}{8} =$ _____

**30.** (a) $\frac{35}{16} =$ _____   (b) $\frac{47}{16} =$ _____   (c) $\frac{55}{16} =$ _____

**31.** (a) $\frac{97}{64} =$ _____   (b) $\frac{139}{64} =$ _____   (c) $\frac{157}{64} =$ _____

**32.** (a) $\frac{14}{2} =$ _____   (b) $\frac{64}{32} =$ _____   (c) $\frac{128}{16} =$ _____

**33.** (a) $\frac{22}{4} =$ _____   (b) $\frac{36}{8} =$ _____   (c) $\frac{44}{16} =$ _____

**34.** (a) $\frac{160}{64} =$ _____   (b) $\frac{500}{32} =$ _____   (c) $\frac{82}{8} =$ _____

**35.** (a) $\frac{17}{10} =$ _____   (b) $\frac{189}{100} =$ _____   (c) $\frac{238}{15} =$ _____

**36.** (a) $\frac{76}{13} =$ _____   (b) $\frac{88}{11} =$ _____   (c) $\frac{165}{21} =$ _____

**Problems 37 through 41.** Write each *mixed number* as an improper fraction.

**37.** (a) $2\frac{1}{2}$     (b) $3\frac{3}{8}$     (c) $5\frac{3}{16}$     (d) $4\frac{5}{8}$

**38.** (a) $3\frac{7}{8}$     (b) $4\frac{9}{16}$     (c) $6\frac{17}{13}$     (d) $10\frac{3}{8}$

**39.** (a) $12\frac{5}{8}$     (b) $15\frac{17}{32}$     (c) $22\frac{1}{4}$     (d) $18\frac{3}{4}$

**40.** (a) $13\frac{25}{64}$     (b) $18\frac{15}{16}$     (c) $6\frac{7}{10}$     (d) $17\frac{23}{100}$

**41.** (a) $34\frac{57}{100}$     (b) $9\frac{23}{100}$     (c) $16\frac{147}{1,000}$     (d) $245\frac{897}{10,000}$

**Problems 42 through 45.** Write each of the measurements in improper fraction form.

   **Example:** $4\frac{3}{4}$ pound $= \frac{19}{4}$ pound

**42.** (a) $5\frac{3}{8}$ inches     (b) $4\frac{7}{16}$ inches     (c) $8\frac{9}{16}$ inches     (d) $7\frac{3}{32}$ inches

**43.**  (a) $7\frac{1}{3}$ feet        (b) $5\frac{3}{4}$ feet        (c) $18\frac{7}{12}$ feet        (d) $45\frac{11}{12}$ feet

**44.**  (a) $3\frac{34}{60}$ hours     (b) $5\frac{3}{7}$ weeks     (c) $14\frac{1}{3}$ yards     (d) $7\frac{34}{52}$ years

**45.**  (a) $5\frac{3}{8}$ gallons    (b) $2\frac{54}{100}$ cm     (c) $9\frac{15}{24}$ days     (d) $2\frac{36}{365}$ years

## SEC. 3   COMMON FRACTIONS AND RATIOS

What does a machinist mean when he says: "For proper operation of this lathe, you must be sure the *gear ratio* is three to one"?

How does the auto mechanic interpret the engine specification information:

Chevrolet Corvette: 425 horsepower

*Compression Ratio:* 9 : 1 (nine to one)?

What does a carpenter mean when he says: "The roof has a pitch (ratio) of one to three"?

Statements like these are made everyday by men and women in all trades and vocations. Since you use the word *ratio* in your work, you must be sure of its meaning.

To find out what the word *ratio* means, look at the statement made by the machinist: ". . . the *gear ratio* is three to one." If you could see the gears involved, they might look like those in Fig. 7-21.

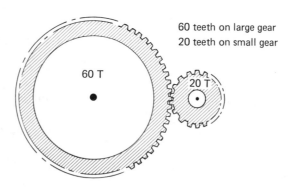

Figure 7-21

The phrase "*gear ratio is three to one*" refers to a *comparison* of the number of teeth on the large gear to the number of teeth on the small gear. To say the gear *ratio* is three to one means that there are *three times as many teeth on the large gear than there are on the small gear*. The numbers 3 and 1 result from using the numbers 60 and 20 as numerator and denominator of a fraction $\left(\frac{60}{20}\right)$ and then reducing the fraction to lowest terms;

$$\frac{60 \text{ teeth on large gear}}{20 \text{ teeth on small gear}} = \frac{60}{20} = \frac{3}{1}.$$

Here 60 compares to 20 as 3 compares to 1. A *ratio is a comparison of two like quantities and is written in fraction form.* (Sometimes ratios are written as decimals or percent and you will study this in Chapters 8 and 9.) If the machinist replaces the large gear with one having 40 teeth, the resulting *gear ratio* becomes

$$\frac{40 \text{ teeth on large gear}}{20 \text{ teeth on small gear}}$$

When this fraction is written in lowest terms it is

$$\frac{2}{1}$$

and the gear ratio is *two to one.*

In the second statement the auto mechanic is talking about the *compression ratio* of an automobile engine. The *compression ratio* of an engine compares the *volume of expanded gases* in a cylinder to the *volume of compressed gases* in the cylinder.

In Fig. 7-22(a) the piston is at the bottom of the stroke. In (b) the piston is at the top of the stroke. If the volume of expanded gases (a) is 45 cubic inches and the compressed volume (b) is 5 cubic inches, then the *compression ratio* is:

$$\text{Compression Ratio} = \frac{\text{Expanded volume} = 45 \text{ cu. in.}}{\text{Compressed volume} = 5 \text{ cu. in.}}$$

$$= \frac{45}{5}$$

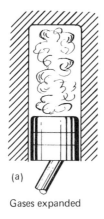

(a)

Gases expanded

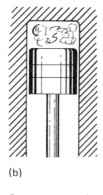

(b)

Gases compressed

**Figure 7-22**

$$\text{Compression Ratio} = \frac{9}{1} \text{ (nine to one)}$$

Sometimes a colon (:) is used to write ratios, and $9:1$ means the same as the fraction $\frac{9}{1}$. In either form it is read "nine to one."

In the third statement the carpenter is talking about the steepness of the roof. He uses the word *pitch* to describe the ratio of the *rise* of a rafter to the *span* of the roof. (The *rise* is the vertical distance between the ridge and the supporting plate. The *span* is the distance between the outer edges of the supporting walls.)

$$\text{Pitch} = \frac{\text{Rise of rafter}}{\text{Span of roof}}$$

In Fig. 7-23 the roof has a rise of 6 feet and a span of 18 feet. The pitch of the roof is:

$$\text{Pitch} = \frac{6 \text{ ft. rise}}{18 \text{ ft. span}}$$

$$\text{Pitch} = \frac{1}{3} \text{ (one to three)}$$

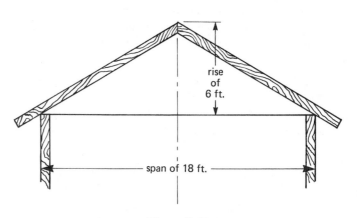

rise
of
6 ft.

span of 18 ft.

*Figure 7-23*

From these examples, you can see that *ratio* is an important concept and worthy of your study. Ratios are used in all technical fields. Some of the more important applications are explained in the examples.

When you write a ratio in the form of a fraction, always write the fraction in its lowest terms. To say that a gear ratio is "twenty to five" is unacceptable. Reduce the fraction $\frac{20}{5}$ to $\frac{4}{1}$, then say the ratio is "four to one."

When writing answers to problems, try to be clear and use correct terms. Apply this principle to all of your work and you will soon be recognized as a first-class tradesman. Study the following ideas and examples to improve your skills.

The *mechanical efficiency* of a motor, steam engine, shop tool, and other mechanical or electrical devices that use and deliver useful energy is the ratio:

$$\text{Mechanical Efficiency (ME)} = \frac{\text{(Useful) output}}{\text{(Total) input}}$$

The *output* is the useful energy delivered *by* the machine, and the *input* is the amount of energy delivered *to* the machine. This ratio is important to the auto mechanic, diesel mechanic, aircraft engine mechanic, electrician, and everyone who works with motors. (The ratio is usually written as a percentage rather than a common fraction. You will study this in Chapter 10.)

**Example 10:** An auto mechanic wants to find the efficiency of an automobile engine before a tune-up. From the engine specification chart he finds that the engine is rated to deliver 315 horsepower at 4,400 rpm. When he tests the engine, he finds that it delivers only 270 horsepower at 4,400 rpm. Find the efficiency.

*Solution:* Compute the ratio of the actual output (270) to the rated input (315).

$$\text{Efficiency} = \frac{270 \text{ hp.}}{315 \text{ hp.}} = \frac{6}{7}$$

*Answer:* The engine is producing six-sevenths of its rated horsepower.

**Example 11:** A class D citizen's band transceiver (CB set) is rated at 5 watts of power. (This is the maximum allowed by the FCC.) The amount of energy delivered to the antenna is usually 3 watts. The loss is due to resistance within the set, the coaxial cable, the antenna, etc. What is the efficiency of a 5 watt CB set?

*Solution:* The efficiency is:

$$\text{Efficiency} = \frac{3 \text{ watts output (delivered)}}{5 \text{ watts input (used)}} = \frac{3}{5}$$

*Answer:* The CB set is three-fifths efficient.

No engine, motor, or other mechanical or electrical device is perfectly efficient. The ratio of useful energy to input energy is always a fraction that is less than one. The principal losses of efficiency are caused by frictional forces within the device.

The concept of *mechanical advantage* is closely related to mechanical and electrical efficiency. *Mechanical advantage* is the ratio:

$$\text{Mechanical advantage (MA)} = \frac{\text{Resistance to effort (output)}}{\text{Input effort}}$$

For instance, if a machine has a mechanical advantage of $4:1$, it means that an (input) effort of 20 pounds results in a resistance (output) of $4 \times 20 = 80$ pounds. $\left(\text{NOTE: } \frac{4}{1} = \frac{80}{20}\right)$ There are many machines that have a mechanical advantage. They range from the simplest tool, such as a claw hammer, to the complicated hydraulic production equipment used in manufacturing. All of them use the principle of *mechanical advantage*. A few examples are discussed in this section. There are more applications in the problems.

The first, and probably one of the best examples of the idea of mechanical advantage, is the lever-fulcrum which you studied in Chapter 6. To find the mechanical advantage of this system, use the *total* length of the lever. That portion of the lever from fulcrum to the end to which a force can be applied is called the *force arm*. The portion of the lever from the fulcrum to the other end is called the *load arm*. The load arm supports the load. (See Fig. 7-24.)

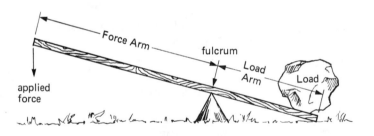

**Figure 7-24**

If the force arm is twice as long as the load arm, the mechanical advantage is $2:1$. The mechanical advantage is written in terms of the *ratio of lengths* of the bar. It is independent of the load or applied force. For example, if the length of the force arm is 8 feet and the length of the load arm is 2 feet, the mechanical advantage is *four to one*. A 100-pound load at the *end* of the load arm requires a 25-pound force at the end of the force arm to move the load. If the force is applied between the fulcrum and the end, the total mechanical advantage is not used.

The wheel-axle combination is also an important use of the principle of mechanical advantage. In Fig. 7-25 a weight $W$ (100 pounds) is lifted by applying a force $F$ to the rope wound around the large wheel. For this system the mechanical advantage is the ratio:

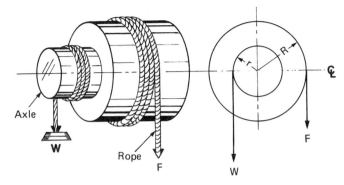

**Figure 7-25**

$$\text{Mechanical advantage} = \frac{\text{Radius of large wheel } (R)}{\text{Radius of axle } (r)}$$

(The radius is the distance from the center line to the edge.)

If the radius $(r)$ of the axle is 2 inches and the radius $(R)$ of the large wheel is 10 inches, the system has a mechanical advantage of 10: 2, which reduces to 5: 1. Therefore, the 100-pound weight is lifted by a force of 20 pounds. A mechanical advantage of *five to one* means that five times as much load can be lifted as there is force applied.

$$(\text{Mechanical advantage}) \times (\text{Force}) = \text{Weight lifted}$$

$$5 \times 20 = 100$$

Another familar system which uses the idea of mechanical advantage is the hydraulic press. There are many uses of the hydraulic press in today's technology. The familiar hydraulic jack, the auto hoist at a service station, and the hydraulic brakes and power accessories on cars are just a few of the important uses of the hydraulic press.

Basically, the hydraulic press has two pistons (one small and one large), a fluid, and tubing to move the fluid under pressure. Figure 7-26 shows the basic parts of any hydraulic system.

Engineers use two ways to find the mechanical advantage of a hydraulic press. Both ways use a ratio.

$$\text{1. Mechanical advantage} = \frac{\text{Resulting force}}{\text{Applied force}}$$

$$\text{2. Mechanical advantage} = \frac{\text{Area of large piston}}{\text{Area of small piston}}$$

In most cases the first formula is not easy to use because forces are hard to measure on the job. But it is easy to find the areas of the two pistons since there are formulas for this. (In Chapter 11 areas of plane figures are discussed.)

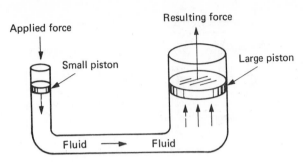

***Figure 7-26***    *Hydraulic press*

When the necessary information is given, use one of these ratios to find the mechanical advantage.

**Example 12:** A force of 50 pounds is applied to the small piston in a hydraulic jack. This force lifts an auto that weighs 4,000 pounds. What is the MA?

*Solution:* Mechanical advantage (MA) $= \dfrac{4{,}000 \text{ pounds lifted}}{50 \text{ pounds applied}}$

*Answer:*                              $\text{MA} = \dfrac{80}{1}$ (eighty to one)

**Example 13:** The face of a hydraulic ram is 500 square inches in area. The area of the small piston is 10 square inches. What is the MA of the system?

*Solution:* $\text{MA} = \dfrac{500 \text{ sq. in.}}{10 \text{ sq. in.}}$

*Answer:* $\text{MA} = 50:1$ (fifty to one)

**Example 14:** Use the MA from Example 7-13. How many pounds of force are at the ram face if 150 pounds of force is applied to the small piston?

*Solution:* $\text{MA} \times (\text{applied force}) = \text{result force}$

*Answer:* $(50) \times (150) = 7{,}500$ pounds

These examples show why the hydraulic press is widely used in autos, airplanes, and other mechanical equipment. Engineers combine the hydraulic press and the lever-fulcrum to get an extremely large mechanical advantage.

# PROBLEMS: Sec. 3
## Common Fractions and Ratio

1. Find each of these ratios. Write the answers in reduced fractional form.

   **Example:** Find the ratio of 6 to 18. Answer: $\frac{6}{18} = \frac{1}{3}$

   Therefore: 6 to 18 is equal to the ratio of 1 to 3 or $\frac{1}{3}$.

   (a) 4 to 6      (b) 9 to 12      (c) 8 to 32      (d) 16 to 64
   (e) 12 to 40      (f) 15 to 25      (g) 45 to 180      (h) 90 to 450

2. Find each of these ratios. State the quantities being compared in the same units.

   **Example:** 14 days to 6 weeks. (6 weeks $= 6 \times 7$ days)

   $$\frac{14 \text{ days}}{42 \text{ days}} = \frac{1}{3} \text{ which is also written } 1:3$$

   (a) 2 in. to 8 in.      (b) 8 in. to 1 ft.      (c) 14 in. to 1 yd.
   (d) 36 min. to 2 hr.      (e) 12 oz. to 3 lb. (1 lb. $= 16$ oz.)
   (f) 120 in. to 20 ft.      (g) 3 quarters to 1 dollar
   (h) 3 nickels to 4 dimes

3. Find each of the ratios and write the answers in simplest form.

   **Example:** $\frac{36 \text{ teeth}}{144 \text{ teeth}} = \frac{1}{4}$

   or write the answer this way:

   36 teeth: 144 teeth $= 1:4$

   (a) $\frac{7 \text{ days}}{28 \text{ days}}$      (b) $\frac{18 \text{ inches}}{72 \text{ inches}}$      (c) $\frac{75 \text{ pounds}}{125 \text{ pounds}}$

   (d) $\frac{45 \text{ square inches}}{9 \text{ square inches}}$      (e) $\frac{3 \text{ ft. 6 in.}}{4 \text{ ft. 8 in.}}$ (change to inches)

   (f) $\frac{1 \text{ hr. 20 min.}}{3 \text{ hr. 0 min.}}$ (change to minutes)      (g) $\frac{5 \text{ cents}}{25 \text{ cents}}$

   (h) $\frac{10 \text{ cents}}{\$2 \text{ and 60 cents}}$

4. Find the gear ratios. Write the answers: $\frac{\text{larger}}{\text{smaller}}$

   **Example:** Find the ratio of two gears if one gear has 48 teeth and the other has 36 teeth.

   *Answer:* The ratio is $\frac{48}{36} = \frac{4 \times 12}{3 \times 12} = \frac{4}{3}$ or $4:3$.

   (a) Larger gear, 56 teeth. Smaller gear, 48 teeth.
   (b) Larger gear, 186 teeth. Smaller gear, 31 teeth.
   (c) Larger gear, 72 teeth. Smaller gear, 24 teeth.
   (d) Larger gear, 288 teeth. Smaller gear, 24 teeth.
   (e) Larger gear, 96 teeth. Smaller gear, 24 teeth.

5. Find the compression ratios.

> **Example:** What is the compression ratio, if the volume of expanded gases is 80 cubic inches and the volume of compressed gases is 20 cubic inches?
>
> *Answer:* Compression ratio $= \dfrac{\text{Expanded Volume}}{\text{Compressed volume}} = \dfrac{80}{20} = \dfrac{4}{1}$

    (a) Expanded volume = 125 cubic inches; compressed volume = 25 cubic inches.

    (b) Expanded volume = 160 cubic centimeters; compressed volume = 48 cubic centimeters.

    (c) Expand volume = 280 cubic inches; compressed volume = 14 cubic inches.

    (d) Expanded volume = 12 liters; compressed volume = 2 liters.

6. An automobile engine is rated at 375 horsepower. When tested, the engine produces only 350 horsepower. What is the mechanical efficiency?

7. An electric motor uses 12 volts to produce an equivalent output of 10 volts. What is the efficiency of the motor?

8. A machinist earns $180 per week. His apprentice earns $100 per week. What is the ratio of their earnings?

9. One share of International Stump costs $54. The stock paid $3 in earnings. What is the cost: earning ratio?

10. The span of a roof is 32 feet. The rise of the roof is 8 feet. What is the pitch?

11. Find the pitch for these.
    (a) Rise = 4 feet    Span = 16 feet
    (b) Rise = 6 feet    Span = 24 feet
    (c) Rise = 8 feet    Span = 48 feet
    (d) Rise = 9 feet    Span = 72 feet
    (e) Rise = 3 feet    Span = 24 feet
    (f) Rise = 12 feet   Span = 24 feet

12. A lever 6 feet long has the fulcrum placed 2 feet from the end. What is the mechanical advantage (MA)?

13. If the force arm for a lever is 4 feet 8 inches long and the load arm is 1 foot 2 inches long, what is the MA?

14. A torque wrench has a mechanical advantage of 25: 1. How much torque will result if 15 pounds is applied at the end of the force arm?

15. In a hydraulic jack the area of the large driven piston is 150 square inches, and the area of the small driving piston is 3 square inches. What is the MA?

16. A force of 2,500 pounds is applied to the small piston of a hydraulic ram. If the MA is 45: 1, how much force is measured at the face of the ram?

17. What is the MA of a wheel-axle combination, if the radius of the axle is 2 inches and the radius of the wheel is 1 foot, 6 inches?

18. There are 155 students enrolled in the Transporation Division at the College of Alameda. The Division has 5 instructors. What is the student: teacher ratio?

**19.** The radiator of a car holds 24 quarts of fluid. If the radiator has 4 quarts of antifreeze, what is the ratio of antifreeze to water?

**20.** The State of Alaska has an area of 158,693 square miles. The population of Alaska is 402,173. Round off both values to two significant digits and find the ratio, population: area.

## SEC. 4   DIRECT PROPORTIONS

Ratios are written as fractions. For instance, a gear tooth ratio of $\frac{4}{1}$ (four to one) means that one gear has four times as many teeth as the other. What the ratio does *not* say is how many teeth either gear has. It is possible for several sets of gears with different numbers of teeth to have the same final gear ratio.

**Example 15:** Find the gear ratios of these:
 (a) 80 teeth on large gear; 20 teeth on small gear

*Solution:* Ratio $= \dfrac{80}{20}$

*Answer:* Ratio $= \dfrac{4}{1}$

**Example 15:** (b) 40 teeth on large gear; 10 teeth on small gear

*Solution:* Ratio $= \dfrac{40}{10}$

*Answer:* Ratio $= \dfrac{4}{1}$

**Example 15:** (c) 144 teeth on large gear; 36 teeth on small gear.

*Solution:* Ratio $= \dfrac{144}{36}$

*Answer:* Ratio $= \dfrac{4}{1}$

In each example, completely different sets of gears show the same ratio of 4:1. When two ratios have the same final value, such as $\dfrac{80}{20}$ and $\dfrac{144}{36}$, the ratios form a *proportion*.

*A proportion means that two ratios are equal.*
The three gear ratios in Example 7-15 are proportional since they have the same final ratio. To say that two ratios are porportional is another way of saying that the fractions used for the ratios are equal.

Proportions are used tó find the value of unknown quantities which involve ratios. To solve problems in proportions use this method:

*When BOTH sides of an equality are* **multiplied** *or* **divided** *by the* **same** *nonzero number, the results are the same.*

1. If $\dfrac{x}{a} = b$, then **multiplying** both sides by $a$ gives $a\dfrac{(x)}{(a)} = ab$

   and $\quad x = ab$ since $\dfrac{a}{a} = 1$

2. If $a \cdot x = b$, then **dividing** both sides by $a$ gives $\dfrac{a \cdot x}{a} = \dfrac{b}{a}$

   and $\quad x = \dfrac{b}{a} \quad$ since $\quad \dfrac{a}{a} = 1.$

Two kinds of proportions are used by technicians—*direct proportions* and *indirect proportions*. In this section you will study *direct proportions*. In the next section you will study *indirect proportions*.

Here are two ways to find a direct proportion:

> *When an* **increase** *in the terms of one ratio in a proportion* **produces an increase** *in like terms of the other ratio of a proportion, then the ratios are in* **direct proportion.**

> *When a* **decrease** *in the terms of one ratio in a proportion* **produces a decrease** *in like terms of the other ratio in the proportion, then the ratios are in* **direct proportion.**

**Example 16:** The oil-to-gas ratio of a Mercury 3.9 outboard motor is 1 to 50. This ratio means that for each 1 pint of oil used, 50 pints of gasoline are needed to lubricate the motor. How many pints of oil are needed for 25 gallons (200 pints) of gasoline?

*Solution:* This is a direct proportion problem because an *increase* in the amount of gasoline used will require an *increase* in the amount of oil needed. To solve this problem let $x$ represent the amount of oil needed. Write the two ratios:

$$\frac{x \text{ pints of oil}}{200 \text{ pints of gas}} \qquad \frac{1 \text{ pint of oil}}{50 \text{ pints of gas}}$$

Set these two ratios equal to each other to form a direct proportion:

$$\frac{x \text{ pints of oil}}{200 \text{ pints of gas}} = \frac{1 \text{ pint of oil}}{50 \text{ pints of gas}}$$

To find the value for $x$, multiply both sides by 200:

$$200\left(\frac{x}{200}\right) = 200\left(\frac{1}{50}\right)$$

Since $200\left(\dfrac{x}{200}\right)$ is equal to $\dfrac{x}{1}$ and $200\left(\dfrac{1}{50}\right)$ is equal to $\dfrac{4}{1}$ you see that $x = 4$.

*Answer:* 4 pints of oil are needed for 25 gallons of gas.

**Example 17:** The label on a 50-pound bag of Evergreen Lawn fertilizer show that it will treat 2,500 square feet of lawn. How many pounds would you use on a golf putting green which is 500 square feet?

*Solution:* This is a direct proportion problem because a *decrease* in the amount of lawn treated will result in a *decrease* in the amount of fertilizer used. To solve this problem, let $x$ equal the amount of fertilizer needed. Write the two ratios:

$$\frac{x \text{ pounds of fertilizer}}{500 \text{ square feet of lawn}} \qquad \frac{50 \text{ pounds of fertilizer}}{2,500 \text{ square feet of lawn}}$$

Set these two ratios equal to each other to form a direct proportion:

$$\frac{x \text{ pounds}}{500 \text{ sq. ft.}} = \frac{50 \text{ pounds}}{2,500 \text{ sq. ft.}}$$

To solve for $x$, multiply both sides by 500:

$$500\left(\frac{x}{500}\right) = 500\left(\frac{50}{2,500}\right)$$

Simplify both sides to get:

$$\frac{x}{1} = \frac{10}{1}$$

*Answer:* You need 10 pounds of fertilizer.

Ratios and proportions are used to solve a wide variety of problems. All top technicians, mechanics, machinists, or other people who work with numbers are familiar with sound, practical methods of problem solving.

# PROBLEMS: Sec. 4   Direct Proportions

1. Which of these ratios are proportional?

   **Example:** (a) The ratios $5:3$ and $35:21$ *are* proportional because

   $$\frac{5}{3} = \frac{35}{21}$$

(b) The ratios 27 : 36 and 30 : 45 *are not* proportional because

$\frac{27}{36} \neq \frac{30}{45}$. (The symbol $\neq$ means *not equal to*.)

(a) 14 : 16 and 21 : 24

(b) 16 : 15 and 48 : 64

(c) 18 : 27 and 12 : 36

(d) 35 : 45 and 40 : 50

(e) 16 : 64 and 1 : 4

(f) 36 : 48 and 4 : 3

(g) 72 : 60 and 288 : 240

2. Find the value for $x$ which will make the two ratios proportional.

**Example:** $\frac{x}{6} = \frac{18}{36}$

*Solution:* Solve for $x$ by multiplying BOTH sides by 6:

$$6\left(\frac{x}{6}\right) = 6\left(\frac{18}{36}\right) \qquad x = \frac{(6)18}{36} = 3$$

*Answer:* $x = 3$

(a) $\frac{x}{27} = \frac{70}{135}$                  (b) $\frac{x}{100} = \frac{3}{20}$

(c) $\frac{x}{75} = \frac{16}{15}$                  (d) $\frac{x}{250} = \frac{270}{750}$

Problems 3–11. Do these *direct proportion* problems.

3. Two gears are in the ratio of 3 : 5. If the larger gear has 40 teeth, how many teeth does the smaller gear have?

4. Two gears have 48 teeth and 132 teeth. If the smaller gear is replaced by a gear with 28 teeth, how many teeth must the larger gear have to keep the same gear ratio?

5. The power-to-weight ratio for a certain automobile engine is 7 : 8. If the engine weighs 400 pounds, how much power does it produce?

6. An unlimited AA-fuel class supercharged dragster can deliver 1,700 horsepower. If the engine weighs 450 pounds, what is the power-to-weight ratio?

7. If the power-to-weight ratio of a racing engine is 5 : 2, what is the weight of the engine if it produces 750 horsepower?

8. The operating instructions for a certain 4-cycle engine call for the fuel-to-oil mix to be 35 : 1. How many *pints* of oil are needed for 70 *gallons* of fuel?

9. The odds of getting a *pair* of cards (two cards with the same face value) in a single deal in a poker game are 3 : 4. (In four deals you can *expect* to get a pair of cards three times.) How many pairs can you expect to get in 24 deals of the cards?

10. The odds of getting "snake eyes" (a total sum of two) in a throw of dice are 1 : 35. If the dice are thrown 280 times, how many times can you expect "snake eyes"?

**11.** In one year the win-loss ratio of the California Seals hockey team was 2 : 5. If they lost 45 games, how many did they win?

## SEC. 5   INDIRECT PROPORTIONS

The proportions you studied in Section 4 were examples of *direct proportions.*

The second type of proportion is *indirect proportion.* Here are two ways to find an indirect proportion:

> *When an **increase** in the terms of one ratio in a proportion **produces a decrease** in like terms of the other ratio of a proportion, the ratios are in **indirect proportion.***

> *When a **decrease** in the terms of one ratio in a proportion **produces an increase** in like terms of the other ratio of a proportion, the ratios are in **indirect proportion.***

You must know when two ratios are in *direct* or *indirect* proportion. Stop and apply reason to the problem. Avoid "jumping in" with a formula or rule that may result in loss of time, waste of material, or loss of your job!

Look at a problem that results in an *indirect proportion.* In Fig. 7-27 the two pulleys are connected by a belt. Suppose the pulleys rotate in the direction indicated and that the larger pulley has completed one revolution. How many revolutions has the smaller pulley completed?

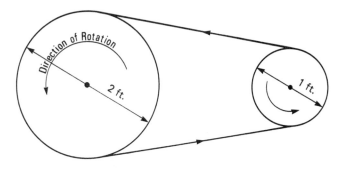

**Figure 7-27**

Reason tells you: "When the big pulley goes around once, the little pulley must go around several times." What you want to know is exactly how many times the small pulley goes around during one revolution of the large pulley.

It seems reasonable to believe that the number of revolutions completed by the pulleys is in some way related to their sizes (diameters). You expect to find a proportion is involved. Set up a direct proportion and see what happens. Will the result of a direct proportion be consistent with what you know *must* happen? A direct proportion is:

$$\frac{\text{Diameter of small pulley}}{\text{Diameter of large pulley}} = \frac{\text{Number of revolutions of small pulley}}{\text{Number of revolutions of large pulley}}$$

Put in the values that you know:

$$\frac{1 \text{ ft.}}{2 \text{ ft.}} = \frac{X = \text{unknown number of revolutions of small pulley}}{1 \text{ revolution of large pulley}}$$

Therefore, $X = \frac{1}{2}$ revolution. This means that the small pulley turns one-half of a revolution when the large pulley turns one complete revolution. Look at the answer. Does it seem reasonable that the small pulley completes only one-half a revolution when the large pulley makes *one* complete revolution? The answer is obviously NO. This answer is not consistent with the reasoning that the smaller pulley goes around *more* times than the larger pulley. The ratios are NOT a direct proportion.

What do you do now? Again set up a proportion. This time, reverse the numerator and denominator of the first ratio:

$$\frac{\text{Diameter of large pulley}}{\text{Diameter of small pulley}}$$

$$= \frac{X = \text{unknown number of revolutions of small pulley}}{1 \text{ revolution of large pulley}}$$

Note: Only *one* of the ratios is reversed.
Put in the values known:

$$\frac{2 \text{ ft.}}{1 \text{ ft.}} = \frac{X \text{ revolutions of small pulley}}{1 \text{ revolutions of large pulley}}$$

Answer: $X = 2$ revolutions

The answer of 2 revolutions is consistent, because when two pulleys are connected together, the smaller pulley turns faster than the larger pulley. *The ratio of the number of revolutions of each is in* **indirect proportion** *to their respective sizes.*

Belt and pulley systems are closely related to gear systems. When two gears of different sizes mesh together, the ratio of their sizes is in *indirect proportion* to the ratio of speeds of revolution. The effect of meshing two gears together changes their direction of rotation and produces different rates of revolution. In designing gears that mesh, engineers have found that: *To find the speeds of the two gears it is necessary to know only the number of teeth on each gear.* (The number of teeth on a gear is directly proportional to the diameter of a gear.) When studying gear systems, use the number of teeth on a gear instead of its size to compute ratios.

**Example 18:** Figure 7-28 shows two gears that are meshed together. If the smaller gear rotates at 150 rpm, what is the rate of rotation of the larger gear?

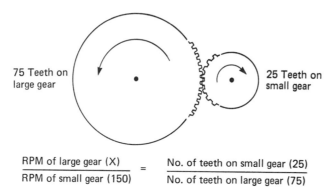

$$\frac{\text{RPM of large gear (X)}}{\text{RPM of small gear (150)}} = \frac{\text{No. of teeth on small gear (25)}}{\text{No. of teeth on large gear (75)}}$$

**Figure 7-28**

*Solution:* This is an *indirect proportion* problem since a small gear with 25 teeth turning at 150 rpm will drive a large gear with 75 teeth at a speed less than 150 rpm. *An increase in size results in a decrease in speed.* Let $x$ equal the speed (in rpm) of the larger gear. Write the ratios:

$$\frac{x = \text{rpm of large gear}}{150 \text{ rpm of small gear}} \quad \frac{25 \text{ teeth on small gear}}{75 \text{ teeth on large gear}}$$

Notice that the terms *small* and *large* are reversed in the second ratio. This is because the ratios are indirect. Set the ratios equal:

$$\frac{x = \text{rpm large gear}}{150 \text{ rpm small gear}} = \frac{25 \text{ teeth on small gear}}{75 \text{ teeth on large gear}}$$

To solve for $x$, multiply both sides by 150:

$$150\left(\frac{x}{150}\right) = 150\left(\frac{25}{75}\right)$$

$$x = \frac{50}{1}$$

*Answer:* $x = 50$ rpm (speed of large gear).

In many mechanical systems gears are meshed in series to reduce the loss of energy caused by friction and slippage. In a two-gear system, one of the gears is called the *driver* (supplies energy), and the other is called the *driven* (receives the energy). When three or more gears are meshed in series, the in-between gears are called *idlers*. (They transmit energy.) You will see that:

> Idlers have NO effect on the speeds of the *driver* and *driven* gears. The only effect an idler has is to change the direction of the adjacent gear and to transmit energy.

The gears in Fig. 7-29 show that the direction of the driven gear is the same as the driver when one idler is used. To prove the statement that idlers do not affect speeds, find the speed of the driven gear with the idler. (Note: The speeds are in *indirect proportion* to the number of teeth.)

$$\frac{\text{rpm of idler }(X)}{\text{rpm of driver }(1)} = \frac{\text{Number of teeth on driver }(60)}{\text{Number of teeth on idler }(20)}$$

This gives: $X = \dfrac{60}{20} = 3$ rpm for the idle gear.

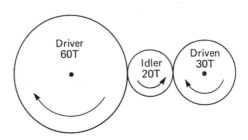

*Figure 7-29*

Now find the speed of the driven gear:

$$\frac{\text{rpm of driven }(X)}{\text{rpm of idler }(3)} = \frac{\text{Number of teeth on idler }(20)}{\text{Number of teeth on driven }(30)}$$

This gives: $X = \dfrac{(3)(20)}{30}$

Therefore, $X = \dfrac{60}{30} = 2$ rpm (driven)

If the idler is omitted from all computations:

$$\frac{\text{rpm of driven }(X)}{\text{rpm of driver }(1)} = \frac{\text{Number of teeth on driver }(60)}{\text{Number of teeth on driven }(30)}$$

$$X = \frac{60}{30} = 2 \text{ rpm (driven)}$$

The effect of the idler is to change direction and transmit energy only. It does NOT affect the speeds. In finding the speeds of either driver or driven gears, *all* idlers are omitted from the computations.

## Gear Trains and Compound Gearing

Many automobiles, particularly sports cars and most light-duty trucks and tractors, use a three- or four-speed gear transmission. In these transmissions, power from the engine is delivered to the transmission through the clutch and clutch gear.

It takes a great deal more power to start a car or truck in motion than it does to keep it in motion. Also, it takes more power for a vehicle to climb a hill than it does to go down hill. With the transmission the engine can get a mechanical advantage over the drive train to the rear wheels.

Most automobile transmissions have three forward speed combinations and one reverse gearing. Basically, the transmission has several gears each with different numbers of teeth. The gear teeth are meshed together by moving the gear shift into different positions. The forward speeds, the reverse speed, and a neutral position are all obtained by engaging or disengaging different sets of gears. Figure 7-30 shows the general four positions of a gear-driven transmission.

When the transmission is in the first-speed, second-speed, or reverse-speed positions, the engine power is transmitted from the clutch gear (driver) to the countershaft drive gear (driven). The power is then transmitted to the drive line through the countershaft gears—all of which are *driver* gears. When power is transferred from one gear to another, the gear *with the power* is called the *driver*. The gear *receiving the power* is called the *driven* gear.

The final gear ratio that results from connecting driving and driven gears can be found by using this formula:

$$\text{Ratio (final)} = \frac{\text{Product of the number of teeth on all DRIVEN gears}}{\text{Product of the number of teeth on all DRIVER gears}}$$

The formula does NOT include any idler gears since they serve only to change direction of rotation of the drive shaft.

**Example 19:** Figure 7-31 shows a three-speed auto transmission. Pay particular attention to which gears are *drivers* and which are *driven*. Notice that they go in sequence:

Driver $\longrightarrow$ Driven $\longrightarrow$ Driver $\longrightarrow$ Driven

Use the values on the gears and find:
(a) The first-speed gear ratio.
(b) The second-speed gear ratio.
(c) The reverse-speed gear ratio.

*Solution* (a): First-speed ratio

$$= \frac{(28T \text{ on countershaft}) \times (24T \text{ on first-speed})}{(16T \text{ on clutch gear}) \times (14T \text{ on first-speed countershaft gear})}$$

$$= \frac{28 \times 24}{16 \times 14} = \frac{3}{1}$$

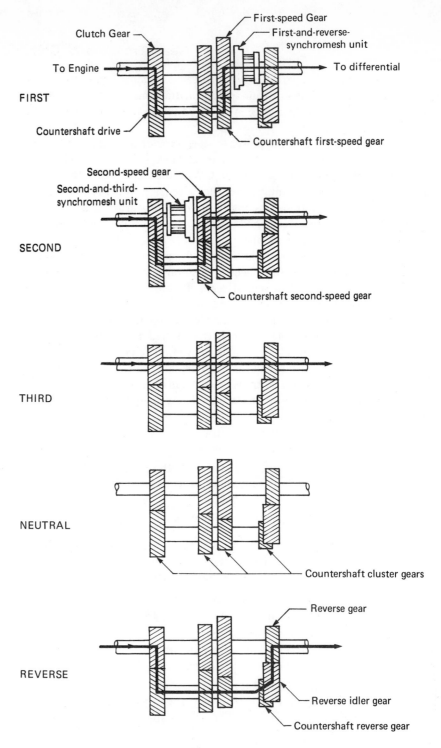

**FIRST**

Clutch Gear

First-speed Gear

First-and-reverse-
synchromesh unit

To Engine

To differential

Countershaft drive

Countershaft first-speed gear

**SECOND**

Second-speed gear

Second-and-third-
synchromesh unit

Countershaft second-speed gear

**THIRD**

**NEUTRAL**

Countershaft cluster gears

**REVERSE**

Reverse gear

Reverse idler gear

Countershaft reverse gear

*Figure 7-30*

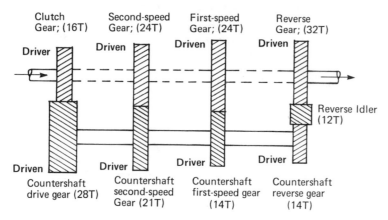

**Figure 7-31**

*Answer* (a): The first-speed gear ratio is 3:1.

*Solution* (b): Second-speed ratio

$$= \frac{(28T \text{ on countershaft}) \times (24T \text{ on second-speed})}{(16T \text{ on clutch gear}) \times (21T \text{ on second-speed countershaft gear})}$$

$$= \frac{28 \times 24}{16 \times 21} = \frac{2}{1}$$

*Answer* (b): The second-speed gear ratio is 2:1.

*Solution* (c): Reverse-speed ratio

$$= \frac{(28T \text{ on countershaft}) \times (32T \text{ on reverse gear})}{(16T \text{ on clutch gear}) \times (14T \text{ on reverse-speed countershaft gear})}$$

$$= \frac{28 \times 32}{16 \times 14} = \frac{4}{1}$$

*Answer* (c): The reverse-speed gear ratio is 4:1.

Sometime you may need to know the speed of the gears instead of the gear ratio. If so, remember that the ratio of the speeds of gears is *inversely* proportional to the number of teeth on the gears.

Here are two formulas for finding the desired speeds:

1. Speed of driver $= \dfrac{(\text{Speed of driven}) \times (\text{Teeth on driven})}{\text{Teeth on driver}}$

2. Speed of driven $= \dfrac{(\text{Speed of driver}) \times (\text{Teeth on driver})}{\text{Teeth on driven}}$

These formulas can be more easily remembered in the form:

1. Speed of driver $= \dfrac{\text{Product of All driven values}}{\text{Product of All teeth on driver gears}}$

2. Speed of driven $= \dfrac{\text{Product of All driver values}}{\text{Product of All teeth on driven gears}}$

**Example 20:** (a) For the transmission in Fig. 7-31, what is the speed of the drive shaft in rpm? The clutch gear is rotating at 900 rpm and the transmission is in first-speed.

*Solution:* (a) Speed of driven $= \dfrac{(900 \text{ rpm}) \times (16T) \times (14T)}{(28T) \times (24T)}$

$$= \frac{900 \times 16 \times 14}{28 \times 24}$$

*Answer:* (a)         $= 300$ rpm

**Example 20:** (b) How fast must the clutch gear rotate in order for the driveshaft to rotate at 600 rpm with the transmission in second-speed?

*Solution:* (b) Speed of driver $= \dfrac{(600 \text{ rpm}) \times (28T) \times (24T)}{(16T) \times (21T)}$

$$= \frac{600 \times 28 \times 24}{16 \times 21}$$

*Answer:* (b)         $= 1{,}200$ rpm

You may use the principles developed for gears to find the ratios and speeds of pulley trains. Change the formulas and use the diameters of the pulleys instead of the number of teeth.

**Example 21:** (a) Figure 7-32 shows a pulley train with pulleys A, B, C, and D. Pulleys $B$ and $C$ are keyed to the same shaft and rotate at the same speed. The diameters of the pulleys are:

A $= 7$ inches   B $= 20$ inches   C $= 10$ inches   D $= 14$ inches

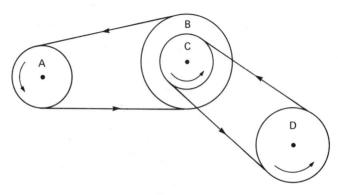

*Figure 7-32*

If the driver pulley $A$ rotates at a speed of 360 rpm, what is the speed of D?

*Solution:* (a)

$$\text{Speed of driven} = \frac{(360 \text{ rpm}) \times (7 \text{ in.}) \times (10 \text{ in.})}{(20 \text{ in.}) \times (14 \text{ in.})} = \frac{(\text{Driver})}{(\text{Driven})}$$

$$= \frac{360 \times 7 \times 10}{20 \times 14}$$

*Answer:* (a)    $D = 90$ rpm

**Example 21:** (b) If power is supplied to pulley $A$, what is the pulley ratio between $A$ and $D$ in the pulley train?

*Solution:* (b)

$$\text{Pulley ratio} = \frac{\text{Product of diameters of driven pulleys}}{\text{Product of diameters of driver pulleys}}$$

$$= \frac{(20 \text{ in.}) \times (14 \text{ in.})}{(7 \text{ in.}) \times (10 \text{ in.})}$$

$$= \frac{20 \times 14}{7 \times 10} = \frac{4}{1}$$

*Answer:* (b) The pulley ratio is 4:1.

It is important that you understand the next statement thoroughly before you go to Sec. 5. Study this statement carefully.

*The problem involves **direct proportion** if an **increase** in one quantity produces an **increase** in another. The problem involves **indirect proportion** if an **increase** in one quantity produces a **decrease** in another.*

When you analyze and set up problems involving proportions, a little common sense will save you embarrassment, wrong answers, and costly mistakes.

## PROBLEMS: Sec. 5 Indirect Proportions

1. Find the speed (rpm) of the small pulley (Fig. 7-33).

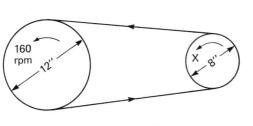

**Figure 7-33**

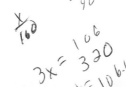

HINT: $\dfrac{x \text{ rpm}}{160 \text{ rpm}} = \dfrac{12 \text{ in.}}{8 \text{ in.}}$

Solve for $x$.

2. Find the speed (rpm) of the large pulley (Fig. 7-34).

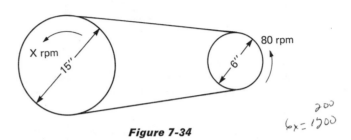

**Figure 7-34**

$6x = 1200$    $200$

HINT: $\dfrac{x \text{ rpm}}{80 \text{ rpm}} = \dfrac{6 \text{ in.}}{15 \text{ in.}}$

Solve for $x$.

$15x = 480$

3. In Fig. 7-35 pulley $A$ is 10 inches in diameter and turns at 150 rpm. What is the diameter of pulley $B$ for a required speed of 300 rpm?

**Figure 7-35**

4. In Fig. 7-36 pulley $A$ is 8 inches in diameter. Pulley $B$ is 20 inches in diameter, pulley $C$ is 10 inches in diameter, and pulley $D$ is 17 inches in diameter. Pulleys $B$ and $C$ are connected to the same shaft and turn at the same speed. If pulley $A$ turns at 170 rpm, find the speed of pulleys $B$ and $D$.

$\dfrac{8}{20} = \dfrac{X}{170}$

$\dfrac{5}{425}$    $2\overline{)850}$

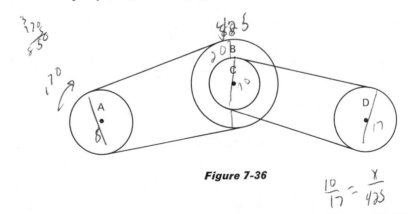

**Figure 7-36**

$\dfrac{10}{17} = \dfrac{X}{425}$

5. Four pulleys are connected as shown in Fig. 7-37. Pulley *A* has a diameter of 14 inches and is turning at 192 rpm. Pulley *B* is 8 inches in diameter. Pulley *C* is 6 inches in diameter and must turn at 1,400 rpm. Find the diameter of pulley *D*, fastened to the same shaft as *B*, so that *C* will turn at the required speed.

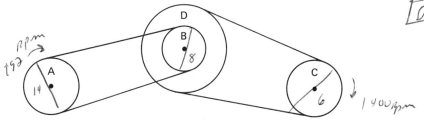

**Figure 7-37**

6. A counter shaft is turning at 630 rpm (Fig. 7-38). A cone pulley is attached to the shaft with steps of 12, 8, and 6 inches in diameter. The cone pulley is attached to a grinder with a pulley 9 inches in diameter. What range of speeds are available at the grinder?

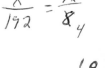

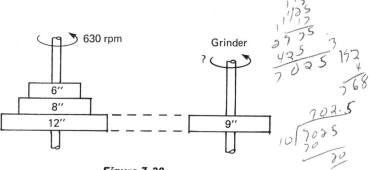

**Figure 7-38**

7. The driver of two cone pulleys turns at 1,500 rpm. Find the three speeds of the driven pulleys (Fig. 7-39).

8. The shaft of a generator is to turn at 1,800 rpm. The shaft has a 4-inch pulley and is belted to a driver shaft that turns at 450 rpm. What is the diameter of the pulley on the driver shaft?

9. The governor on a diesel engine turns at 1,400 rpm. The diameter of the governor pulley is 6 inches. The speed of the engine is 2,100 rpm. Find the diameter of the pulley on the engine shaft.

10. The pulley on the headstock of a wood lathe is 4 inches in diameter. The pulley is belted to a pulley cone with pulleys 3, 7, and 10 inches in diameter. If the shaft of the pulley cone turns at 380 rpm, what speeds are available at the headstock?

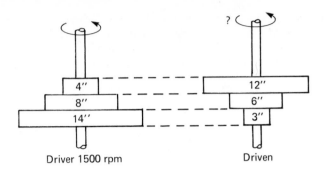

**Figure 7-39**

11. The crankshaft of an automobile engine has two 8-inch diameter pulleys connected to it. One pulley is belted to the alternator. The other is belted to the cooling fan and power steering. The alternator pulley is 4 inches in diameter. The cooling fan pulley is 6 inches in diameter. The power steering pulley is 10 inches in diameter. What are the speeds of all pulleys if the engine is turning at 4,200 rpm (Fig. 7-40)?

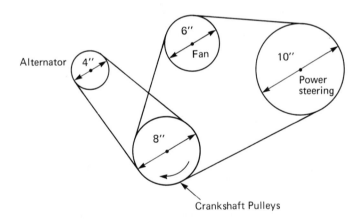

**Figure 7-40**

12. A centrifugal water pump is driven by a shaft with a 5-inch pulley. The pulley is belted to a stationary engine that drives the pump. A 16 inch pulley on the engine is turning at 2,750 rpm. What is the speed of the pump shaft?

13. Find the speed of the gear indicated.

$$\text{Note:} \quad \frac{\text{Teeth on driven}}{\text{Teeth on driver}} = \frac{\text{Speed of driver (rpm)}}{\text{Speed of driven (rpm)}}$$

(The number of teeth are *inversely* proportional to the speeds of the gears.)

(a) Large gear has 70 teeth; small gear has 20;
speed of large gear = 100 rpm;
speed of small gear = $x$ rpm.
(b) Large gear has 160 teeth; small gear has 55 teeth;
speed of large gear = 2,200 rpm;
speed of small gear = $x$ rpm.
(c) Large gear has 180 teeth; small gear has 60 teeth;
speed of small gear = 450 rpm;
speed of large gear = $x$ rpm.
(d) Large gear has 250 teeth; small gear has 80 teeth;
speed of small gear = 550 rpm;
speed of large gear = $x$ rpm.

**14.** Find the missing values in this table:

| | Teeth on large gear | Teeth on small gear | Speed of large gear | Speed of small gear |
|---|---|---|---|---|
| (a) | 100 | 40 | 750 | _____ |
| (b) | 450 | _____ | 200 | 600 |
| (c) | _____ | 75 | 375 | 800 |
| (d) | 175 | 98 | _____ | 150 |

**15.** A gear that has 120 teeth is meshed with a smaller gear that has 24 teeth. If the smaller gear makes 10 revolutions, how many revolutions does the larger gear make?

**16.** A gear with 56 teeth is meshed with a gear which has 160 teeth. If the smaller gear makes 240 rpm, what is the speed of the larger gear?

**17.** A 48-tooth gear is turning at 640 rpm. It is required to drive another gear at 120 rpm. How many teeth are on the driven gear?

**18.** In Fig. 7-41 three gears are meshed. Gear $A$ has 24 teeth, gear $B$ has 36 teeth, and gear $C$ has 80 teeth. If gear $A$ turns at 200 rpm, what is the speed of gear $C$? (NOTE: The speed of the driven gear does NOT depend on the speed of the idler between the driver and driven gears.)

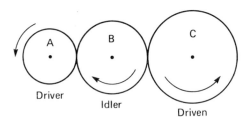

*Figure 7-41*

19.  Two gears have a speed ratio of 7:3. If the smaller gear has 96 teeth, how many teeth does the larger gear have?

20.  The speeds of two gears are in the ratio of 5:2. If the faster gear turns at 180 rpm, what is the speed of the slower gear? (Use *direct proportion*.)

21.  The ratio of the number of teeth of two gears is 12:5. If the larger gear is turning at 65 rpm, find the speed of the smaller gear.

22.  Figure 7-42 shows compound gearing. Gear *A* has 24 teeth. Gear *B* has 36 teeth. Gear *C*, which is on the same shaft as gear *B*, has 20 teeth. Gear *D* has 48 teeth. If gear *A* (the driver) makes 72 rpm, what is the speed of gear *D*? (NOTE: Gears *B* and *C* are *not* idlers.)

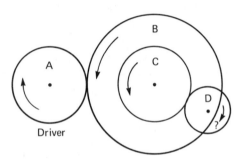

**Figure 7-42**

23.  A five-speed bicycle has a driver gear with 52 teeth. The rear wheel has five gears on the same shaft with 13, 16, 19, 22, and 25 teeth. If the bicyclist pedals the driver at 60 rpm, find (to the nearest rpm) the five speeds of the rear wheel.

24.  Gears *A*, *B*, *C*, *D*, *E*, and *F* are meshed as shown in Fig. 7-43. If gear *A* turns at 160 rpm, find the speed of gear *F*.

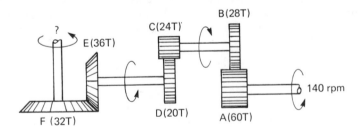

**Figure 7-43**

## SEC. 6   AVERAGES

Are you an *average* person? Do you have *average* intelligence? If you are working, do you make an *average* salary? Do you expect to live an *average* life in an *average* home with an *average* husband or wife and have $2\frac{1}{4}$ *average* children? What does it mean to *be average*?

According to the dictionary, *average* means fair, mediocre, medium, run-of-the-mill, so-so, and passable. Certainly, no one wants to be classified as mediocre, run-of-the-mill, or just so-so. Before you use the word *average* in technology, be sure you have a clear understanding of what it means.

When the word *average* is used in mathematics and technology, it refers to a number that represents a set of numbers.

### Rule for Finding the Arithmetic Average of a Set of Numbers

Step 1. *Find the sum of all numbers of the set.*
(Be sure to take into account the signs of the numbers.)

Step 2. *Divide the sum by the number of numbers in the set.*
*The quotient is the average for the set of numbers.*

Study the following examples:

**Example 22:** Before starting a motor overhaul, an auto mechanic wanted to know the *average* compression in an eight-cylinder engine. He used a compression gauge to get this data:

| Cylinder number | Compression (lbs./sq. in.) |
|:---:|:---:|
| 1 | 134 |
| 2 | 136 |
| 3 | 128 |
| 4 | 130 |
| 5 | 138 |
| 6 | 140 |
| 7 | 136 |
| 8 | 122 |
| 8 = Number of samples | 1,064 = Sum of readings |

*Solution:* Average compression $= \dfrac{1,064}{8} = 133$ pounds per square inch. The *average* (133 psi) is not equal to *any* of the measured values. The number 133 is the *typical* compression of the eight measurements. Do not be surprised by the value of an average. The number that represents a set of values may not be equal to any one of them.

*Answer:* Average compression $= 133$ psi.

**Example 23:** The foreman in a machine shop needs to estimate the time needed for the completion of a machined part. The part goes through four separate operations during its manufacture. Find the *average* time at each stage and obtain a time estimate for the complete manufacture.

| Sample number | Time at lathe | Time at drill press | Time at milling machine | Time at grinding machine |
|---|---|---|---|---|
| 1 | 14 min. 10 sec. | 4 min. 30 sec. | 17 min. 20 sec. | 8 min. 30 sec. |
| 2 | 15 min. 20 sec. | 4 min. 40 sec. | 16 min. 40 sec. | 9 min. 15 sec. |
| 3 | 13 min. 30 sec. | 4 min. 15 sec. | 18 min. 10 sec. | 8 min. 25 sec. |
| 4 | 14 min. 20 sec. | 4 min. 15 sec. | 16 min. 50 sec. | 7 min. 20 sec. |
| 5 | 15 min. 15 sec. | 4 min. 50 sec. | 17 min. 25 sec. | 9 min. 30 sec. |
| 6 | 14 min. 50 sec. | 4 min. 10 sec. | 17 min. 50 sec. | 8 min. 45 sec. |
| 7 | 15 min. 05 sec. | 4 min. 15 sec. | 17 min. 10 sec. | 8 min. 30 sec. |
| 8 | 12 min. 50 sec. | 4 min. 25 sec. | 18 min. 35 sec. | 7 min. 45 sec. |
| Totals | 112 min. 200 sec. | 32 min. 200 sec. | 136 min. 240 sec. | 64 min. 240 sec. |

*Solution:*

$$\text{Average time at lathe} = \frac{112 \text{ min. } 200 \text{ sec.}}{8 \text{ samples}} = 14 \text{ min. } 25 \text{ sec.}$$

$$\text{Average time at drill press} = \frac{32 \text{ min. } 200 \text{ sec.}}{8 \text{ samples}} = 4 \text{ min. } 25 \text{ sec.}$$

$$\text{Average time at milling machine} = \frac{136 \text{ min. } 240 \text{ sec.}}{8 \text{ samples}}$$
$$= 17 \text{ min. } 30 \text{ sec.}$$

$$\text{Average time at grinding machine} = \frac{64 \text{ min. } 240 \text{ sec.}}{8 \text{ samples}}$$
$$= 8 \text{ min. } 30 \text{ sec.}$$

Estimate of manufacture time = sum of averages
$$= 43 \text{ min. } 110 \text{ sec.}$$
$$= 44 \text{ min. } 50 \text{ sec.}$$

*Answer:*     Estimate = 45 min (nearest minute)

NOTE: Using this estimate, the shop foreman can expect one part to be completed every 45 minutes. Also, using this estimate, he can find the total shop production, use of materials, and production costs, etc.

## PROBLEMS: Sec. 6 Averages

**1.** Find the average of each set of numbers.
  (a) 6, 8, 5, 8, 7, 6, 7, 8, 8.
  (b) 68, 59, 75, 65, 72, 74, 69, 70.
  (c) 345, 350, 340, 365, 355.
  (d) 1,650; 1,665; 1,646; 1,635; 1,654.

**2.** Find the average length of these measurements.
  4 ft. 8 in.     4 ft. 8 in.     5 ft. 2 in.
  5 ft. 3 in.     4 ft. 11 in.     4 ft. 10 in.

3. Find the average weight of these measurements. (NOTE: 16 oz. = 1 lb.)
   12 lb. 14 oz.   13 lb. 6 oz.   11 lb. 13 oz.
   12 lb. 7 oz.   12 lb. 15 oz.

4. A surveyor needs a very accurate base line. He measures the base line ten
   times and uses the average. What is this average to the nearest inch?
   874 ft. 9 in.   876 ft. 2 in.   876 ft. 4 in.   875 ft. 11 in.
   875 ft. 10 in.   876 ft. 1 in.   875 ft. 6 in.   874 ft. 11 in.
   875 ft. 3 in.   874 ft. 10 in.

5. To find an average working time for a machined part, four machinists are
   given the same job. Their working times are:
   1 hr. 36 min.   1 hr. 46 min.   1 hr. 26 min.   1 hr. 24 min.
   What is the average working time?

6. The Mid-Nite Auto Supply has a delivery truck for servicing their accounts.
   To find the average distance traveled per day, the driver made this record:
   Monday, 46 miles        Tuesday, 54 miles        Wednesday, 38 miles
   Thursday, 65 miles       Friday, 59 miles         Saturday, 44 miles
   To the nearest mile, what is the average distance per day for this driver?

7. Find the average compression, to the nearest pound, in psi for a six-cylinder
   engine with compression readings of:
   142 psi   136 psi   138 psi
   144 psi   140 psi   139 psi

8. A parts salesman earns commissions of $65, $78, $92, $80, $72, $128 for
   a six-day week. What is his average commission per day, to the nearest
   dollar?

9. The Brittany Bus Line needs a time schedule for a daily run from San
   Francisco to Sacramento, a distance of 80 miles. Each day, for five days,
   the bus leaves San Francisco at 1:25 P.M. for Sacramento. Arrival times
   for the five runs are:
       Monday, 2:45 P.M.   Tuesday, 2:55 P.M.   Wednesday, 2:50 P.M.
       Thursday, 3:05 P.M.   Friday, 2:50 P.M.
   (a) Find the average time for the run from S.F. to Sacramento.
   (b) If the bus is scheduled to leave S. F. each week day at 1:25 P.M., what
   time is the bus *expected* to be in Sacramento?

10. In one week a weather station recorded the following 12:00 P.M. tempera-
    tures:
    Monday   64°F   Tuesday   66°F   Wednesday 61°F   Thursday   58°F
    Friday   58°F   Saturday 62°F   Sunday   65°F
    What is the average daily noon temperature for the week, to the nearest
    degree?

11. The weights of the seven linemen on the offensive team for the Honeydew
    Hornets are 195, 216, 234, 228, 255, 220, 205 pounds. The weights of the
    quarterback and the other three backs are 195, 236, 248, 208 pounds. Find
    the following values to the nearest pound.
    (a) What is the average weight per man on the line?
    (b) What is the average weight per man in the backfield?
    (c) What is the average weight per man for the offensive team?

**12.** In six home games Piedmont High School's basketball team scored 66, 58, 79, 82, 45, and 54 points. What is its average score per game?

## Practice Test for Chapter 7

**1.** Name the fraction that shows the shaded portion of Figs. 7-44 and 7-45.

(a)

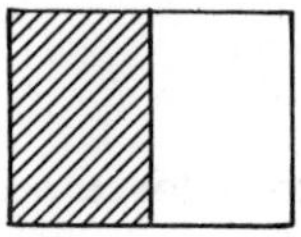

(b)

Answer: _____          Answer: _____

**Figure 7-44**              **Figure 7-45**

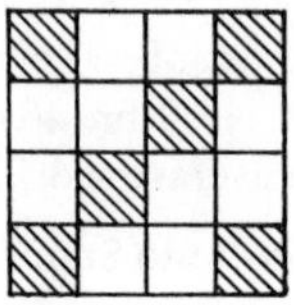

**2.** Which pair of fractions are equal?

(a) $\frac{5}{8}$ and $\frac{15}{24}$          (b) $\frac{7}{16}$ and $\frac{15}{32}$    *Answer:* _____

**3.** Write each of the mixed numbers as improper fractions.

(a) $2\frac{1}{4}$ = _____          (b) $3\frac{5}{8}$ = _____

**4.** Write each improper fraction as a mixed number.

(a) $\frac{23}{4}$ = _____          (b) $\frac{53}{16}$ = _____

**5.** Write each fraction in simplest form.

(a) $\frac{6}{8}$ = _____          (b) $\frac{56}{64}$ = _____

**6.** Write these fractional parts in simplest form.
(a) Six hours is what fractional part of a day?    *Answer:* (a) _____
(b) Six quarts is what fractional part of 2 gallons?
*Answer:* (b) _____
(c) Eight millimeters is what fractional part of a centimeter?
*Answer:* (c) _____

**7.** Write each of these as a ratio.
(a) 18 inches to 4 feet.                    *Answer:* (a) _____
(b) 8 dimes to 12 nickels.                  *Answer:* (b) _____

**8.** Find the gear ratios. Write: $\frac{\text{larger}}{\text{smaller}}$.

(a) 48 teeth to 36 teeth                    *Answer:* (a) _____
(b) 256 teeth to 64 teeth.                  *Answer:* (b) _____

**9.** A force of 500 pounds is applied to the small piston of a hydraulic jack. If the MA (mechanical advantage) is 35:1, how much force is at the face of the large piston?                    *Answer:* _____

10.   The ratio of *pounds per horsepower*, which is written lb/hp, is a fairly reliable index of a car's performance. A low lb/hp ratio means a high performance capability. The Ford-powered Pantera weighs approximately 2,790 lbs and the engine can develop 310 hp. What is the lb/hp ratio?

*Answer:* _____

11.   Find the speeds of pulleys *B, C, D, E,* and *F* in the belt-pulley system in Fig. 7-46.

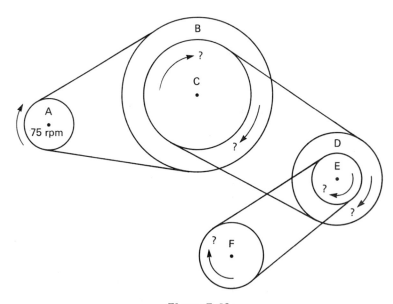

**Figure 7-46**

$A = 12$ inches
$B = 20$ inches          *Answer:*   $B =$ _____ rpm
$C = 8$ inches                        $C =$ _____ rpm
$D = 15$ inches                       $D =$ _____ rpm
$E = 10$ inches                       $E =$ _____ rpm
$F = 12$ inches                       $F =$ _____ rpm

# 8

# Adding,
# Subtracting,
# Multiplying,
# and Dividing
# Common Fractions

## SEC. 1  ADDING FRACTIONS

In the shop, out in the field, or wherever your job takes you, you will make important measurements. These measurements will be made with a ruler, compass, micrometer, tape measure, or some other instrument that will require careful and accurate reading. After making careful measurements, you may have to carry out a computation, such as addition, subtraction, or some other operation in order to get a final result. Be neat and strive for accuracy in whatever you do.

In earlier chapters you learned to add, subtract, multiply, and divide whole numbers. In this section you will learn the methods for adding common fractions. Working with fractions is more difficult than working with whole numbers. But you can master fractions by studying the examples, paying attention to the rules, and working the problems.

When two or more fractions have the *same* denominator, it is easy to add the fractions together.

To see how to add fractions with the same denominator, add $\frac{5}{16}$ to $\frac{7}{16}$. The fraction $\frac{5}{16}$ means that 5 of 16 equal subdivisions of a unit are represented. The fraction $\frac{7}{16}$ means that 7 of 16 equal subdivisions of a unit are represented. To add the fractions $\frac{5}{16}$ and $\frac{7}{16}$ means to find the total number of 16ths represented by both fractions. From this you see that:

$$\frac{5}{16} + \frac{7}{16} = \frac{(5 + 7)}{16}$$
$$= \frac{12}{16}$$

To be a bit more general, let the letter $c$ represent a common denominator for two fractions with numerators $a$ and $b$. To add the fractions $\frac{a}{c}$ and $\frac{b}{c}$, add the two numerators $a$ and $b$ together. Use this sum as the numerator of the fraction with denominator $c$:

$$\frac{a}{c} + \frac{b}{c} = \frac{a + b}{c}$$

This result is stated as the

### Rule for Adding Fractions with a Common Denominator

*To add two or more fractions with the **same** denominator, add all numerators together. Use this sum as the numerator of the fraction. Use the denominator as the common denominator.*

**Example 1:** Find the sum of $\frac{1}{8} + \frac{3}{8} + \frac{5}{8}$.

*Solution:* Add the numerators: $1 + 3 + 5 = 9$. Use this as the numerator of the fraction. The common denominator is 8.

$$\frac{1}{8} + \frac{3}{8} + \frac{5}{8} = \frac{(1 + 3 + 5)}{8} = \frac{9}{8}$$

*Answer:* $\frac{9}{8}$

**Example 2:** Find the sum of $\frac{3}{32} + \frac{7}{32} + \frac{11}{32} + \frac{5}{32}$.

*Solution:* All of these fractions have the common denominator 32. Add the numerators:

$$3 + 7 + 11 + 5 = 26$$

and write the fraction $\frac{26}{32}$

*Answer:* $\frac{26}{32}$

NOTE: You may simplify this result to $\frac{13}{16}$ since $\frac{26 \div 2}{32 \div 2} = \frac{13}{16}$

Use the rule for adding fractions with a common denominator to prove these answers:

1. $\dfrac{3}{8} + \dfrac{5}{8} + \dfrac{7}{8} + \dfrac{1}{8} = \dfrac{(3 + 5 + 7 + 1)}{8} = \dfrac{16}{8} = 2$

2. $\dfrac{5}{16} + \dfrac{7}{16} + \dfrac{9}{16} + \dfrac{11}{16} = \dfrac{(5 + 7 + 9 + 11)}{16} = \dfrac{32}{16} = 2$

3. $\dfrac{7}{64} + \dfrac{15}{64} + \dfrac{19}{64} = \dfrac{(7 + 15 + 19)}{64} = \dfrac{41}{64}$

In each case the numerators are added together. The result is written as a fraction with the common denominator. The final result is always written in the simplest form.

When adding two or more fractions that have different denominators you must first find a common denominator. To try to add two fractions such as $\dfrac{2}{3}$ and $\dfrac{3}{4}$ without first finding a common denominator is like trying to add feet and inches, yards and miles, or nickels and dimes. *Fractions can only be added together when they are all written with the same denominator.* The fractions $\dfrac{2}{3}$ and $\dfrac{3}{4}$ can both be written with the common denominator 12:

$$\frac{2}{3} + \frac{3}{4} = \frac{8}{12} + \frac{9}{12}$$

$$= \frac{8 + 9}{12}$$

$$= \frac{17}{12}$$

From this result it is possible to state the

### Rule for Adding Fractions with Different Denominators

*To add two or more fractions with different denominators, write all fractions using a common denominator. Add the fractions by adding their numerators together. Use this sum as the numerator. Use the common denominator as the denominator.*

Study Examples 3 —— 6.

**Example 3:** Find the sum of $\dfrac{1}{2} + \dfrac{1}{4}$.

*Solution:* Use 4 as the common denominator because $\dfrac{1}{2} = \dfrac{2}{4}$ and $\dfrac{1}{4} = \dfrac{1}{4}$.

*Answer:* $\dfrac{1}{2} + \dfrac{1}{4} = \dfrac{2}{4} + \dfrac{1}{4} = \dfrac{3}{4}$

**Example 4:** Find the sum of $\frac{1}{8} + \frac{5}{16} + \frac{3}{32}$.

*Solution:* Find a common denominator for the three fractions. You need a number that is divisible by 8, 16, and 32. Use 32 as the common denominator because $\frac{1}{8} = \frac{4}{32}$, $\frac{5}{16} = \frac{10}{32}$, and $\frac{3}{32} = \frac{3}{32}$.

*Answer:* $\frac{4}{32} + \frac{10}{32} + \frac{3}{32} = \frac{17}{32}$

**Example 5:** Add the fractions $\frac{1}{3}, \frac{1}{4}$, and $\frac{5}{6}$.

*Solution:* Here you have three denominators, 3, 4, and 6. Find a number that has 3, 4, and 6 as factors. Use 12 because $3 \times 4 = 12$, $4 \times 3 = 12$, and $2 \times 6 = 12$. Write each fraction with the common denominator 12.

$$\frac{1}{3} = \frac{1 \times 4}{3 \times 4} = \frac{4}{12}$$

$$\frac{1}{4} = \frac{1 \times 3}{4 \times 3} = \frac{3}{12}$$

$$\frac{5}{6} = \frac{5 \times 2}{6 \times 2} = \frac{10}{12}$$

Add the fractions by adding the numerators 4, 3, and 10.

*Answer:* $\frac{17}{12} = 1\frac{5}{12}$

**Example 6:** Find the sum of the mixed numbers $4\frac{1}{3} + 3\frac{1}{2} + 5\frac{3}{4}$.

*Solution:* Use 12 as the common denominator for the three fractions. Add the whole numbers and fractions separately. Reduce to simplest form.

$$4\frac{4}{12} + 3\frac{6}{12} + 5\frac{9}{12} = 12\frac{19}{12}$$

*Answer:* $\qquad 12\frac{19}{12} = 12 + 1\frac{7}{12} = 13\frac{7}{12}$

**Example 7:** Find the sum of $2\frac{1}{2} + 3\frac{1}{5} + 3\frac{2}{3}$.

*Solution:* Use 30 as the common denominator.

$$2\frac{15}{30} + 3\frac{6}{30} + 3\frac{20}{30} = 8\frac{41}{30}$$

*Answer:*
$$8\frac{41}{30} = 8 + 1\frac{11}{30} = 9\frac{11}{30}$$

Any number may be used as a common denominator as long as it is divisible by all of the denominators. It is best to use as the common denominator the smallest number into which all of the denominators will divide. The smallest common denominator is also called the *Least Common Denominator* (LCD). Study Examples 8-10 to learn how to find the LCD.

**Example 8:** Figure 8-1 shows a brace plate. Find the distance around it.

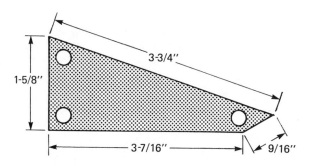

**Figure 8-1**

*Solution:* The distance $D$ around the figure is the sum of the four measurements:

$$D = 3\frac{7''}{16} + \frac{9''}{16} + 3\frac{3''}{4} + 1\frac{5''}{8}$$

Add the whole numbers:

$$3 + 3 + 1 = 7$$

Then add the fractions. Use 16 as the *least common denominator* because it is the smallest number into which 4, 8, and 16 will divide.

$$\frac{7}{16} + \frac{9}{16} + \frac{12}{16} + \frac{10}{16} = \frac{38}{16} = 2\frac{3}{8}$$

Add the whole number 7 to this sum.

*Answer:* $D = 7 + 2\frac{3}{8} = 9\frac{3''}{8}$

**Example 9:** Find the length of the bolt shown in Fig. 8-2.

*Solution:* Length of bolt $= 1\frac{3''}{16} + 2\frac{1''}{2} + \frac{5''}{8}$

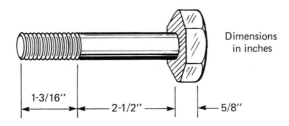

Dimensions
in inches

1-3/16" | 2-1/2" | 5/8"

**Figure 8-2** *Dimensions in inches*

$$\text{(Change to 16ths)} = 1\frac{3''}{16} + 2\frac{8''}{16} + \frac{10''}{16}$$

*Answer:* $\qquad \text{Length} = 3\frac{21''}{16} = 4\frac{5''}{16}$

**Example 10:** Find the length of the piston and connecting rod shown in Fig. 8-3. All dimensions are in inches.

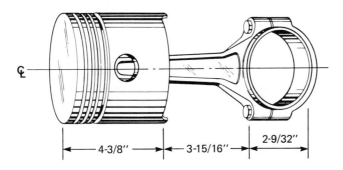

₵

4-3/8" | 3-15/16" | 2-9/32"

**Figure 8-3**

*Solution:* Length of piston and connecting rod = sum of all values.

$$\text{Length} = 4\frac{3''}{8} + 3\frac{15''}{16} + 2\frac{9''}{32}$$

Change all fractions to 32nds.

$$\text{Length} = 4\frac{12''}{32} + 3\frac{30''}{32} + 2\frac{9''}{32}$$

$$= 9\frac{51''}{32} = 9'' + 1\frac{19''}{32}$$

*Answer:* $\qquad \text{Length} = 10\frac{19''}{32}$

Most fractions used in technology have denominators of 2, 4, 8, 16, 32, 64, or a multiple of 10. Sometimes when you use a formula or a rule, you may get fractions with other denominators however. In Section 4 of this chapter you will learn to work with all kinds of fractions. To be expert in your field you must know and understand fractions.

## PROBLEMS:  Sec. 1  Adding Fractions

**1.** Do these additions. Write the answers in the simplest form.

(a) $\frac{1}{10} + \frac{2}{10} + \frac{3}{10}$  (b) $\frac{1}{8} + \frac{3}{8} + \frac{3}{8}$

(c) $\frac{3}{16} + \frac{5}{16}$  (d) $\frac{3}{8} + \frac{1}{8}$

(e) $\frac{5}{10} + \frac{3}{10}$  (f) $\frac{1}{12} + \frac{5}{12} + \frac{7}{12}$

(g) $\frac{5}{16} + \frac{7}{16} + \frac{3}{16}$  (h) $\frac{3}{10} + \frac{5}{10}$

(i) $\frac{5}{32} + \frac{7}{32} + \frac{9}{32}$  (j) $\frac{7}{64} + \frac{15}{64}$

(k) $\frac{7}{16} + \frac{5}{16} + \frac{9}{16}$  (l) $\frac{15}{32} + \frac{17}{32} + \frac{9}{16} + \frac{15}{16}$

(m) $\frac{21}{32} + \frac{19}{32} + \frac{15}{32}$  (n) $\frac{1}{4} + \frac{3}{4}$

(o) $\frac{3}{8} + \frac{5}{8}$

**2.** Find the sums. Add whole numbers and fractions separately. Write answers as mixed numbers with fractions in the simplest form.

(a) $1\frac{1}{4} + 2\frac{1}{4} + 5\frac{1}{4}$  (b) $7\frac{1}{8} + 3\frac{1}{8} + 5\frac{3}{8}$

(c) $3\frac{5}{16} + 7\frac{3}{16} + 6\frac{7}{16}$  (d) $8\frac{3}{32} + 10\frac{5}{32} + 6\frac{7}{32}$

(e) $10\frac{1}{10} + 3\frac{3}{10} + 7\frac{7}{10}$  (f) $4\frac{5}{32} + 5\frac{4}{32} + 3\frac{3}{32}$

(g) $7\frac{17}{64} + 6\frac{13}{64} + 9\frac{27}{64}$  (h) $4\frac{5}{12} + 1\frac{7}{12} + 3\frac{2}{12}$

(i) $5\frac{17}{100} + 7\frac{45}{100} + 9\frac{35}{100}$  (j) $5\frac{153}{1,000} + 18\frac{237}{1,000}$

**3.** Add the measurements. Write answers in simplest form.

(a) $3\frac{1}{2}$ in. $+ 4\frac{1}{2}$ in. $+ 7\frac{1}{2}$ in.

(b) $1\frac{1}{4}$ lb. $+ 4\frac{3}{4}$ lb. $+ 6\frac{3}{4}$ lb.

(c) $4\frac{3}{8}$ in. $+ 8\frac{7}{8}$ in. $+ 5\frac{5}{8}$ in.

(d) $\frac{5}{12}$ ft. $+ \frac{7}{12}$ ft. $+ 3\frac{1}{12}$ ft.

(e) $3\frac{5}{36}$ yd. $+ 7\frac{19}{36}$ yd. $+ 4\frac{6}{36}$ yd.

(f) $1\frac{5}{16}$ mi. $+ 6\frac{4}{16}$ mi. $+ 7\frac{3}{16}$ mi.

(g) $19\frac{7}{100}$ cm. $+ 4\frac{16}{100}$ cm. $+ 6\frac{54}{100}$ cm.

**4.** Find a common denominator for these fractions. (A common denominator is any number that ALL other denominators will divide into.)

**Example:** Find a common denominator for the fractions $\frac{1}{2}, \frac{1}{4}, \frac{1}{6}$.

*Solution:* Use 12 as the common denominator since each of the denominators (2, 4, 6) will divide 12.

*Answer:* The common denominator is 12.

(a) $\frac{1}{2}$ and $\frac{1}{4}$    $\frac{1}{2}$ and $\frac{1}{8}$    $\frac{1}{4}$ and $\frac{1}{8}$    $\frac{1}{4}$ and $\frac{1}{16}$

(b) $\frac{1}{2}$ and $\frac{1}{3}$    $\frac{1}{2}$ and $\frac{1}{16}$    $\frac{1}{2}$ and $\frac{1}{12}$    $\frac{1}{4}$ and $\frac{1}{16}$

(c) $\frac{1}{4}$ and $\frac{3}{16}$    $\frac{3}{4}$ and $\frac{5}{8}$    $\frac{5}{16}$ and $\frac{3}{8}$    $\frac{7}{12}$ and $\frac{3}{4}$

(d) $\frac{7}{8}$ and $\frac{3}{4}$    $\frac{1}{2}$ and $\frac{13}{16}$    $\frac{1}{32}$ and $\frac{3}{8}$    $\frac{5}{16}$ and $\frac{7}{32}$

(e) $\frac{1}{2}, \frac{1}{4}$, and $\frac{1}{8}$

(f) $\frac{1}{2}, \frac{1}{4}$, and $\frac{1}{6}$

(g) $\frac{3}{4}, \frac{1}{2}$, and $\frac{5}{8}$

(h) $\frac{5}{16}, \frac{3}{8}$, and $\frac{7}{32}$

(i) $\frac{5}{8}, \frac{17}{64}$, and $\frac{3}{32}$

(j) $\frac{1}{4}, \frac{1}{6}$, and $\frac{1}{8}$

(k) $\frac{1}{2}, \frac{1}{3}$, and $\frac{1}{4}$

(l) $\frac{1}{4}, \frac{1}{5}$, and $\frac{3}{20}$

(m) $\frac{2}{3}, \frac{4}{5}$, and $\frac{1}{4}$

(n) $\frac{5}{6}, \frac{3}{4}$, and $\frac{7}{12}$

(o) $\frac{7}{8}, \frac{5}{6}$, and $\frac{3}{4}$

(p) $\frac{7}{8}, \frac{11}{12}$, and $\frac{5}{6}$

**5.** Change each set of fractions to equal fractions with the indicated denominator.

**Example:** Change $\frac{3}{4}$ and $\frac{5}{6}$ each to 12ths.

*Solution:* $\frac{3}{4} = \frac{9}{12}$ since $\frac{3 \times 3}{4 \times 3} = \frac{9}{12}$

$\frac{5}{6} = \frac{10}{12}$ since $\frac{5 \times 2}{6 \times 2} = \frac{10}{12}$

(a) $\frac{1}{2}$ and $\frac{1}{4}$ to 8ths

(b) $\frac{1}{2}$ and $\frac{1}{3}$ to 6ths

(c) $\frac{3}{4}$ and $\frac{1}{6}$ to 12ths

(d) $\frac{3}{4}$ and $\frac{5}{8}$ to 16ths

(e) $\frac{1}{4}$ and $\frac{1}{3}$ to 12ths

(f) $\frac{1}{6}$ and $\frac{5}{8}$ to 24ths

(g) $\frac{1}{5}$ and $\frac{1}{4}$ to 20ths

(h) $\frac{3}{5}$ and $\frac{1}{3}$ to 15ths

(i) $\frac{11}{16}, \frac{3}{8}$, and $\frac{1}{4}$ to 32nds

(j) $\frac{3}{8}, \frac{5}{16}$, and $\frac{7}{64}$ to 64ths

(k) $\frac{1}{3}, \frac{1}{2},$ and $\frac{3}{4}$ to 12ths     (l) $\frac{4}{5}, \frac{3}{10},$ and $\frac{3}{4}$ to 20ths

(m) $\frac{11}{16}, \frac{5}{8},$ and $\frac{2}{3}$ to 48ths     (n) $\frac{1}{6}, \frac{3}{8},$ and $\frac{5}{12}$ to 24ths

(o) $\frac{5}{6}, \frac{4}{9},$ and $\frac{7}{12}$ to 36ths     (p) $\frac{9}{16}, \frac{5}{8},$ and $\frac{7}{12}$ to 48ths

**Problems 6 through 11.** Add each pair of numbers. Write fractions in simplest form. Don't forget that fractions can be added only when they have the same denominator.

**Example:** (a) $3\frac{1}{2} = 3\frac{2}{4}$     (b) $5\frac{1}{4} = 5\frac{3}{12}$

$+4\frac{1}{4} = 4\frac{1}{4}$          $4\frac{5}{12} = 4\frac{5}{12}$

$\overline{\phantom{+4\frac{1}{4}}}$          $\overline{\phantom{4\frac{5}{12}}}$

$7\frac{3}{4}$          $9\frac{8}{12} = 9\frac{2}{3}$

**6.** (a) $4\frac{1}{2}$     (b) $5\frac{1}{4}$     (c) $4\frac{1}{4}$     (d) $8\frac{1}{8}$     (e) $9\frac{1}{2}$

$5\frac{1}{4}$          $9\frac{3}{8}$          $7\frac{1}{16}$          $2\frac{3}{16}$          $1\frac{7}{16}$

**7.** (a) $3\frac{1}{4}$     (b) $11\frac{3}{4}$     (c) $10\frac{3}{16}$     (d) $14\frac{11}{32}$     (e) $6\frac{17}{32}$

$4\frac{3}{8}$          $12\frac{5}{16}$          $5\frac{3}{8}$          $5\frac{1}{4}$          $9\frac{1}{2}$

**8.** (a) $6\frac{3}{16}$     (b) $4\frac{5}{8}$     (c) $6\frac{5}{16}$     (d) $10\frac{3}{5}$     (e) $8\frac{17}{32}$

$5\frac{1}{4}$          $\frac{7}{32}$          $6\frac{7}{32}$          $19\frac{1}{10}$          $16\frac{3}{64}$

**9.** (a) $\frac{5}{16}$     (b) $7\frac{1}{2}$     (c) $24\frac{5}{6}$     (d) $28\frac{1}{3}$     (e) $2\frac{1}{3}$

$4\frac{11}{64}$          $3\frac{9}{64}$          $6\frac{7}{12}$          $18\frac{1}{2}$          $4\frac{3}{4}$

**10.** (a) $9\frac{1}{4}$     (b) $19\frac{1}{6}$     (c) $7\frac{1}{3}$     (d) $35\frac{5}{12}$     (e) $3\frac{7}{12}$

$7\frac{3}{5}$          $2\frac{1}{8}$          $1\frac{2}{5}$          $21\frac{19}{36}$          $5\frac{13}{60}$

**11.** (a) $3\frac{7}{10}$     (b) $4\frac{7}{25}$     (c) $9\frac{7}{20}$     (d) $3\frac{5}{6}$     (e) $8\frac{1}{6}$

$9\frac{34}{100}$          $6\frac{5}{10}$          $5\frac{11}{25}$          $7\frac{15}{36}$          $3\frac{5}{9}$

**Problems 12 through 19.** Find the sum in each problem.

**12.** (a) $\frac{3}{4}$ and $\frac{5}{8}$     (b) $3\frac{1}{2}$ and $7\frac{1}{8}$     (c) $3\frac{5}{8}$ and $2\frac{3}{16}$

**13.** (a) 4 and $5\frac{1}{8}$     (b) $6\frac{3}{5}$ and $8\frac{3}{10}$     (c) $7\frac{3}{8}$ and $9\frac{5}{64}$

**14.** (a) $\frac{1}{2}$, $\frac{3}{8}$, and $\frac{5}{16}$        (b) $\frac{2}{3}$, $\frac{1}{16}$, and $\frac{5}{12}$

**15.** (a) $3\frac{1}{4}$, $7\frac{2}{5}$, and $9\frac{7}{10}$        (b) $6\frac{1}{3}$, $7\frac{1}{4}$, and $3\frac{5}{8}$

**16.** (a) $23\frac{5}{8}$, $8\frac{7}{16}$, and $5\frac{35}{64}$        (b) $9\frac{1}{5}$, $6\frac{3}{4}$, and $8\frac{9}{25}$

**17.** $5\frac{3}{8}$, $4\frac{7}{16}$, $8\frac{9}{32}$, and $5\frac{3}{32}$

**18.** $23\frac{5}{16}$, $45\frac{7}{32}$, $9\frac{1}{2}$, and $14\frac{5}{8}$

**19.** $\frac{1}{2}$, $\frac{1}{3}$, $\frac{1}{4}$, and $\frac{1}{8}$

**20.** Two pieces of metal are welded together. What is the total thickness if one piece is $1\frac{1}{8}$ inches thick and the other is $\frac{5}{16}$ inch thick?

**21.** Three pieces of plywood $\frac{3}{4}$ inch, $\frac{3}{8}$ inch, and $\frac{1}{4}$ inch are glued together. What is the total thickness of the three pieces?

**22.** A cabinet maker needs moldings of 1 ft. $6\frac{1}{2}$ in., 2 ft. $4\frac{1}{8}$ in., 5 ft. $9\frac{3}{4}$ in., and 4 ft. $7\frac{5}{16}$ in. What is the total length of molding needed?

**23.** A maintenance technician for an air conditioning company needs the following lengths of copper tubing to repair a cooling unit:

$$4\frac{11}{16} \text{ in.} \qquad 6\frac{5}{8} \text{ in.} \qquad 8\frac{7}{32} \text{ in.} \qquad 23\frac{3}{4} \text{ in.}$$

What is the total length of tubing he needs?

**24.** Four standard distance blocks are used to measure the height of a machined part. The four blocks are:

$$\frac{5}{32} \text{ in.} \qquad \frac{7}{64} \text{ in.} \qquad \frac{1}{8} \text{ in.} \qquad \frac{1}{2} \text{ in.}$$

How high is the part?

**25.** What is the length of the bolt shown in Fig. 8-4?

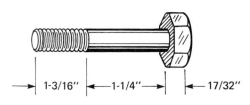

1-3/16"  1-1/4"  17/32"

*Figure 8-4*

26. Figure 8-5 shows an alignment bracket. Find the dimensions marked $A$ and $B$.

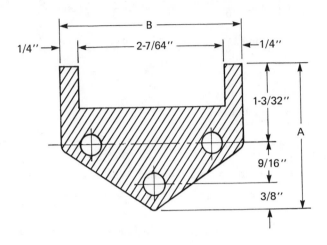

**Figure 8-5**

27. A machinist wants to make five taper pins from a piece of bar stock that is 56 inches long. The taper pins are to be:

$$5\frac{3}{4} \text{ in.} \qquad 8\frac{7}{8} \text{ in.} \qquad 12\frac{9}{16} \text{ in.} \qquad 15\frac{7}{32} \text{ in.} \qquad 6\frac{7}{16} \text{ in.}$$

Allow $\frac{1}{32}$ inch waste for each of the five cuts. Will the machinist be able to make the five taper pins from the piece of bar stock?

28. A finished machine part is shown in Fig. 8-6. What is the dimension marked $A$?

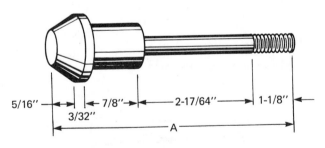

**Figure 8-6**

29. Figure 8-7 shows a double pulley. Find values for the dimensions marked $A$, $B$, and $C$.

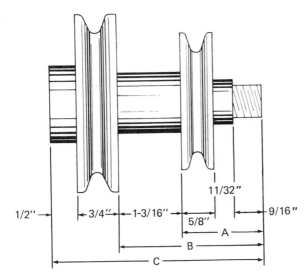

**Figure 8-7**

**30.** A brass rod is cut into lengths of $4\frac{1}{2}$, $3\frac{5}{8}$, $7\frac{9}{16}$, $8\frac{1}{2}$, and $2\frac{5}{32}$ inches. If $\frac{1}{16}$ inch is wasted on each cut, what was the length of the rod? (Note: To get the five pieces, it is necessary to make only *four* cuts—not five.)

**31.** Four low-resistance lamps are connected in series. The resistances are:

$$1\frac{1}{3} \text{ ohms} \qquad 3\frac{1}{2} \text{ ohms} \qquad 2\frac{1}{4} \text{ ohms} \qquad 4\frac{5}{6} \text{ ohms}$$

If the total resistance is the sum of the individual resistances, what is the total resistances (in ohms) for the four lamps?

**32.** A panel door $1\frac{3}{8}$ inches thick is to have a piece of vinyl-surfaced plywood $\frac{3}{16}$ inch thick glued to one side. A piece of prefinished plywood $\frac{5}{32}$ inch thick is to be glued to the other side. What will be the thickness of the finished door?

**33.** An auto-body man at Brownie's Body Shop is making a repair estimate for an insurance company. His estimates are:

Replace rear door     $3\frac{1}{2}$ hours

Outer panel          $4\frac{1}{4}$ hours

Weatherstrip          $\frac{1}{2}$ hour

Paint          $1\frac{1}{3}$ hours

What is the total estimated time for the repair job?

**34.** In one week the weather station at Oakland International Airport recorded the following rainfall:

| | | | |
|---|---|---|---|
| Monday | $\frac{7}{8}$ inch | Tuesday | $\frac{1}{16}$ inch |
| Wednesday | $\frac{3}{32}$ inch | Thursday | $\frac{1}{4}$ inch |
| Friday | $\frac{15}{16}$ inch | Saturday | $\frac{1}{32}$ inch |
| Sunday | no rainfall | | |

What was the rainfall for the week?

## SEC. 2   SUBTRACTING FRACTIONS

Subtracting fractions is similar to adding fractions. If the fractions have the same denominators, you get the difference by subtracting the numerators.

**Example 11:** Do the subtraction: $\frac{17}{32} - \frac{9}{32}$.

*Solution:* $\frac{17}{32} - \frac{9}{32} = \frac{17-9}{32} = \frac{8}{32}$

Simplify $\frac{8}{32} = \frac{8 \div 8}{32 \div 8} = \frac{1}{4}$

*Answer:* Difference $= \frac{1}{4}$

NOTE: Always write your answers in the simplest form.

If one or both of the numbers are mixed numbers, find the difference of the fractions. Then find the difference of the whole numbers. You may want to change mixed numbers to improper fractions and then find the difference.

**Example 12:** Find the difference: $4\frac{1}{2} - 3\frac{1}{8}$.

*Solution:* Find the difference between the fractions:

$$\frac{1}{2} - \frac{1}{8} = \frac{4}{8} - \frac{1}{8} = \frac{3}{8}$$

The difference of the whole numbers is 1.

*Answer:* $4\frac{1}{2} - 3\frac{1}{8} = 1\frac{3}{8}$

**Example 13:** Find the difference: $3\dfrac{7}{8} - 2\dfrac{15}{16}$.

*Solution:* When you write the fractions with common denominator 16, you have: $\dfrac{14}{16} - \dfrac{15}{16}$.

Because 14 is smaller than 15, you must borrow a full unit, $\left(\dfrac{16}{16}\right)$ from the whole number 3. The problem now looks like this:

$$2\dfrac{30}{16} - 2\dfrac{15}{16} = \dfrac{15}{16}$$

In this and similar situations, you can simplify the problem by changing the mixed numbers to improper fractions, and then finding a common denominator.

$$3\dfrac{7}{8} - 2\dfrac{15}{16} = \dfrac{31}{8} - \dfrac{47}{16} = \dfrac{62}{16} - \dfrac{47}{16}$$

*Answer:* $\qquad 3\dfrac{7}{8} - 2\dfrac{15}{16} = \dfrac{15}{16}$

**Example 14:** Find the difference between the diameters at the ends of the tapered rod shown in Fig. 8-8.

1-15/16″                    1-29/64″

**Figure 8-8**

*Solution:* The difference is: $1\dfrac{15''}{16} - 1\dfrac{29''}{64}$. Change to 64ths, which gives:

*Answer:* $1\dfrac{60''}{64} - 1\dfrac{29''}{64} = \dfrac{31''}{64}$

In some problems you must do both adding and subtracting. (Generally, it is easier to do the adding before doing any subtracting.)

**Example 15:** A hole is bored off-center as shown in Fig. 8-9. Find the dimension marked D.

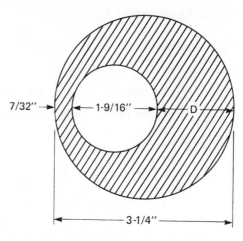

**Figure 8-9**

*Solution:* The value for D is the difference between the overall dimension $\left(3\frac{1}{4}\right)$ and the sum of the two interior dimensions $\left(\frac{7}{32} + 1\frac{9}{16}\right)$.

$$D = 3\frac{1}{4} - \left(\frac{7}{32} + 1\frac{9}{16}\right)$$

Do the adding in the parentheses. Change all denominators to 32nds, and then find the difference.

$$D = \frac{104}{32} - \frac{57}{32}$$

*Answer:* $D = \frac{47}{32} = 1\frac{15}{32}$ inches.

NOTE: Some of the steps are left out. Fill in the missing steps to be sure the answer is correct.

For some problems you will find that *borrowing* is the best way to get the answer. For other problems you will find that *changing to improper fractions* is the best way to get the answer. In any problem you must decide which method to use.

## PROBLEMS: Sec. 2
### Subtracting Fractions

**Problems 1 through 5.** Use a ruler to find the missing numerators.

**Example:** $4\frac{1}{2}$ in. $= 3\frac{?}{8}$ in.

*Solution:* Start at the 3-inch mark on the ruler. Count the number of $\frac{1}{8}$ inch divisions until you reach the $4\frac{1}{2}$ inch mark. From the ruler you find that:

*Answer:* $4\frac{1}{2}$ in. $= 3\frac{12}{8}$ in.

1. (a) 3 in. $= 2\frac{}{4}$ in.  (b) $4\frac{1}{4}$ in. $= 4\frac{}{16}$ in.  (c) 5 in. $= 4\frac{}{16}$ in.

2. (a) $1\frac{3}{4}$ in. $= \frac{}{4}$ in.  (b) $6\frac{7}{16}$ in. $= 5\frac{}{16}$ in.  (c) $8\frac{1}{8}$ in. $= 7\frac{}{8}$ in.

3. (a) $11\frac{1}{4}$ in. $= 10\frac{}{4}$ in.  (b) $4\frac{5}{8} = 3\frac{}{8}$ in.  (c) $6\frac{5}{16}$ in. $= 5\frac{}{16}$ in.

4. $6\frac{1}{2}$ in. $= 6\frac{}{4}$ in. $= 5\frac{}{8}$ in.

5. $9\frac{3}{4}$ in. $= 8\frac{}{4}$ in. $= 8\frac{}{16}$ in.

**Problems 6 through 9.** Find the missing numerator.

**Example:** $4 = 3\frac{?}{2}$

*Answer:* $4 = 3 + 1 = 3 + \frac{2}{2} = 3\frac{2}{2}$

**Example:** $5\frac{1}{4} = 4\frac{?}{8}$

*Answer:* $5\frac{1}{4} = 5 + \frac{1}{4} = 4 + 1 + \frac{1}{4} = 4 + \frac{8}{8} + \frac{2}{8} = 4\frac{10}{8}$

6. (a) $5 = 4\frac{}{16}$  (b) $6 = 5\frac{}{4}$  (c) $8 = 7\frac{}{8}$  (d) $10 = 9\frac{}{16}$

7. (a) $1\frac{1}{2} = \frac{}{2}$  (b) $2\frac{1}{4} = 1\frac{}{16}$  (c) $4\frac{2}{5} = 3\frac{}{10}$  (d) $8\frac{1}{2} = 7\frac{}{10}$

8. (a) $1\frac{15}{16} = \frac{}{16}$  (b) $5\frac{7}{32} = 4\frac{}{64}$  (c) $7\frac{9}{16} = 6\frac{}{64}$  (d) $11\frac{9}{10} = 10\frac{}{100}$

9. (a) $2\frac{3}{7} = 1\frac{}{7}$  (b) $5\frac{11}{12} = 4\frac{}{24}$  (c) $2\frac{1}{3} = 1\frac{}{9}$  (d) $7\frac{3}{5} = 5\frac{}{20}$

**Problems 10 through 15.** Subtract the fractions. Write answers in simplest form.

**Example:** $\frac{5}{8} - \frac{3}{8} = \frac{(5-3)}{8} = \frac{2}{8} = \frac{1}{4}$ (reduced to simplest form)

10. (a) $\frac{5}{16} - \frac{3}{16}$  (b) $\frac{7}{8} - \frac{5}{8}$  (c) $\frac{3}{4} - \frac{1}{4}$  (d) $\frac{15}{16} - \frac{9}{16}$

11. (a) $\frac{3}{8} - \frac{2}{8}$  (b) $\frac{7}{16} - \frac{4}{16}$  (c) $\frac{9}{32} - \frac{5}{32}$  (d) $\frac{17}{32} - \frac{16}{32}$

12. (a) $\frac{7}{10} - \frac{4}{10}$  (b) $\frac{2}{3} - \frac{1}{3}$  (c) $\frac{4}{8} - \frac{2}{8}$  (d) $\frac{8}{16} - \frac{5}{16}$

13. (a) $\frac{15}{32} - \frac{9}{32}$  (b) $\frac{27}{64} - \frac{16}{64}$  (c) $\frac{5}{6} - \frac{3}{6}$  (d) $\frac{55}{64} - \frac{33}{64}$

**14.** (a) $\dfrac{45}{100} - \dfrac{25}{100}$ (b) $\dfrac{13}{10} - \dfrac{9}{10}$ (c) $\dfrac{12}{8} - \dfrac{7}{8}$ (d) $\dfrac{24}{16} - \dfrac{13}{16}$

**15.** (a) $\dfrac{40}{32} - \dfrac{20}{32}$ (b) $\dfrac{7}{4} - \dfrac{3}{4}$ (c) $\dfrac{17}{12} - \dfrac{11}{12}$ (d) $\dfrac{25}{36} - \dfrac{18}{36}$

**Problems 16 through 19.** Find the differences. Write answers in simplest form.

**16.** (a) $4\dfrac{3}{4}$ in. $- 2\dfrac{1}{4}$ in. (b) $6\dfrac{7}{16}$ in. $- 5\dfrac{3}{16}$ in.

**17.** (a) $7\dfrac{7}{8}$ in. $- 5\dfrac{1}{8}$ in. (b) $10\dfrac{8}{10}$ cm. $- 7\dfrac{5}{10}$ cm.

**18.** (a) $13\dfrac{15}{16}$ lb. $- 10\dfrac{7}{16}$ lb. (b) $5\dfrac{11}{32}$ in. $- 5\dfrac{10}{32}$ in.

**19.** (a) $15\dfrac{25}{36}$ yd. $- 12\dfrac{22}{36}$ yd. (b) $4\dfrac{3}{7}$ week $- 3\dfrac{1}{7}$ week

**Problems 20 through 25.** Do the subtractions. Write answers in simplest form. (NOTE: Fractions can be subtracted *only* when the denominators are the same.)

Example: (a) $\begin{aligned} 7\dfrac{1}{2} &= 7\dfrac{2}{4} \\ -3\dfrac{1}{4} &= 3\dfrac{1}{4} \\ \hline &\phantom{=}4\dfrac{1}{4} \end{aligned}$ (b) $8 - 2\dfrac{5}{16} = 7\dfrac{16}{16} - 2\dfrac{5}{16} = 5\dfrac{11}{16}$

**20.** (a) $4\dfrac{5}{8} - 2\dfrac{3}{16}$ (b) $7\dfrac{1}{2} - 5\dfrac{3}{8}$ (c) $8\dfrac{7}{16} - 4\dfrac{1}{4}$ (d) $5\dfrac{5}{32} - 1\dfrac{1}{16}$

**21.** (a) $16\dfrac{7}{8} - 4\dfrac{17}{32}$ (b) $10\dfrac{3}{10} - 5\dfrac{1}{5}$ (c) $4\dfrac{3}{4} - 2\dfrac{5}{32}$ (d) $\dfrac{3}{4} - \dfrac{1}{2}$

**22.** (a) $\dfrac{15}{16} - \dfrac{5}{8}$ (b) $\dfrac{45}{64} - \dfrac{3}{8}$ (c) $4\dfrac{7}{32} - 4\dfrac{1}{64}$ (d) $6\dfrac{5}{12} - 4\dfrac{1}{4}$

**23.** (a) $6\dfrac{1}{2} - 5\dfrac{1}{3}$ (b) $4\dfrac{1}{2} - 3\dfrac{1}{5}$ (c) $5\dfrac{3}{4} - 2\dfrac{1}{6}$ (d) $\dfrac{3}{4} - \dfrac{1}{3}$

**24.** (a) $10 - 5\dfrac{1}{2}$ (b) $19 - 15\dfrac{11}{16}$ (c) $32 - 10\dfrac{11}{32}$ (d) $1 - \dfrac{45}{64}$

**25.** (a) $4 - 3\dfrac{3}{4}$ (b) $5\dfrac{3}{8} - 2$ (c) $4\dfrac{7}{10} - 3\dfrac{7}{100}$ (d) $10 - \dfrac{1}{10}$

**Problems 26 through 31.** Do the subtractions. Write the answers in simplest form.

Example: $4\dfrac{5}{16} - 2\dfrac{3}{4} = 3\dfrac{21}{16} - 2\dfrac{12}{16} = 1\dfrac{9}{16}$

NOTE: Borrow 1 which is equal to $\dfrac{16}{16}$ from the 4 to get $3\dfrac{21}{16}$, then subtract $2\dfrac{12}{16}$ to get the answer.

**26.** (a) $6\dfrac{1}{2} - 4\dfrac{3}{4}$ (b) $9\dfrac{3}{8} - 7\dfrac{9}{16}$ (c) $5\dfrac{1}{4} - 2\dfrac{1}{2}$

**27.** (a) $10\frac{3}{4} - 7\frac{13}{16}$    (b) $22\frac{5}{8} - 10\frac{7}{8}$    (c) $11\frac{1}{3} - 5\frac{2}{3}$

**28.** (a) $8\frac{5}{12} - 5\frac{1}{2}$    (b) $8\frac{15}{16} - 7\frac{31}{32}$    (c) $16\frac{17}{32} - 9\frac{45}{64}$

**29.** (a) $7\frac{1}{4} - 4\frac{3}{5}$    (b) $4\frac{1}{3} - 2\frac{1}{2}$    (c) $7\frac{1}{4} - 5\frac{2}{3}$

**30.** (a) $9\frac{1}{5} - 3\frac{1}{3}$    (b) $4\frac{1}{6} - 2\frac{1}{2}$    (c) $14\frac{1}{8} - 10\frac{1}{3}$

**31.** (a) $17\frac{2}{5} - 11\frac{2}{3}$    (b) $5\frac{1}{3} - 3\frac{7}{12}$    (c) $6\frac{11}{20} - 4\frac{3}{5}$

**32.** From $7\frac{1}{8}$ subtract $3\frac{3}{4}$.

**33.** From $5\frac{1}{2}$ take $2\frac{1}{3}$.

**34.** From $7$ subtract $5\frac{11}{32}$.

**35.** Subtract $10\frac{5}{8}$ from $15\frac{3}{4}$.

**36.** What is the result of $5\frac{1}{4}$ decreased by $3\frac{7}{16}$?

**37.** Find the difference between $4\frac{1}{5}$ and $2\frac{3}{10}$.

**38.** Take $23\frac{1}{2}$ from $45\frac{3}{8}$.

**39.** Take $5\frac{1}{32}$ from $12$.

**40.** Decrease $7\frac{1}{2}$ by $4\frac{3}{8}$.

**41.** How much is left when $17\frac{1}{5}$ is subtracted from $23\frac{9}{10}$?

**42.** By how much does $10\frac{7}{16}$ exceed $7\frac{7}{8}$?

**43.** How much larger is $5\frac{7}{8}$ than $3\frac{7}{16}$?

**44.** How much larger is $11\frac{7}{32}$ than $11\frac{9}{64}$?

**45.** When $6\frac{3}{8}$ is subtracted from $12\frac{7}{16}$ what is the difference?

**46.** What is the difference between a piece of steel plate $1\frac{7}{16}$ inches thick and a piece $2\frac{17}{32}$ inches thick?

**47.** A piece of round bar stock is $\frac{3}{4}$ inch in diameter. How much must be removed to get a diameter of $\frac{37}{64}$ inch?

**48.** A sheet of prefinished hardwood plywood is $\frac{5}{32}$ inch thick. A sheet of vinyl finished plywood is $\frac{3}{16}$ inch thick. What is the difference in their thicknesses?

**49.** A planer makes a cut $\frac{3}{64}$ inch deep in a piece of aluminum $1\frac{5}{32}$ inch thick. What is the remaining thickness?

**50.** A piece of channel iron 4 feet $6\frac{7}{8}$ inches long is cut from a piece 6 feet $5\frac{1}{4}$ inches long. How much is left?

**51.** The diameter of a hole in a casting is $8\frac{5}{8}$ inches. The casting is placed in a vertical boring mill and the hole is bored to a diameter of $9\frac{17}{32}$ inches. By how much is the diameter of the hole increased?

**52.** A piece of tapered stock is $1\frac{7}{32}$ inches in diameter at one end and $\frac{59}{64}$ inches in diameter at the other end. What is the difference in diameters?

**53.** A live chicken weighs $4\frac{1}{4}$ pounds. After dressing, it weighs $3\frac{3}{8}$ pounds. What is the weight loss?

**54.** At the opening bell, one share of International Stump was selling at $45\frac{7}{8}$ points. At the closing bell, the stock was selling at $42\frac{1}{4}$ points. How many points did it lose?

**55.** An experienced mechanic can assemble an engine in $6\frac{1}{3}$ hours. An apprentice can do the same job in $9\frac{1}{4}$ hours. How much faster is the experienced mechanic than the apprentice?

**56.** A nail $2\frac{3}{4}$ inches long is driven through a piece of wood $1\frac{3}{16}$ inch thick. How much of the nail is exposed?

**57.** A casting weighed $34\frac{7}{8}$ pounds. After machining, it weighed $29\frac{1}{2}$ pounds. How much metal was removed by machining?

**58.** The cavity for a sand casting is $3\frac{11}{32}$ inches deep. After casting, the metal shrank to a thickness of $3\frac{17}{64}$ inches. How much did the casting shrink?

**59.** A tank is $\frac{3}{4}$ full of gasoline. If $\frac{1}{8}$ is drawn off, what fraction of the tank of gas remains?

**60.** Last month Bruce weighed $234\frac{1}{4}$ pounds. Now he weighs $227\frac{5}{16}$ pounds. How much weight has he lost?

**61.** A plumber needs pieces of pipe measuring $5\frac{1}{2}$ feet, $4\frac{1}{4}$ feet, $2\frac{1}{3}$ feet, and $3\frac{2}{3}$ feet. If all pieces are cut from a piece $18\frac{1}{2}$ feet long, how much is left? Disregard any waste in cutting.

**62.** In one 8-hour working day a mechanic spent $1\frac{3}{4}$ hours doing a tune-up, $2\frac{1}{2}$ hours doing a brake job, and $3\frac{1}{3}$ hours replacing a clutch. How many hours did he have left to work on a broken axle?

**63.** Find the dimension marked A in Fig. 8-10.

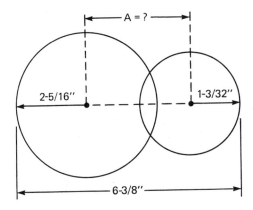

**Figure 8-10**

**64.** Find the dimension marked Y in Fig. 8-11.

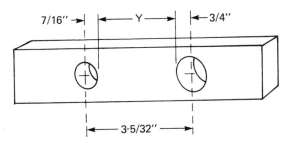

**Figure 8-11**

**65.** Find the dimension marked X in Fig. 8-12.

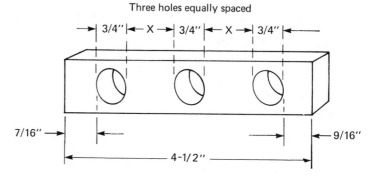

**Figure 8-12**

**66.** Find the distance marked X in Fig. 8-13.

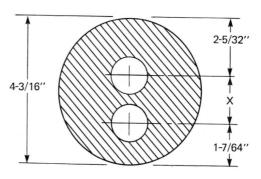

**Figure 8-13**

## FIRST Practice Test—CHAPTER 8

**1.** Do the additions. Write the answers in simplest form.

(a) $\dfrac{3}{4} + \dfrac{1}{4} + \dfrac{2}{4} =$ _____

(b) $\dfrac{3}{8} + \dfrac{5}{8} + \dfrac{1}{8} + \dfrac{7}{8} =$ _____

(c) $\dfrac{5}{16} + \dfrac{7}{16} + \dfrac{1}{16} + \dfrac{11}{16} =$ _____

(d) $\dfrac{7}{32} + \dfrac{11}{32} + \dfrac{9}{32} + \dfrac{15}{32} =$ _____

**2.** Add the mixed numbers. Write answers in simplest form.

(a) $2\dfrac{1}{2} + 3\dfrac{1}{4} + 4\dfrac{1}{4} + 2\dfrac{3}{4} =$ _____

(b) $5\frac{1}{8} + 4\frac{3}{4} + 3\frac{1}{2} + 10\frac{3}{8} = $ _____

(c) $6\frac{3}{16} + 4\frac{1}{8} + 2\frac{1}{4} + 9\frac{1}{2} = $ _____

(d) $10\frac{5}{16} + 8\frac{7}{64} + 9\frac{13}{32} + 11\frac{5}{16} = $ _____

3. Add the measurements. Write the answers in the simplest form.

(a) 2 ft. $3\frac{1}{2}$ in. + 3 ft. $6\frac{1}{8}$ in. + 4 ft. $5\frac{3}{4}$ in. = _____

(b) 2 lb. $6\frac{1}{2}$ oz. + 3 lb. $5\frac{1}{4}$ oz. + 7 lb. $10\frac{3}{4}$ oz. = _____

(c) 3 yd. $2\frac{1}{2}$ ft. + 4 yd. $1\frac{1}{4}$ ft. + 5 yd. $2\frac{1}{3}$ ft. = _____

(d) 1 m. $45\frac{1}{2}$ cm. + 4 m. $36\frac{1}{5}$ cm. + 86 m. $56\frac{7}{10}$ cm. = _____

(1 m = 100 cm)

4. Do the additions. Write all answers in simplest form.

(a) $4\frac{1}{32}$ in.  (b) $5\frac{1}{16}$ lb.  (c) $1\frac{1}{2}$ cm.

   $3\frac{5}{32}$ in.    $4\frac{3}{8}$ lb.     $3\frac{1}{5}$ cm.

   $8\frac{7}{32}$ in.    $7\frac{1}{4}$ lb.     $6\frac{3}{10}$ cm.
   _____         _____         $2\frac{9}{100}$ cm.
                                      _____

*Answer:* (a) _____ (b) _____ (c) _____ .

5. Boards measuring $3\frac{1}{2}$, $4\frac{3}{4}$, $5\frac{7}{12}$, and $2\frac{1}{3}$ feet are needed by a carpenter. Disregard any waste. Can all of the boards be cut from a single board 16 feet long?     *Answer:* _____

6. Pieces of wood with thicknesses of $\frac{1}{16}$, $\frac{3}{32}$, $\frac{5}{64}$, and $\frac{3}{8}$ inch are laminated together. What is the thickness of the result?     *Answer:* _____

7. A laboratory technician weighed samples of $4\frac{1}{5}$, $7\frac{3}{10}$, $18\frac{9}{10}$, $4\frac{39}{100}$, and $54\frac{7}{20}$ grams. Find the total weight of all the samples.     *Answer:* _____

8. Find the length of the bolt in the Fig. 8-14.     *Answer:* _____

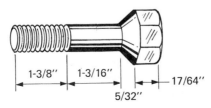

1-3/8''   1-3/16''   17/64''

5/32''

**Figure 8-14**

Use Fig. 8-15 to answer problems 9, 10, 11, 12, and 13.

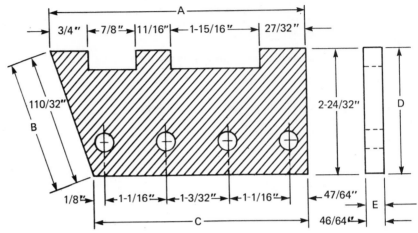

**Figure 8-15**

9. Express the dimension A as a mixed number in simplest form.

                                                *Answer:* _____

10. Express the dimension B as a mixed number in simplest form.

                                                *Answer:* _____

11. What is the value of the dimension C?      *Answer:* _____

12. Express the height D as a mixed number in simplest form.

                                                *Answer:* _____

13. Express the width E as a fraction in simplest form.   *Answer:* _____

14. Do the subtractions. Write all answers in simplest form.

    (a) $5\frac{3}{8} - 3\frac{1}{8} =$ _____         (b) $17\frac{3}{4} - 6\frac{1}{2} =$ _____

    (c) $45\frac{7}{8} - 9\frac{11}{16} =$ _____       (d) $6\frac{5}{64} - 3\frac{1}{32} =$ _____

15. (a) From $5\frac{1}{16}$ subtract $3\frac{7}{16}$.             *Answer:* (a) _____

    (b) Find the difference between $11\frac{3}{5}$ and $8\frac{7}{20}$.     (b) _____

    (c) What is the result of decreasing $15\frac{3}{32}$ by $11\frac{5}{16}$?   (c) _____

    (d) If $56\frac{7}{8}$ is diminished by $34\frac{27}{32}$ what is the result?   (d) _____

16. A board is $1\frac{5}{32}$ inches thick. A piece of marine plywood is $\frac{57}{64}$ inch thick. Find the difference of their thicknesses.       *Answer:* _____

**17.** A piece of tapered stock measures $1\frac{9}{32}$ inches in diameter at one end and $\frac{51}{64}$ inches in diameter at the other. Find the difference in the diameters.

*Answer:* _____

**18.** A cylinder in an automobile engine measures $3\frac{5}{16}$ inches in diameter. After reboring, the cylinder measures $3\frac{23}{64}$ inches in diameter. By how much is the diameter of the cylinder increased? *Answer:* _____

**19.** A cylinder in an engine block measures $3\frac{3}{16}$ inches in diameter. A piston measuring $3\frac{21}{128}$ inches is placed in the cylinder. What is the clearance between piston and cylinder? *Answer:* _____

**20.** Find the dimension marked Y in Fig. 8-16. *Answer:* _____

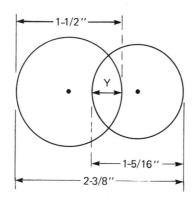

**Figure 8-16**

**21.** A cabinet maker cuts lengths of $5\frac{1}{8}$, $18\frac{3}{16}$, $24\frac{7}{8}$, and $48\frac{11}{16}$ inches from a piece of quarter-round. The original piece of quarter-round was 10 feet $6\frac{1}{2}$ inches long. Allowing $\frac{1}{16}$ inch waste per cut, how much of the original piece is left? *Answer:* _____

**22.** Two pieces of plywood $1\frac{1}{8}$ inches and $\frac{3}{4}$ inch are nailed together with an 8d common nail measuring $2\frac{1}{2}$ inches long. How much of the nail goes through and must be bent over? *Answer:* _____

**23.** A machinist starts with a piece of bar stock $2\frac{5}{32}$ inches thick. He makes cuts of $\frac{3}{32}$, $\frac{5}{64}$, and $\frac{3}{64}$ inch on a milling machine, and then removes $\frac{17}{128}$ inch on the grinder. How thick is the finished part?

*Answer:* _____

## SEC. 3   MULTIPLYING FRACTIONS

You know that multiplying is a shortcut for adding the same quantity to itself several times. For example, if a carpenter cuts 12 boards each 2 feet long, then you find the total length of all the boards by multiplying:

$$\text{Total length} = 12 \times 2 = 24 \text{ feet}$$

If the carpenter cuts 6 boards each $2\frac{1}{2}$ feet long, what is the total length of all the boards? You find the total length by multiplying:

$$\text{Total length} = 6 \times 2\frac{1}{2}$$

This multiplication is a shortcut for the addition:

$$2\frac{1}{2} + 2\frac{1}{2} + 2\frac{1}{2} + 2\frac{1}{2} + 2\frac{1}{2} + 2\frac{1}{2}$$

Add these mixed numbers to find the total length of all 6 boards:

$$6 \times 2\frac{1}{2} = (6 \times 2) + \left(6 \times \frac{1}{2}\right)$$
$$= 12 + 3$$
$$= 15 \text{ feet}$$

Suppose the carpenter has a board 6 feet long and he wants to cut it in half. What is the length of one piece? Look at Fig. 8-17 and you can see that:

$$\frac{1}{2} \text{ of 6 feet} = 3 \text{ feet}$$

The answer is the same as if the numerator of the fraction $\frac{1}{2}$ were multiplied by 6 and the answer simplified.

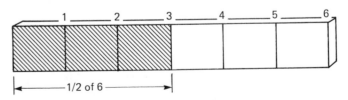

**Figure 8-17**

$$\frac{1}{2} \text{ of } 6 = \frac{1}{2} \times 6 = \frac{1 \times 6}{2}$$

$$= \frac{6}{2}$$

$$\frac{1}{2} \times 6 = 3 \qquad \left( \frac{1}{2} \text{ of } 6 \text{ is } 3 \right)$$

Now suppose the carpenter has a board $\frac{3}{4}$ foot long and he wants to cut it in half. That is, he wants $\frac{1}{2}$ of $\frac{3}{4}$ foot. What is the length of one piece? Look at Fig. 8-18 which shows a board 1 foot long that has been

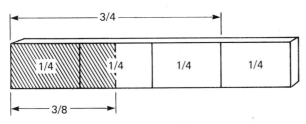

**Figure 8-18**

divided into fourths. The carpenter wants to cut the $\frac{3}{4}$ foot length into two equal parts. From your experience with equivalent fractions, you know that one-fourth can be divided into two equal parts—each called one-eighth.

$$\frac{1}{2} \text{ of } \frac{1}{4} = \frac{1}{8}$$

From this you reason that: "Since one-half of one-fourth is one-eighth, then one-half of three-fourths is three-eights."

*Answer:* $\frac{1}{2}$ of $\frac{3}{4}$ foot $= \frac{3}{8}$ foot

NOTE: The fraction $\frac{3}{8}$ is the result of the multiplication:

$$\frac{1}{2} \times \frac{3}{4} = \frac{1 \times 3}{2 \times 4} = \frac{3}{8}$$

The numerators of the two fractions are multiplied together and the denominators are multiplied together separately. From these examples, the rule for multiplying fractions is:

### *Rule for Multiplying Fractions*

*To find the product of two or more fractions, multiply all numerators together to get the numerator of the product. Multiply all denominators together to get the denominator of the product. Simplify the result if possible.*

$$\text{Product of fractions} = \frac{\text{Product of all numerators}}{\text{Product of all denominators}}$$

For the fractions $\frac{a}{b}$ and $\frac{c}{d}$ their product is:

$$\frac{a}{b} \times \frac{c}{d} = \frac{a \times c}{b \times d}$$

When multiplying fractions together, you do not need a common denominator.

**Example 16:** Find the product $\frac{3}{4} \times \frac{5}{8}$.

*Solution:* Multiplying the numerators gives:

$$3 \times 5 = 15 \text{ (numerator of answer)}$$

Multiplying the denominators gives:

$$4 \times 8 = 32 \text{ (denominator of answer)}$$

*Answer:* $\frac{3}{4} \times \frac{5}{8} = \frac{3 \times 5}{4 \times 8} = \frac{15}{32}$

If any of the factors in a multiplication problem are mixed numbers, you *must* change them to *improper* fractions *before* multiplying.

**Example 17:** Find the product: $3\frac{1}{2} \times 4\frac{1}{2} \times \frac{3}{8}$.

*Solution:* Change *all* mixed numbers to *improper fractions*. Then multiply according to the rule for multiplying fractions.

*Answer:* $\frac{7}{2} \times \frac{9}{2} \times \frac{3}{8} = \frac{189}{32} = 5\frac{29}{32}$

**Example 18:** A fuel tank has a capacity of $35\frac{1}{2}$ gallons. If the fuel weighs $6\frac{1}{4}$ pounds per gallon, what is the weight of the fuel in a full tank?

*Solution:* Weight of fuel = (weight per gallon) × (number of gallons)

$$\text{Weight} = \left(6\frac{1}{4}\right) \times \left(35\frac{1}{2}\right) \left(\text{Change } 35\frac{1}{2} \text{ to the improper fraction } \frac{71}{2}\right)$$

$$= \frac{25}{4} \times \frac{71}{2}$$

$$= \frac{1{,}775}{8}$$

*Answer:* Weight = $221\frac{7}{8}$ pounds

In the following examples you will see how fractions are used in problems of unit conversions.

**Example 19:** In each part of this example, use conversion factors to change a given quantity to an equivalent quantity. (You should know standard conversion factors by now.)

(a) Change $2\frac{1}{2}$ yards to inches. $2\frac{1}{2}$ yards = ___?___ inches.

*Solution* (a): Change yards to inches by multiplying by 36 (36 inches = 1 yard)

$$2\frac{1}{2} \text{ yd.} = 36 \times 2\frac{1}{2} = 36 \times \frac{5}{2}$$

$$= \frac{180}{2} = 90$$

*Answer* (a): $2\frac{1}{2}$ yards = 90 inches

**Example 19:** (b) Change $3\frac{2}{3}$ feet to inches.

*Solution* (b): Change feet to inches by multiplying by 12. (12 inches = 1 foot)

$$3\frac{2}{3} \text{ ft.} = 12 \times 3\frac{2}{3} \text{ inches}$$

$$= 12 \times \frac{11}{3} \text{ inches}$$

$$= \frac{132}{3} \text{ inches}$$

*Answer* (b): $3\frac{2}{3}$ ft. = 44 inches

**Example 19:** (c) How many pints are there in $8\frac{3}{16}$ gallons?

*Solution* (c): Change gallons to pints by multiplying by 8.
     (8 pts. = 1 gal.)

$$8\frac{3}{16} \text{ gal.} = 8 \times 8\frac{3}{16} \text{ pints}$$

$$= 8 \times \frac{131}{16} \text{ pints}$$

$$= \frac{8 \times 131}{16} \text{ pints}$$

$$= \frac{\cancel{8} \times 131}{\cancel{8} \times 2} \quad \text{(Divide numerator and denominator by 8.)}$$

*Answer* (c): $8\frac{3}{16}$ gal. $= 65\frac{1}{2}$ pints

**Example 19:** (d) One gallon of water weighs $8\frac{1}{3}$ pounds. Find the weight of $\frac{1}{2}$ gallon.

*Solution* (d): Weight $= \frac{1}{2}$ of $8\frac{1}{3}$

$$= \frac{1}{2} \times \frac{25}{3} = \frac{25}{6}$$

*Answer* (d): $\frac{1}{2}$ gallon $= \frac{25}{6} = 4\frac{1}{6}$ pounds

**Example 19:** (e) There are $2\frac{54}{100}$ centimeters in 1 inch. How many centimeters are there in $12\frac{1}{2}$ inches?

*Solution* (e): Find the answer by multiplying.

$$2\frac{54}{100} \times 12\frac{1}{2} = \frac{254}{100} \times \frac{25}{2}$$

$$= \frac{254 \times 25}{100 \times 2}$$

$$= \frac{127 \times 50}{2 \times 50 \times 2} \quad \text{(Divide numerator and denominator by 50.)}$$

$$= \frac{127}{4} = 31\frac{3}{4}$$

*Answer* (e): $12\frac{1}{2}$ inches $= 31\frac{3}{4}$ centimeters

# PROBLEMS: Sec. 3
## Multiplying Fractions

**Multiply the fractions in problems 1 through 4.**

Example: $\dfrac{1}{2} \times \dfrac{1}{4} = \dfrac{1 \times 1}{2 \times 4} = \dfrac{1}{8}$

1. (a) $\dfrac{1}{2} \times \dfrac{1}{8}$   (b) $\dfrac{1}{2} \times \dfrac{1}{2}$   (c) $\dfrac{1}{4} \times \dfrac{1}{8}$   (d) $\dfrac{1}{4} \times \dfrac{1}{16}$   (e) $\dfrac{1}{2} \times \dfrac{1}{10}$

2. (a) $\dfrac{1}{2} \times \dfrac{3}{5}$   (b) $\dfrac{1}{8} \times \dfrac{3}{16}$   (c) $\dfrac{3}{8} \times \dfrac{3}{4}$   (d) $\dfrac{2}{3} \times \dfrac{1}{5}$   (e) $\dfrac{7}{8} \times \dfrac{3}{4}$

3. (a) $\dfrac{1}{3} \times \dfrac{3}{5}$   (b) $\dfrac{1}{4} \times \dfrac{4}{5}$   (c) $\dfrac{3}{10} \times \dfrac{1}{3}$   (d) $\dfrac{2}{3} \times \dfrac{1}{2}$   (e) $\dfrac{3}{8} \times \dfrac{8}{10}$

4. (a) $\dfrac{5}{8} \times \dfrac{4}{5}$   (b) $\dfrac{7}{16} \times \dfrac{8}{10}$   (c) $\dfrac{7}{10} \times \dfrac{5}{14}$   (d) $\dfrac{15}{16} \times \dfrac{4}{5}$   (e) $\dfrac{3}{10} \times \dfrac{5}{12}$

**Do the multiplications for problems 5 through 8.** Remember to divide the numerators and denominators by common factors before multiplying.

Example: $\dfrac{15}{16} \times \dfrac{4}{10} = \dfrac{3 \times \cancel{5} \times 1 \times \cancel{4}}{4 \times \cancel{4} \times 2 \times \cancel{5}} = \dfrac{3}{8}$

5. (a) $\dfrac{2}{5} \times \dfrac{5}{12}$   (b) $\dfrac{4}{5} \times \dfrac{5}{8}$   (c) $\dfrac{1}{3} \times \dfrac{6}{10}$   (d) $\dfrac{5}{8} \times \dfrac{4}{15}$   (e) $\dfrac{7}{16} \times \dfrac{16}{21}$

6. (a) $\dfrac{21}{10} \times \dfrac{15}{14}$   (b) $\dfrac{10}{9} \times \dfrac{15}{16}$   (c) $\dfrac{7}{2} \times \dfrac{12}{14}$   (d) $\dfrac{15}{4} \times \dfrac{6}{5}$   (e) $\dfrac{7}{8} \times \dfrac{16}{21}$

7. (a) $\dfrac{9}{32} \times \dfrac{7}{15}$   (b) $\dfrac{21}{16} \times \dfrac{8}{15}$   (c) $\dfrac{14}{15} \times \dfrac{30}{21}$   (d) $\dfrac{7}{5} \times \dfrac{10}{14}$   (e) $\dfrac{3}{16} \times \dfrac{6}{15}$

8. (a) $\dfrac{5}{8} \times \dfrac{16}{15}$   (b) $\dfrac{15}{64} \times \dfrac{3}{10}$   (c) $\dfrac{25}{48} \times \dfrac{3}{20}$   (d) $\dfrac{15}{32} \times \dfrac{12}{45}$   (e) $\dfrac{7}{8} \times \dfrac{16}{35}$

**Do the multiplications for problems 9 through 12.** Write answers in simplest form.

Example: Multiply $5 \times \dfrac{7}{8} = \dfrac{35}{8} = 4\dfrac{3}{8}$

9. (a) $2 \times \dfrac{1}{2}$   (b) $3 \times \dfrac{1}{3}$   (c) $4 \times \dfrac{1}{4}$   (d) $8 \times \dfrac{1}{8}$   (e) $5 \times \dfrac{1}{5}$

10. (a) $2 \times \dfrac{3}{4}$   (b) $5 \times \dfrac{3}{10}$   (c) $\dfrac{11}{16} \times 8$   (d) $\dfrac{5}{6} \times 6$   (e) $\dfrac{7}{16} \times 16$

11. (a) $6 \times \dfrac{2}{3}$   (b) $\dfrac{3}{4} \times 8$   (c) $\dfrac{7}{12} \times 4$   (d) $\dfrac{11}{16} \times 12$   (e) $\dfrac{17}{32} \times 64$

12. (a) $45 \times \dfrac{7}{15}$   (b) $\dfrac{17}{12} \times 8$   (c) $24 \times \dfrac{3}{8}$   (d) $30 \times \dfrac{5}{6}$   (e) $32 \times \dfrac{15}{64}$

**Do the multiplications for problems 13 through 16.** Write the answers as mixed numbers with the fractions in simplest form.

13. (a) $4 \times \dfrac{2}{3}$   (b) $6 \times \dfrac{5}{8}$   (c) $7 \times \dfrac{3}{4}$   (d) $9 \times \dfrac{5}{8}$   (e) $11 \times \dfrac{3}{8}$

14. (a) $\frac{7}{8} \times 9$    (b) $\frac{5}{16} \times 11$    (c) $\frac{7}{16} \times 9$    (d) $\frac{4}{5} \times 7$    (e) $7 \times \frac{5}{12}$

15. (a) $11 \times \frac{15}{32}$    (b) $13 \times \frac{15}{32}$    (c) $\frac{17}{64} \times 27$    (d) $\frac{13}{4} \times 29$    (e) $45 \times \frac{8}{9}$

16. (a) $9 \times \frac{17}{64}$    (b) $\frac{5}{12} \times 25$    (c) $17 \times \frac{23}{8}$    (d) $11 \times \frac{15}{36}$    (e) $19 \times \frac{25}{12}$

**Do the multiplications for problems 17 through 20.** Write all mixed numbers as improper fractions before multiplying.

**Example:** Multiply: $2\frac{1}{12} \times 4\frac{3}{4} = \frac{25}{12} \times \frac{19}{4} = \frac{475}{48} = 9\frac{43}{48}$

17. (a) $1\frac{1}{2} \times \frac{2}{3}$    (b) $2\frac{1}{3} \times \frac{5}{7}$    (c) $2\frac{1}{4} \times \frac{2}{5}$    (d) $1\frac{3}{8} \times \frac{1}{2}$

    (e) $\frac{1}{8} \times 2\frac{1}{4}$

18. (a) $1\frac{1}{2} \times 2\frac{1}{4}$    (b) $2\frac{3}{8} \times 1\frac{1}{4}$    (c) $4\frac{5}{16} \times 2\frac{1}{2}$    (d) $3\frac{1}{5} \times 1\frac{3}{4}$

    (e) $5\frac{1}{3} \times 3\frac{3}{4}$

19. (a) $2\frac{1}{8} \times 3$    (b) $4 \times 1\frac{5}{16}$    (c) $7 \times 4\frac{3}{4}$    (d) $5 \times 6\frac{3}{10}$    (e) $8 \times 4\frac{3}{16}$

20. (a) $2\frac{1}{2} \times 4\frac{1}{8} \times 1\frac{1}{4}$    (b) $1\frac{5}{16} \times 2\frac{2}{3} \times 3\frac{1}{7}$    (c) $4\frac{1}{6} \times 3\frac{1}{5} \times 1\frac{3}{10}$

21. Simplify each of the fractions.

    **Example:** $\dfrac{\cancel{22}^{11} \times \cancel{21}^{3} \times \cancel{2}}{\cancel{7} \times \cancel{2} \times \cancel{2}} = \dfrac{11 \times 3 \times 1}{1 \times 1 \times 1} = 33$

    (a) $\dfrac{1{,}760 \times 20}{60}$     (b) $\dfrac{75 \times 20 \times 60}{30 \times 25 \times 10}$     (c) $\dfrac{5{,}280 \times 100}{3{,}600}$

    (d) $\dfrac{12 \times 18 \times 24}{4 \times 36 \times 10}$     (e) $\dfrac{120 \times 40 \times 22}{33{,}000}$

22. Do the multiplications.

    (a) Multiply $14\frac{5}{8}$ by $1\frac{1}{2}$

    (b) Find the product of $2\frac{1}{6}$ and $3\frac{1}{2}$

    (c) Find $\frac{1}{2}$ of $\frac{3}{8}$

    (d) What is the product when $1\frac{7}{16}$ is multiplied by $2\frac{3}{4}$?

**Find the fractional parts of the given numbers in problems 23 through 26.** Find fractional parts by multiplying.

    **Example:** What is $\frac{2}{3}$ of 9 feet?

    *Solution:* Multiply: $\frac{2}{3} \times \frac{9}{1} = \frac{18}{3} = 6$

    *Answer:* Two-thirds of 9 feet is 6 feet.

23. (a) $\frac{1}{2}$ of 36 in.  (b) $\frac{1}{4}$ of 12 ft.  (c) $\frac{1}{3}$ of 24 in.  (d) $\frac{3}{4}$ of 16 oz.

24. (a) $\frac{1}{5}$ of $100  (b) $\frac{2}{3}$ of 87¢  (c) $\frac{3}{5}$ of 15 yd.  (d) $\frac{5}{8}$ of 136 cm.

25. (a) $\frac{1}{4}$ of $3\frac{1}{2}$ ft.  (b) $\frac{5}{10}$ of $7\frac{1}{5}$ mi.  (c) $\frac{3}{8}$ of $9\frac{1}{3}$ ft.  (d) $\frac{4}{5}$ of $6\frac{3}{4}$ qt.

26. (a) $1\frac{1}{2}$ of $5\frac{3}{4}$ ft.  (b) $2\frac{1}{4}$ of $3\frac{1}{2}$ doz.  (c) $1\frac{1}{4}$ of $1\frac{1}{4}$ day.

27. What is the total length of 9 taper pins, if each pin is $5\frac{3}{8}$ inches long? $\left(\text{H\small{INT}: Multiply } 9 \times 5\frac{3}{8} \text{ inches.}\right)$

28. A piece of brass rod is to be used for making spacers $\frac{3}{8}$ inch thick. Disregard any waste in cutting. How long must the rod be to make 75 spacers?

29. A cabinetmaker needs 18 pieces of quarter-round. Each piece is $17\frac{3}{8}$ inches long. What is the total length of quarter-round needed?

30. A number 7 standard taper pin is $\frac{13}{32}$ inch in diameter at the large end. It has a maximum length of $3\frac{3}{4}$ inches. Allow $\frac{1}{32}$ inch waste per cut. How much bar stock is needed to make 15 pins? $\left(\text{N\small{OTE}: The } \frac{13}{32} \text{ inch is not used in the problem.}\right)$

31. It takes a machinist $8\frac{1}{2}$ minutes to mill a key-way in a piece of round stock. It takes him $2\frac{1}{4}$ minutes to set up and remove the finished piece from the milling machine. How long will it take him to mill 45 key-ways? $\left(\text{N\small{OTE}: The total time for 1 key-way is } \left(8\frac{1}{2} + 2\frac{1}{4}\right) \text{ minutes.}\right)$

32. Figure 8-19 is a drawing of a No. 24 American Standard Wood screw. Allowing $\frac{1}{32}$ inch waste per cut, how much stock is needed for 375 screws?

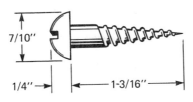

7/10″    1/4″→    1-3/16″

***Figure 8-19***

33. Bob wants to buy a new four-barrel carburator and intake manifold for his Pantera. The total price is $186. If he pays one-third down, what is the balance?

34. In the fourth year of apprenticeship, a machinist apprentice can expect to make $10,800. He budgets one-fourth for food, three-tenths for housing, one-eighth for clothing, three-twentieths for investment and savings, and the rest for miscellaneous expenses. How much can he budget for each item, including miscellaneous expenses?

35. A fuel tank contains 360 gallons of gasoline. A pump removes $\frac{1}{6}$ of the fuel in 1 hour. How much fuel is left in the tank at the end of 5 hours?

36. If an engine is turning at 2,100 rpm, how many revolutions will it turn in two-thirds of a minute?

37. A painter mixed $7\frac{3}{4}$ gallons of paint. Of the total number of gallons, $\frac{3}{16}$ is linseed oil. How much of the total is oil?

38. Sam's Super Service is featuring a special on motor oil. At the beginning of the month Sam had 120 cases of oil. During the first week Sam sold $\frac{3}{5}$ of the 120 cases. During the second week he sold $\frac{1}{2}$ of the remainder. If each case holds 24 quarts, how many quarts does Sam now have?

39. When a 26-inch bicycle wheel makes one revolution, the bike moves forward $81\frac{3}{5}$ inches. How far will the bicycle move when the wheel makes $1\frac{1}{4}$ revolutions?

40. A patternmaker is paid $7\frac{1}{2}$ per hour. How much does he earn by working $6\frac{1}{2}$ hours on Monday and $5\frac{3}{4}$ hours on Tuesday?

41. Parts-Unlimited gives special discounts to independent auto and diesel mechanics. On certain items the parts house gives discounts of *one-fourth and one-fifth*. This means that a discount of $\frac{1}{4}$ is subtracted from the original price, and the remainder is further discounted by $\frac{1}{5}$.

> **Example:** A short block is listed at $280. Using the *one-fourth—one-fifth* discount method, how much will a mechanic pay for the block?

*Solution:* Find $\frac{1}{4}$ of $280.

$$\frac{1}{4} \times \$280 = \$70$$

To get the first discount price subtract $280 - $70 = $210.

Now, discount $210 by $\frac{1}{5}$

$$\frac{1}{5} \text{ of } \$210 = \frac{1}{5} \times \$210 = \$42.$$

*Answer:* Final price = $280 − $70 − $42 = $168

Use the *one-fourth—one-fifth* discount method to find the final list price of the following items:

(a) A rebuilt automatic transmission that lists for $320.
(b) A rebuilt VW engine that lists for $560.
(c) A parts order of $480 to overhaul a Cummings diesel engine.
(d) A transmission, drive shaft, and differential that lists for $1,160.
(e) A Keep-Kool-With-Koolege air conditioner with a retail list price of $368.
(f) An exhaust manifold, tail pipe assembly, and mufflers with a retail list of $112.

## SEC. 4   DIVIDING FRACTIONS

When a quantity is divided, it is separated into a number of parts. When a foot is divided into inches, it is separated into 12 equal parts— each 1 inch long. When a yard is divided into feet, it is separated into 3 equal parts—each part 1 foot long. Divisions such as these are easy, and you do them all the time. But do you know how to divide $4\frac{1}{2}$ feet into pieces so that each piece is $\frac{1}{4}$ inch long? How many pieces will result? To answer this and similar questions, study this section on dividing fractions.

When a ruler 12 inches long is divided in half, there are 2 pieces. Each piece is $12 \div 2 = 6$ inches long.

When a ruler 12 inches long is divided in thirds, there are 3 pieces. Each piece is $12 \div 3 = 4$ inches long.

If you divide a 12-inch ruler into $\frac{1}{2}$ inch pieces, how many pieces will there be? Write this as $12 \div \frac{1}{2} = ?$ Look at the ruler in Fig. 8-20. Because each inch is made up of 2 pieces, each $\frac{1}{2}$ inch long, there are 2 halves per inch. Therefore, there are 24 half-inch pieces in 12 inches:

$$12 \div \frac{1}{2} = 24$$

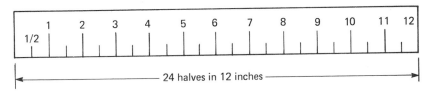

**Figure 8-20**

If you divide the 12-inch ruler into $\frac{1}{4}$ inch pieces, how many pieces will there be? Write this as $12 \div \frac{1}{4} = \underline{?}$ Look at the ruler in Fig. 8-21. Because there are 4 one-fourths in 1 inch, there are $12 \times 4 = 48$ one-fourths in 12 inches. Therefore,

$$12 \div \frac{1}{4} = 48.$$

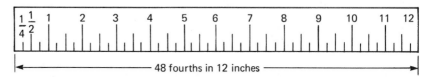

**Figure 8-21**

You found the answers to each of the above questions by looking at a picture of a ruler 12 inches long and counting the result. This is one way of finding answers to division problems. In addition to this visual-counting approach, you need to learn another method of solving division problems. This is the method of *multiplying by the* **reciprocal** *of a fraction*. To understand a *reciprocal* of a fraction, look at these multiplications:

$$\frac{2}{3} \times \frac{3}{2} = \frac{2 \times 3}{3 \times 2} = \frac{6}{6} = 1$$

$$\frac{3}{8} \times \frac{8}{3} = \frac{3 \times 8}{8 \times 3} = \frac{24}{24} = 1$$

$$\frac{1}{4} \times \frac{4}{1} = \frac{1 \times 4}{4 \times 1} = \frac{4}{4} = 1$$

$$\frac{5}{1} \times \frac{1}{5} = \frac{5 \times 1}{1 \times 5} = \frac{5}{5} = 1$$

For each fraction there is a *companion* fraction (or whole number) and their product is the number 1. The companion of $\frac{2}{3}$ is $\frac{3}{2}$ because $\frac{2}{3} \times \frac{3}{2}$ $= 1$. The companion of $\frac{3}{8}$ is $\frac{8}{3}$ because $\frac{3}{8} \times \frac{8}{3} = 1$. The companion of $\frac{1}{4}$ is $\frac{4}{1}$ because $\frac{1}{4} \times \frac{4}{1} = 1$. The companion of 5 is $\frac{1}{5}$ because $5 \times \frac{1}{5} = 1$. In mathematics these companion numbers are called *reciprocals*. Thus, *the reciprocal of a number N is another number R so that their product is 1*.

$$\text{(Number)} \times \text{(Reciprocal)} = 1$$

Here are some more numbers and their reciprocals:

$\dfrac{3}{4}$ and $\dfrac{4}{3}$ are *reciprocals* because $\dfrac{3}{4} \times \dfrac{4}{3} = \dfrac{3 \times 4}{4 \times 3} = \dfrac{12}{12} = 1$

$\dfrac{7}{16}$ and $\dfrac{16}{7}$ are *reciprocals* because $\dfrac{7}{16} \times \dfrac{16}{7} = \dfrac{112}{112} = 1$

Do you understand how to get the reciprocal of a fraction? Notice that the reciprocal of $\dfrac{3}{4}$ is $\dfrac{4}{3}$. You get the reciprocal $\dfrac{4}{3}$ when you switch the numerator and denominator of the fraction.

*When the given fraction is $\dfrac{N}{D}$, then the reciprocal is $\dfrac{D}{N}$*

There is only one number that does not have a reciprocal. Zero does not have a reciprocal. Since $0 = \dfrac{0}{1}$ the reciprocal would look like $\dfrac{1}{0}$. Division by zero is not possible. (See page 80.)

Be careful when you try to find the reciprocal of a mixed number. You must write the mixed number as an improper fraction before you can write its reciprocal.

*The reciprocal of $5\dfrac{1}{4} = \dfrac{21}{4}$ is $\dfrac{4}{21}$ NOT $4\dfrac{1}{5}$*

You will now learn to divide a number by a fraction using the idea of the reciprocal.

**Example 20:** Divide 12 by $\dfrac{3}{8}$.

*Solution:* This is written:

$$12 \div \dfrac{3}{8} = \dfrac{12}{\dfrac{3}{8}} \longleftarrow \text{Division bar}$$

Recall from Section 1 of this chapter the rule that says you can *multiply numerator and denominator of ANY fraction by the same nonzero number and the result is the same.* Apply this rule to the problem:

$$\dfrac{12}{\dfrac{3}{8}}$$

Multiply the numerator (12) and the denominator $\left(\dfrac{3}{8}\right)$ by the RECIPROCAL of $\dfrac{3}{8}$, which is $\dfrac{8}{3}$.

$$\dfrac{12 \times \dfrac{8}{3}}{\dfrac{3}{8} \times \dfrac{8}{3}} = \dfrac{\dfrac{12}{1} \times \dfrac{8}{3}}{\dfrac{3 \times 8}{8 \times 3}} = \dfrac{\dfrac{96}{3}}{1} = \dfrac{\dfrac{32}{1}}{1} = 32$$

NOTE: Any number divided by 1 is that same number.

*Answer:* 32

Here is the way the problem looks when letters are used instead of numbers:

$$\frac{\dfrac{A}{N}}{\dfrac{N}{D}} = \frac{\dfrac{A}{1} \times \dfrac{D}{N}}{\dfrac{N}{D} \times \dfrac{D}{N}} = \frac{\dfrac{A \times D}{N}}{1} = \frac{A \times D}{N}$$

*Multiply numerator and denominator by the reciprocal of the divisor*

Study the following example. Multiply the number above the division line by the *reciprocal* of the fraction below the division line.

**Example 21:** (a) $\dfrac{\dfrac{4}{1}}{\dfrac{3}{8}} = \dfrac{4 \times \dfrac{8}{3}}{\dfrac{3}{8} \times \dfrac{8}{3}} = \dfrac{\dfrac{32}{3}}{1} = \dfrac{32}{3} = 10\dfrac{2}{3}$

(b) $\dfrac{\dfrac{4}{5}}{\dfrac{2}{3}} = \dfrac{\dfrac{4}{5} \times \dfrac{3}{2}}{\dfrac{2}{3} \times \dfrac{3}{2}} = \dfrac{\dfrac{12}{10}}{1} = \dfrac{6}{5} = 1\dfrac{1}{5}$

Take a moment and look at the process of multiplying by the reciprocal. You probably noticed that when you multiplied the fraction below the division line by its reciprocal, the product was always 1. The multiplication above the division line is a product of *the reciprocal of the denominator fraction* and the original number. The product of a number and its reciprocal is always 1.

NOTE:

**Rule for Dividing a Number by a Fraction**

*To **divide** a number **by a fraction**, multiply it by **the reciprocal of the fraction**.*

$$\frac{A}{\dfrac{N}{D}} = A \times \frac{D}{N}$$

**Practice Problems**

Use the Rule for Dividing by a Fraction to prove these answers.

**1.** $4 \div \dfrac{2}{3} = 4 \times \dfrac{3}{2} = \dfrac{12}{2} = 6$

**2.** $\dfrac{\dfrac{7}{8}}{\dfrac{3}{4}} = \dfrac{7}{8} \times \dfrac{4}{3} = \dfrac{28}{24} = \dfrac{7}{6} = 1\dfrac{1}{6}$

**3.**  $\dfrac{5}{6} \div \dfrac{3}{4} = \dfrac{5}{6} \times \dfrac{4}{3} = \dfrac{20}{18} = \dfrac{10}{9} = 1\dfrac{1}{9}$

**4.**  $\dfrac{3}{8} \div 4 = \dfrac{3}{8} \times \dfrac{1}{4} = \dfrac{3}{32}$

**5.**  $\dfrac{2\frac{1}{2}}{1\frac{1}{4}} = \dfrac{\frac{5}{2}}{\frac{5}{4}} = \dfrac{5}{2} \times \dfrac{4}{5} = \dfrac{20}{10} = 2$  (Change mixed numbers to improper fractions *before* doing the division.)

Examples 22, 23, and 24 are typical of the type of problems that involve dividing by fractions.

**Example 22:** A piece of sheet metal is 33 inches wide. How many strips $1\dfrac{3}{8}$ inches wide can be cut from it?

*Solution:* Find the number of $1\dfrac{3}{8}$ inch strips by dividing:

$$33 \text{ inches} \div 1\dfrac{3}{8} \text{ inches per strip}$$

Write the whole number 33 as $\dfrac{33}{1}$.

$$\dfrac{33}{1} \div 1\dfrac{3}{8} = \dfrac{33}{1} \div \dfrac{11}{8}$$

$$= \dfrac{33}{1} \times \dfrac{8}{11} = 24 \text{ strips}$$

*Answer:* 24 strips

**Example 23:** An automatic screw machine makes brass washers which are $\dfrac{3}{32}$ inch thick. Allow $\dfrac{1}{64}$ inch waste for each cut-off operation. How many washers can be made from a piece of bar stock 8 feet 2 inches long? (See Fig. 8-22.)

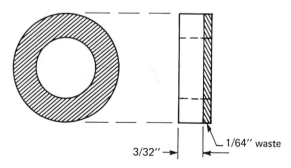

3/32″ →    1/64″ waste

**Figure 8-22**

*Solution:* Each washer uses $\left(\dfrac{3}{32} + \dfrac{1}{64}\right)$ inches of stock which is equal to (washer + waste).

$$\frac{3}{32} + \frac{1}{64} = \frac{6}{64} + \frac{1}{64} = \frac{7}{64} \text{ inch used for each washer.}$$

Find out how many times $\dfrac{7}{64}$ inch will divide into 8 feet 2 inches.

Change 8 ft. 2 in. to inches.

$$8 \text{ ft. } 2 \text{ in.} = (12 \times 8) + 2 \text{ inches}$$
$$= 98 \text{ inches}$$

Divide: $98$ inches $\div \dfrac{7}{64}$ inches per washer

$$= \frac{98}{1} \times \frac{64}{7} \text{ (Multiply by reciprocal)}$$

$$= \frac{7 \times 14 \times 64}{1 \times 7} = 896 \quad \text{(Simplify fraction)}$$

*Answer:*          896 washers

**Example 24:** A fuel storage tank has a capacity of 3,000 gallons. If one cubic foot is equal to $7\dfrac{1}{2}$ gallons, how many cubic feet are there in the tank.

*Solution:* Find the answer by dividing.

$$3,000 \text{ gallons} \div 7\frac{1}{2} \text{ cubic feet per gallon}$$

$$3,000 \div 7\frac{1}{2} = \frac{3,000}{1} \times \frac{2}{15} \quad \text{(Multiply by reciprocal)}$$

$$= \frac{6,000}{15} = 400 \quad \text{(Simplify fraction)}$$

*Answer:*          400 cubic feet

## PROBLEMS : Sec. 4
### Dividing Fractions

**Problems 1 through 4.** Find the reciprocal of each number. If the number is a mixed number, change it to an improper fraction *before* finding the reciprocal.

**Example:** (a) Find the reciprocal of $\dfrac{7}{8}$. *Answer:* $\dfrac{8}{7}$ since $\dfrac{7}{8} \times \dfrac{8}{7} = 1$

(b) Find the reciprocal of $4\dfrac{1}{2}$. *Answer:* Change the improper fraction $\dfrac{9}{2}$ to get the reciprocal $\dfrac{2}{9}$.

1. (a) $\frac{1}{2}$  (b) $\frac{3}{4}$  (c) $\frac{4}{5}$  (d) $\frac{3}{8}$  (e) $\frac{5}{16}$  (f) $\frac{7}{8}$  (g) $\frac{9}{10}$  (h) $\frac{15}{16}$

2. (a) $\frac{5}{32}$  (b) $\frac{19}{64}$  (c) $\frac{5}{2}$  (d) $\frac{7}{4}$  (e) $\frac{25}{64}$  (f) $\frac{17}{8}$  (g) $\frac{28}{5}$  (h) $\frac{19}{10}$

3. (a) $1\frac{1}{2}$  (b) $3\frac{3}{4}$  (c) $5\frac{3}{8}$  (d) $8\frac{1}{4}$  (e) $9\frac{7}{16}$  (f) $5\frac{1}{3}$  (g) $4\frac{7}{12}$

4. (a) $2\frac{3}{8}$  (b) $6\frac{1}{4}$  (c) $8\frac{7}{32}$  (d) $5\frac{7}{10}$  (e) $4\frac{11}{12}$  (f) $9\frac{1}{5}$  (g) $5\frac{17}{36}$

**Problems 5 through 12.** Do the divisions.

**Example:** Divide: $\dfrac{3}{4} \div \dfrac{7}{8} = \dfrac{3}{4} \times \dfrac{8}{7} = \dfrac{3 \times \overset{2}{\cancel{8}}}{7 \times \cancel{4}} = \dfrac{6}{7}$

Divide: $2\dfrac{1}{2} \div 1\dfrac{1}{4} = \dfrac{5}{2} \div \dfrac{5}{4} = \dfrac{5}{2} \times \dfrac{4}{5} = 2$

5. (a) $\frac{1}{4} \div \frac{1}{3}$  (b) $\frac{3}{4} \div \frac{1}{8}$  (c) $\frac{7}{8} \div \frac{3}{2}$  (d) $\frac{9}{16} \div \frac{3}{8}$  (e) $\frac{5}{2} \div \frac{3}{4}$

6. (a) $\frac{1}{2} \div \frac{3}{4}$  (b) $\frac{5}{4} \div \frac{3}{8}$  (c) $\frac{7}{16} \div \frac{5}{32}$  (d) $\frac{5}{4} \div \frac{9}{2}$  (e) $\frac{6}{5} \div \frac{17}{10}$

7. (a) $2\frac{1}{2} \div \frac{3}{4}$  (b) $4\frac{2}{3} \div \frac{5}{6}$  (c) $4\frac{3}{8} \div 2\frac{1}{4}$  (d) $4\frac{1}{5} \div 2\frac{3}{10}$

8. (a) $3\frac{1}{8} \div \frac{3}{16}$  (b) $7\frac{1}{2} \div \frac{5}{4}$  (c) $3\frac{1}{3} \div 4\frac{5}{12}$  (d) $15\frac{1}{4} \div 2\frac{3}{8}$

9. (a) $\frac{3}{2} \div \frac{1}{5}$  (b) $\frac{7}{16} \div \frac{3}{5}$  (c) $4\frac{1}{6} \div 2\frac{3}{5}$  (d) $7\frac{7}{8} \div \frac{1}{3}$  (e) $\frac{3}{4} \div \frac{4}{3}$

10. (a) $4 \div \frac{1}{2}$  (b) $6 \div \frac{2}{3}$  (c) $8 \div 1\frac{1}{4}$  (d) $3 \div \frac{3}{4}$  (e) $10 \div 1\frac{3}{5}$

11. (a) $\frac{3}{4} \div 4$  (b) $\frac{1}{2} \div 6$  (c) $\frac{7}{8} \div 5$  (d) $1\frac{3}{4} \div 6$  (e) $4\frac{7}{10} \div 12$

12. (a) $14 \div \frac{1}{4}$  (b) $85 \div 1\frac{1}{8}$  (c) $40 \div 2\frac{1}{5}$  (d) $60 \div \frac{3}{16}$  (e) $24\frac{1}{3} \div 1\frac{5}{6}$

**Problems 13 through 18.** Do the divisions.

**Example:** $4\dfrac{1}{8}$ ft. $\div 6 = \dfrac{33}{8} \div 6 = \dfrac{33}{8} \times \dfrac{1}{6} = \dfrac{11}{16}$ ft.

13. (a) $4\frac{1}{2}$ in. $\div 3$  (b) $6\frac{2}{3}$ ft. $\div 5$  (c) $2\frac{1}{4}$ lb. $\div 3$

14. (a) $5\frac{1}{4}$ ft. $\div \frac{1}{2}$  (b) $7\frac{1}{10}$ cm. $\div \frac{1}{5}$  (c) $4\frac{5}{16}$ lb. $\div \frac{1}{8}$

15. (a) $\frac{15}{16}$ in. $\div \frac{1}{2}$  (b) $\frac{11}{32}$ in. $\div \frac{3}{8}$  (c) $6\frac{2}{3}$ ft. $\div \frac{3}{4}$

16. (a) $4\frac{1}{2}$ lb. $\div 2\frac{1}{2}$  (b) $6\frac{1}{3}$ yd. $\div 4\frac{1}{6}$  (c) $10\frac{1}{2}$ hr. $\div \frac{3}{4}$

17. (a) $4\frac{1}{4}$ qt. $\div 1\frac{1}{8}$  (b) $22\frac{3}{5}$ °F $\div 2\frac{1}{2}$  (c) $42$ mi. $\div 2\frac{5}{8}$

**18.**  (a) $50¢ \div 2\frac{1}{2}$    (b) $\$58 \div 3\frac{5}{8}$    (c) $252 \text{ mi.} \div 5\frac{1}{4}$

(d) $28 \text{ ft.} \div 3\frac{1}{2}$

**Problems 19 and 20.** Find the answer to each problem. Follow these steps: (1) Find the value of the numerator. (2) Find the value of the denominator. (3) Do the division. (4) Reduce the result to simplest terms when possible.

**Example:** Find the value of $\dfrac{4\frac{1}{2} - 2\frac{1}{4}}{3\frac{1}{2} + 1\frac{1}{2}}$

Step 1.  $4\frac{1}{2} - 2\frac{1}{4} = 2\frac{1}{4} = \frac{9}{4}$

Step 2.  $3\frac{1}{2} + 1\frac{1}{2} = 5 = \frac{5}{1}$

Step 3.  $\dfrac{\frac{9}{4}}{\frac{5}{1}} = \frac{9}{4} \times \frac{1}{5} = \frac{9}{20}$

**19.** (a) $\dfrac{4\frac{1}{4} + 1\frac{1}{2}}{2\frac{1}{4} - 1\frac{1}{8}}$  (b) $\dfrac{3\frac{1}{2} + 1\frac{1}{8}}{2\frac{1}{4} - 1\frac{3}{8}}$  (c) $\dfrac{3\frac{1}{8} + 4\frac{3}{8}}{5\frac{1}{4} + 1\frac{1}{4}}$  (d) $\dfrac{3\frac{1}{2} + \frac{7}{8}}{4\frac{3}{4} + 3\frac{1}{2}}$

**20.** (a) $\dfrac{2\frac{1}{2} \times 3\frac{1}{4}}{4\frac{1}{2} \times 2\frac{1}{8}}$  (b) $\dfrac{1\frac{3}{4} + 4}{3\frac{1}{2} \times 2\frac{1}{8}}$  (c) $\dfrac{\frac{2}{3} \times 3\frac{1}{16}}{\frac{5}{16} + \frac{3}{4}}$  (d) $\dfrac{3\frac{1}{2} \times 4\frac{1}{2}}{5\frac{1}{2} \times 6\frac{1}{4}}$

**21.**  How many pieces of pipe $1\frac{3}{4}$ feet long can be cut from a piece 14 feet long? (Ignore waste in cutting.)

**22.**  A two-by-eight plank $10\frac{1}{4}$ feet long is cut into 6 pieces of equal length. Disregard waste in cutting and determine the length of each piece.

**23.**  The city of Trenton is 70 miles from New York. It takes a commuter $1\frac{3}{4}$ hours to drive this distance. What is the average speed?
(Speed = distance ÷ time)

**24.**  Bronze spacers $1\frac{1}{16}$ inches thick are machined from a piece of stock 72 inches long. For each spacer, $\frac{3}{64}$ inch is wasted in the cut-off operation. How many spacers are made from the piece of stock?

**25.**  A production control technician finds that it takes 3 hours and 20 minutes $\left(3\frac{1}{3} \text{ hours}\right)$ to make a certain part. He also finds that it costs the company $\$13\frac{1}{4}$ for each part.

(a) How many parts are completed in a 40-hour week?

(b) What does it cost to make all the parts?

26. A stack of 11-gauge sheet metal is 1 foot, $4\frac{1}{2}$ inches high. A single sheet is $\frac{1}{8}$ inch thick. Disregard any space between the sheets. How many sheets are in the stack?

27. A workbench in an assembly plant measures 26 feet, 3 inches. An electrician wants to install 14 electrical outlets evenly spaced along the bench. How far apart are the outlets? (Allow 1 foot 9 inches end distances.)

28. A single page of the classified section of the *San Francisco Chronicle* is $14\frac{5}{8}$ inches wide. The left margin is $\frac{5}{16}$ inch wide. The right margin is $\frac{1}{4}$ inch wide. If there are nine columns of print, what is the width of one column?

29. Find the average for the following five measurements:

$$5\frac{1}{8} \qquad 5\frac{3}{16} \qquad 5\frac{1}{16} \qquad 5\frac{3}{32} \qquad 5\frac{7}{64} \qquad \text{inches}$$

30. A gallon of water weighs about $8\frac{1}{3}$ pounds. How many gallons of water are there in 850 pounds?

31. A single piece of elm hardwood flooring measures $2\frac{1}{16}$ inches wide. If the pieces are laid side-by-side, how many pieces are needed for a floor $16\frac{1}{2}$ feet wide?

32. A quality control inspector inspected 24 parts in $2\frac{1}{2}$ hours. Working at the same rate, how many will he inspect in $6\frac{1}{4}$ hours? (HINT: Work this problem by using direct proportions.)

$$\frac{24 \text{ parts inspected}}{2\frac{1}{2} \text{ hours}} = \frac{x \text{ parts inspected}}{6\frac{1}{4} \text{ hours}}$$

$$x = \frac{\left(6\frac{1}{4}\right)(24)}{2\frac{1}{2}} \quad \left(\text{Multiply both sides of equation by } 6\frac{1}{4}\right)$$

Find $x$.

33. If pine flooring boards are $2\frac{1}{4}$ inches wide, how many boards are needed for a hallway $56\frac{1}{4}$ inches wide?

34. A series of lamps are connected to a circuit that is fused for 20 amps. Each lamp draws $\frac{3}{4}$ amp. How many lamps can be safely placed in the circuit?

35. The pace of a man of average size is about $2\frac{1}{2}$ feet. How many paces are in 1 mile? (1 mile = 5,280 feet)

36. Six metal workers are given the same job to do. Their working times are listed below. Find the *average* time for the job.

    Working times:   $24\frac{1}{2}$, $25\frac{3}{4}$, $22\frac{1}{2}$, 26, $23\frac{1}{4}$, and $24\frac{1}{4}$ minutes.

## SECOND Practice Test—CHAPTER 8

1. Do the multiplications. Write answers in simplest form.

   (a) $\frac{5}{16} \times \frac{3}{8}$ 　　　　　　　　　　　　　　　 *Answer:* (a) _____

   (b) $\frac{9}{32} \times \frac{8}{3}$ 　　　　　　　　　　　　　　　 *Answer:* (b) _____

2. Do the multiplications.

   (a) $1\frac{1}{2} \times 2\frac{3}{8}$ 　　　　　　　　　　　　　　 *Answer:* (a) _____

   (b) $2\frac{3}{8} \times 6 \times 5\frac{1}{3}$ 　　　　　　　　　　　 *Answer:* (b) _____

3. How much is $\frac{7}{8}$ of $4\frac{1}{2}$? 　　　　　　　　　 *Answer:* _____

4. Find $2\frac{1}{2}$ times $4\frac{3}{4}$ inches. 　　　　　　　 *Answer:* _____

5. One sheet of vinyl-surfaced plywood is $\frac{5}{16}$ inches thick. How high is a stack of 176 sheets? Write answer in feet and inches. 　 *Answer:* _____

6. A fuel storage tank has a capacity of 560 gallons. A pump fills the tank at the rate of $\frac{1}{16}$ of the tank per hour. How many gallons are in the tank at the end of $3\frac{1}{5}$ hours? 　　　　　　 *Answer:* _____

7. Use the *one-fourth—one-fifth* discount rule to the find the discount price on the following items: cam shaft, rocker arm assembly, valves, valve springs, and rings with a total retail list price of $220.
   　　　　　　　　　　　　　　　　　　　　　 *Answer:* _____

8. Do the divisions.

   (a) $\frac{3}{4} \div \frac{2}{3}$ 　　　　　　　　　　　　　　　 *Answer:* (a) _____

   (b) $\frac{2}{5} \div \frac{3}{10}$ 　　　　　　　　　　　　　　 *Answer:* (b) _____

9. Do the divisions.

   (a) $4 \div 2\frac{1}{2}$ 　　　　　　　　　　　　　　　 *Answer:* (a) _____

   (b) $4\frac{5}{16} \div 2\frac{1}{2}$ 　　　　　　　　　　　　 *Answer:* (b) _____

10. In order to get an accurate measure for a certain part, a technician took the average of these values:

$$3\frac{3}{4} \text{ in.}, \qquad 3\frac{7}{8} \text{ in.}, \qquad 3\frac{13}{16} \text{ in.}, \qquad \text{and} \qquad 3\frac{27}{32} \text{ in.}$$

What value did he use for the measure?      *Answer:* _____

11. A dressed and finished common two-by-four measures $1\frac{3}{4}$ inches thick.

How many are in a stack $3\frac{1}{2}$ feet high?      *Answer:* _____

12. Ordinary 8d common nails are $2\frac{1}{2}$ inches long and weigh about $\frac{1}{7}$ of an ounce each. How many nails are there in a 10-pound package? (1 pound = 16 ounces)      *Answer:* _____

13. An automatic high speed screw machine makes flat head machine screws. Each finished screw uses $1\frac{3}{32}$ inches of material including waste. How many screws can be made from 3 pieces of bar stock 4 ft. $3\frac{1}{2}$ in., 4 ft. 8 in., and 5 ft. $6\frac{3}{4}$ in.?      *Answer:* _____

# 9

# Decimals

## SEC. 1  READING, WRITING, AND ROUNDING OFF DECIMALS

The evening paper, a pay check, the reading of a micrometer, and the dimensions on a drawing for a precision-machined part all have a great deal in common. To see this, look at the following examples.

The classified section in a newspaper reads:

### HELP WANTED

Journeyman machinist. Minimum 5 years experience in aerospace metals. Ability to read blueprints and do complete set up necessary. Experience required in *controlled atmosphere* production where parts have tolerances of $\pm 0.0001$ inch. Starting salary $8.78 per hour, excellent fringe benefits. Apply: Box 123.

A weekly paycheck stub shows the take home pay after deductions. A glance at the deductions on the paycheck shows:

| Gross Earnings | Fed. Tax | F.I.C.A. | Union Dues |
|:---:|:---:|:---:|:---:|
| $325.00 | $65.40 | $12.00 | $14.75 |
| | Net Earnings: $232.85 | | |

The micrometer in Fig. 9-1 shows a measurement of 0.128 inches. The drawing in Fig. 9-2 is used in the manufacture of a precision steel drill.

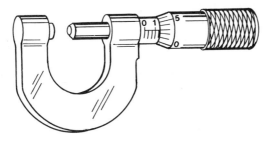

**Figure 9-1**

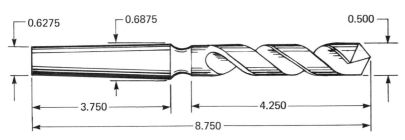

**Figure 9-2**

What do the examples have in common? In each case, the numbers, such as $8.78, $65.40, 0.128, and 0.6275, are all *decimal numbers*. A *decimal number* is a special kind of number. You must be familiar with decimals in order to understand and carry out the jobs assigned to you.

The number system we use today is called the *decimal system* because the number 10 and multiples of 10 are used to find place value. For example, the whole number 1,658 is a *decimal numeral* in which the digit 1 represents 1,000. The digit 6 represents 600. The digit 5 represents 50, and the digit 8 represents 8 units. Recall from Chapter 1 that you can find the place value in a number by using multiples of ten. That is, in the number 15,674 the 4 has the *place value* of **units**. The 7 has the *place value* of **ten**. The 6 has the *place value* of **hundred**. The 5 has the *place value* of **thousand**. The 1 has the *place value* of **ten-thousand**.

Some fractions can be written by a similar use of place value. These fractions are called *decimal fractions*. A *decimal fraction* is a fraction with a *denominator* of 10; 100; 1,000; 10,000; or some higher multiple of 10. For instance, the fractions:

$$\frac{2}{10} \qquad \frac{17}{100} \qquad \frac{457}{1,000} \qquad \frac{758}{10,000}$$

are *decimal fractions*.

You can write every decimal fraction by using a decimal point (.)—which is just a period. When you use the decimal point, you do not need to write the denominator. Here is how you use the decimal point to write decimal fractions. *Look at the decimal fraction and count the number of zeros following the digit 1 in the denominator. Now look at the numerator. Start from the unit's place and count to the **left** as many places as there are zeros in the denominator. Put a decimal point (.) to the left of the last place **you** counted.* Leave out the denominator. As a result you get the fraction written as a decimal.

**Example 1:** Write the decimal fractions

$$\text{(a) } \frac{2}{10} \qquad \text{(b) } \frac{17}{100} \qquad \text{(c) } \frac{457}{1,000} \qquad \text{(d) } \frac{758}{10,000}$$

as decimals. Use the decimal point.

*Solution* (a): The decimal fraction $\frac{2}{10}$ has a denominator of 10. There is *one* zero after the digit 1. Start at unit's place in the numerator (2) and count **left** *one* place. Put a decimal point in *front* of the 2. Write the decimal .2.

*Answer* (a): $\frac{2}{10} = .2$ (1 place)

*Solution* (b): The decimal fraction $\frac{17}{100}$ has a denominator of 100. There are *two* zeros after the digit 1. Start at the digit 7 (the unit's place) in the numerator and count **left** *two* places. Put a decimal point in front of the 1. Write the decimal .17.

*Answer* (b): $\frac{17}{100} = .17$    (2 places)

*Solution* (c): The decimal fraction $\frac{457}{1,000}$ has a denominator of 1,000. There are *three* zeros after the digit 1. Start at the digit 7 in the numerator and count **left** *three* places. Put a decimal point in front of the 4. Write the decimal .457.

*Answer* (c): $\frac{457}{1,000} = .457$    (3 places)

*Solution* (d): The decimal fraction $\frac{758}{10,000}$ has a denominator of 10,000. There are *four* zeros after the digit 1. Start at the digit 8 in the numerator and count **left** *four* places. Put a zero in front of the 7 in order to be able to count four places. This zero is a place holder. Write the decimal .0758.

*Answer* (d): $\frac{758}{10,000} = .0758$    (4 places)

From these examples it is possible to state

### The Rule for Writing a Decimal Fraction as a Decimal

*When a **decimal** fraction is written as a decimal, there must be as many digits to the **right** of the decimal point as there are zeros in the denominator.*
  Study these decimals:

| (*Decimal*) *fraction* | *Decimal number* | *Read as:* |
|---|---|---|
| $\dfrac{3}{10}$ | .3 | three-tenths |
| $\dfrac{56}{100}$ | .56 | fifty-six hundredths |
| $\dfrac{45}{1,000}$ | .045 | forty-five thousandths |
| $\dfrac{75}{10,000}$ | .0075 | seventy-five ten-thousandths |

NOTE: The decimal number has as many places to the right of the decimal point as there are zeros in the denominator of the decimal fraction.

Each digit to the right of the decimal point represents a certain *place value*. The *first* place to the right is the *tenth's* place. The *second* place is the *hundredth's* place. The *third* place is the *thousandth's* place. The *fourth* place is the *ten-thousandth's* place, and so on. You must be able to distinguish between the *ten's* and *tenth's* place, the *hundred's* and *hundredth's* place, the *thousand's* and *thousandth's* place when you write whole numbers and decimals.

As Fig. 9-3 shows, the place value to the *left* of the decimal point is for *whole* numbers. The place value to the *right* of the decimal point is for decimal *fractions*.

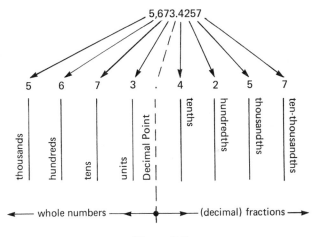

**Figure 9-3**

The FIRST place to the right of the decimal point is the *tenth's* place, the SECOND place is the *hundredth's* place, then the *thousandth's* place, and so on. There is no *unit's* place to the right of the decimal point.

Use a place value chart similar to the one in Chapter 1 to help you locate the correct place value in a decimal. Fill each pocket to the right of the decimal point with one of the digits 0, 1, 2, 3, 4, 5, 6, 7, 8, 9 until the complete decimal fraction is shown. In Fig. 9-4 you see the decimal .0345 (three hundred forty-five ten-thousandths).

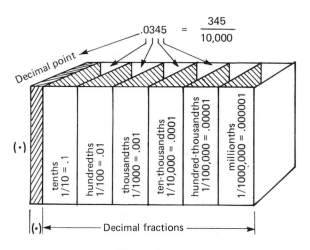

**Figure 9-4**

To read or write a decimal, determine the size of the decimal as to tenths (one place), hundredths (two places), thousandths (three places), etc. Do this by counting the number of digits to the right of the decimal point, starting with tenths. After you find the size, read the numerator in ordinary place-value terms followed by the size of the denominator of the fraction. For instance, read the decimal .0034 "thirty-four ten-thousandths" because there are four places to the right of the decimal point, showing ten-thousandths. The two zeros between the decimal point and the digit 3 are place holders. (They are used for correct location of the decimal point only.)

Study these correct decimal readings:

    (a) For .56 read "fifty-six hundredths."
    (b) For .075 read "seventy-five thousandths."
    (c) For 4.009 read "four *and* nine thousandths."
    (d) For 500.05 read "five hundred *and* five hundredths."

NOTE: In (c) and (d) the word *and* is used to replace the decimal point.

You may find it more convenient to use the word *point* instead of *and* to show the location of the decimal point. For instance, read the decimal number 10.045 "ten *and* forty-five thousandths" or "ten *point* zero-four-five." Either form is acceptable. Use the one that sounds best to you.

When reading and writing decimals, be sure of the size of the number. (See Example 3.)

### Rule for Comparing Decimals

*To compare a set of decimals, be sure that they have the **same** number of decimal places. Then compare the numerators in ordinary numerical fashion.*

**Example 2:** Arrange the decimals

$$.6 \quad .06 \quad .605 \quad .56 \quad .665$$

from smallest to largest.

*Solution:* Write all decimals with three decimal places.

$$.600 \quad .060 \quad .605 \quad .560 \quad .665$$

Ignore decimal points and read .600 as six hundred, .060 as sixty, .605 as six hundred five, .560 as five hundred sixty, and .665 as six hundred sixty-five.

When you arrange these in numerical order and put in the decimal point, you have:

*Answer:* $.060 < .560 < .600 < .605 < .665$

**Example 3:** Arrange the five decimals in increasing size,

$$.7 \quad .77 \quad .07 \quad .707 \quad .077$$

*Solution:* Write each with three decimal places which gives:

$$.7 = .700 \quad .77 = .770 \quad .07 = .070$$
$$.707 = .707 \quad .077 = .077$$

Change the .7 to .700 (tenths to thousandths) by adding two zeros to the right of the digit 7. Change the other decimals by adding as many zeros as necessary to make all decimals read in thousandths (three decimal places).

Ignore the decimal point and compare the numbers:

$$700 \quad 770 \quad 70 \quad 707 \quad 77$$

Obviously, when you arrange these numbers in increasing size you have

$$70 \quad 77 \quad 700 \quad 707 \quad 770$$

*Answer:* Look at the original decimals. This corresponds to the arrangement:

$$.070 < .077 < .700 < .707 < .770$$

When you compare or arrange decimals, you may want to compare them or round them off to the nearest tenth, hundredth, thousandth, ten-thousandth, or some other value.

**Example 4:** For the decimal .88, what is the nearest tenth? Is it .8 or .9?

*Solution:* To see which tenth is the nearest, notice that $.8 = .80$ and $.9 = .90$. You know that .88 (eighty-eight hundredths) is closer to .90 (ninety hundredths) than it is to .80 (eighty-hundredths).

*Answer:* .88 is equal to .9 to the nearest tenth.

NOTE: Use the rules of rounding off to a specific place value that you studied in Chapter 2. These rules apply to decimals as well as whole numbers.

## PROBLEMS: Sec. 1 Reading, Writing, and Rounding Off Decimals

**Problems 1 through 8.** Write the decimals as word statements.

> **Example:**  .35 is written thirty-five hundredths.
> .09 is written nine hundredths.
> 31.45 is written thirty-one *and* forty-five hundredths.

1. (a) .5   (b) .8   (c) .7   (d) .4   (e) .6   (f) .9   (g) .1   (h) .3
2. (a) .50   (b) .08   (c) .07   (d) .04   (e) .06   (f) .09   (g) .01   (h) .03
3. (a) .25   (b) .38   (c) .57   (d) .64   (e) .26   (f) .49   (g) .73
4. (a) 1.2   (b) 2.5   (c) 3.8   (d) 2.4   (e) 7.7   (f) 5.6   (g) 9.3
5. (a) 2.05   (b) 3.04   (c) 5.07   (d) 4.09   (e) 3.08   (f) 8.08   (g) 87.75
6. (a) 3.25   (b) 5.43   (c) 7.68   (d) 9.33   (e) 4.85   (f) 2.49
7. (a) 25.01   (b) 45.05   (c) 50.02   (d) 38.07   (e) 44.06   (f) 99.03
8. (a) 46.38   (b) 89.65   (c) 75.54   (d) 45.54   (e) 34.60   (f) 78.95

**Problems 9 through 14.** Write the word statements as decimal numbers.

> **Example:** Eight hundredths    *Answer:* .08
> Two *and* forty five hundredths    *Answer:* 2.45
> One hundred seventy-five *point* nine-seven
> *Answer:* 175.97

9. (a) Three-tenths   (b) Eight-tenths   (c) Four-tenths   (d) Nine-tenths

**10.** (a) One *and* two-tenths             (b) Three *point* five
  (c) Four *and* six-tenths

**11.** (a) Twenty-five and nine-tenths      (b) Fifty and five-tenths

**12.** (a) Thirty-seven hundredths          (b) *Point* four-six
  (c) Ninety-eight hundredths          (d) Twenty-seven hundredths

**13.** (a) Two *point* three-five           (b) Six and forty-nine hundredths
  (c) Twelve and eighteen hundredths   (d) Thirty and three hundredths

**14.** (a) Eighteen and eighty-one hundredths
  (b) Nineteen *point* zero nine

**Problems 15 through 20.** Write word statements for the decimals.

  **Example:**   4.40 is written four and forty hundredths.
        5.05 is written five and five hundredths.
        10.1 is written ten and one-tenth.
        155.55 is written one hundred fifty-five and fifty-five
        hundredths.

**15.** (a) 4.4   (b) .04   (c) 40.4   (d) 40.04   (e) 44.40 (f) 40.44 (g) 14.41

**16.** (a) .3   (b) 3.0   (c) 3.3   (d) 30.3   (e) 33.03 (f) 31.13 (g) 13.13

**17.** (a) 5.05   (b) 5.50   (c) 5.5   (d) 50.05   (e) 55.50 (f) 50.50 (g) 15.51

**18.** (a) 36.0   (b) 3.60   (c) .36   (d) 30.6   (e) 36.6 (f) 36.36 (g) 36.63

**19.** (a) 70.7   (b) 70.07   (c) 700.7 (d) 700.70 (e) 770.7 (f) 707.7 (g) 707.07

**20.** (a) 8.5   (b) 8.05   (c) 5.8   (d) 8.58   (e) 85.08   (f) 58.85
  (g) 50.85   (h) 858.85

**21.** Write each of the word statements as a decimal number.

  **Example:** Two *point* zero-two    *Answer:* 2.02
        Five hundred and five hundredths    *Answer:* 500.05
        Seven hundred seven and seven hundredths
        *Answer:* 707.07

  (a) Forty-four and four tenths
  (b) Sixty point six-six
  (c) Thirty and thirty-three hundredths
  (d) Fifty-five and five tenths
  (e) Three hundred three point three-three
  (f) Five hundred fifty and fifty-five hundredths
  (g) One hundred and ten hundredths
  (h) One hundred one point zero-one
  (i) Two hundred twenty-two and two hundredths
  (j) Sixty-nine point nine-six

**Problems 22 through 27.** Write word statements for the decimal numbers.

  **Example:**   2.547 is written two and five hundred forty-seven thousandths.

**22.** (a)   .123   (b)   .456   (c)   .789   (d)   .655   (e)   .875
  (f)   .984   (g)   .542

**23.** (a)    1.345    (b)    2.438    (c)    6.784    (d)    19.556    (e)    23.856
(f)    34.855

**24.** (a)    10.465    (b)    12.116    (c)    16.750    (d)    56.655    (e)    80.557
(f)    92.433

**25.** (a)  145.646    (b)  456.243    (c)  789.554    (d)  344.098
(e)  275.750

**26.** (a)  105.055    (b)    2.002    (c)    25.252    (d)  500.005    (e)  505.050

**27.** (a)    40.040    (b)    35.053    (c)    64.640    (d)  475.475    (e)    33.333

**28.** Write each of the word statements as a decimal number.
(a) Two hundred sixty-five thousandths
(b) Point four five-nine
(c) Seven hundred twenty-four thousandths
(d) Three point five-six-seven
(e) Forty-five and three hundred six thousandths
(f) Fifty-five and five thousandths
(g) Nineteen and one hundred nineteen thousandths
(h) One thousand four point four-zero-zero

**Problems 29 through 33.** Write word statements for the decimal numbers.

**Example:** 1.9564 is written one *and* nine thousand five hundred sixty-four *ten-thousandths*.

**29.** (a) .8764        (b) .5678        (c) .3564        (d) .1982
(e) .2375        (f) .1875

**30.** (a) 2.1457        (b) 4.7595        (c) 3.8857        (d) 10.7845
(e) 15.5465        (f) 25.2545

**31.** (a) 23.5056        (b) 56.7780        (c) 45.8070        (d) 140.6065
(e) 234.4320

**32.** (a) 50.0505        (b) 205.0054        (c) 500.0005        (d) 505.0505
(e) 1,595.4578

**33.** (a) 10,000.0905        (b) 10,010.0101        (c) 2,000.0002        (d) 1.0001

**34.** Write each of the word statements as a decimal number.
(a) One thousand five hundred sixty-four ten-thousandths
(b) Point five-two-three-seven
(c) Twelve and four hundred thirty-four ten-thousandths
(d) Twenty-six ten-thousandths
(e) Point zero-zero-two-six
(f) One hundred fifty point zero-four-nine-eight
(g) One thousand ten and one thousand one ten-thousandth

**35.** Name the place value of the digit 6 in each of the given decimal numbers.

**Example:** 5.8640        The digit 6 is in the *hundredths* place.

(a) 4.674        (b) 3.067        (c) 5.9865        (d) 1.0076
(e) 6.108        (f) 56.89        (g) 98.0689        (h) 657.4
(i) 645.78        (j) 45.9565        (k) 6,540.89        (l) 65,789.1

**36.** Name the place value of the digit 5 in the numbers.
(a) 456.897      (b) 145.7683      (c) .1592      (d) 3,478.9502

**37.** Write a decimal number with 5 in the hundred's place, 8 in the ten's place, 1 in the unit's place, 2 in the tenth's place, 4 in the hundredth's place, 9 in the thousandth's place, and 5 in the ten-thousandth's place.

**38.** Write a decimal number with 3 in the thousand's place, 1 in the hundred's place, 8 in the unit's place, 0 in both the tenth's and hundredth's place, 5 in the thousandth's place, and 4 in the ten-thousandth's place.

**39.** For each group of decimals, find the smallest and the largest.

**Example:** For the group 5.5, .55, 5.05, .055, 55.5, the decimal .055 is the smallest, and the decimal 55.5 is the largest.

(a)  .5       .55      .05      .055     .0505
(b)  .8       .89      .98      .90      .098
(c)  1.2      2.1      1.12     2.11     1.21
(d)  .35      3.5      5.3      .53      .053     .035
(e)  1.10     1.01     1.101    1.011    1.110

**Problems 40 through 45.** Round off decimals to the value specified. Review rules and examples of rounding off in Chapter 2 (page 20) before doing these problems.

**40.** Round off to nearest tenth
(a) .16    (b) .47    (c) .34    (d) .49    (e) 1.53    (f) 2.89
(g) 4.95

**41.** Round off to nearest tenth
(a) 2.467    (b) 5.779    (c) 1.0993    (d) 2.0015    (e) .8904
(f) 34.0901

**42.** Round off to nearest hundredth
(a) 1.784    (b) 2.519    (c) .088    (d) 3.126    (e) 4.455
(f) 5.365

**43.** Round off to nearest hundredth
(a) 3.1089    (b) 5.6553    (c) 8.2300    (d) .0095    (e) 45.1500

**44.** Round off to nearest thousandth
(a) .2148    (b) .3459    (c) 2.0833    (d) 17.1555    (e) 36.0094
(f) 44.0099

**45.** Round off to nearest thousandth
(a) .5654    (b) .6705    (c) 1.0076    (d) 35.1234    (e) 1.9985
(f) 2.9996

**Problems 46 through 51.** Round off the decimals to the indicated number of decimal places.

**Example:** Round off 3.4556, 4.9087, 6.75441 to *two places*.

*Solution:* For each number start counting *at the decimal point*, and count two places to the right (which is the hundredth's place), and round off.

*Answer:* 3.4556 is 3.46 to two places.
            4.9087 is 4.91 to two places.
            6.7544 is 6.75 to two places.

**46.** Round off to one place (same as nearest tenth).
(a) 2.36     (b) 5.68     (c) 17.14     (d) 10.69     (e) 12.547
(f) 2.1158

**47.** Round off to one place.
(a) .56     (b) 2.65     (c) 3.389     (d) 6.6856     (e) 10.0504
(f) 23.0479

**48.** Round off to two places (same as nearest hundredth).
(a) 1.468     (b) 2.557     (c) 18.509     (d) .248     (e) .993
(f) 1.897

**49.** Round off to two places.
(a) 3.066     (b) 11.6790     (c) 14.0109     (d) 23.4554
(e) 9.9805     (f) 2.001

**50.** Round off to three places (same as nearest thousandth).
(a) .1786     (b) 2.3543     (c) 4.5568     (d) 2.3442     (e) 15.1905

**51.** Round off to three places.
(a) 1.3050     (b) 3.1556     (c) 2.5439     (d) 8.1175     (e) 5.0067
(f) 4.8995

**Problems 52 through 54.** Round off each measurement to the indicated value.

**52.** Round off to nearest tenth.
(a) 2.54 cm.     (b) .868 mile     (c) .9463 liter     (d) 3.1416 in.
(e) 1.09 yards

**53.** Round off to two places.
(a) 1.0936 yards     (b) 2.2056 pounds     (c) .3937 inch
(d) 1.1515 mile     (e) 6.436 cm.

**54.** Round off to three places.
(a) .3937 inch     (b) 1.5564 mile     (c) 1.093 yards
(d) 2.2046 pounds     (e) $3.0756

**55.** One meter is equal to 1.0936 yards. Round off this conversion factor to:
(a) nearest tenth     (b) nearest hundredth     (c) three places.

**56.** A machinist measures a piece of metal and it is 0.1469 inch thick. What is the thickness, to the nearest hundredth of an inch?

**57.** A surveyor measured the width of a city lot to be 56.84 feet. Show this distance to:
(a) nearest tenth of a foot
(b) nearest foot
(c) one significant digit (Review Chapter 2.)
(d) nearest 10 feet

**58.** An electrical technician measured the resistance in a length of wire. The resistance was 15.9565 ohms. The technician said the resistance was 15.956 ohms, to the nearest thousandth. Was he correct?

**59.** A chemical technician weighed 16.4535 grams of calcium oxide (ordinary lime). What is this weight to the:
   (a) nearest thousandth of a gram?
   (b) to two places?
   (c) nearest gram?
   (d) nearest tenth of a gram?

**60.** A quality control inspector measured four finished machine parts. Round off each measurement to the nearest thousandth. Arrange them in order from the smallest to the largest. The measurements are:
   3.3694 cm.    3.3683 cm.    3.3696 cm.    3.3674 cm.

**61.** Last month Mrs. Campbell spent $87.46 on groceries.
   (a) How much did she spend to the nearest 10 cents?
   (b) How much did she spend to the nearest dollar?
   (c) How much did she spend to the nearest $10.00?

**62.** Brown & Sharp 11-gauge wire is 0.0907 inch in diameter. Write this measurement to the:
   (a) nearest thousandth    (b) nearest hundredth    (c) nearest tenth.

## SEC. 2    ADDING AND SUBTRACTING DECIMALS

To find the answers to the following exercises, you must know how to add and subtract decimal fractions.

Exercise 1. An engine cylinder is 3.855 inches in diameter. It is to be rebored 0.065 inches oversize. What will be the diameter of the rebored cylinder?

Exercise 2. During the rainy season, a weather station recorded the following rainfall for one week:

| | |
|---|---|
| Sunday | 3.57 inches |
| Monday | 2.78 inches |
| Tuesday | 2.89 inches |
| Wednesday | 3.86 inches |
| Thursday | 1.98 inches |
| Friday | 4.19 inches |
| Saturday | 3.88 inches |

What is the total rainfall for the week?

Exercise 3. Mr. Buchanan is a cabinet maker. His wages are $216.45 per week before any deductions. His deductions are:

| | | |
|---|---|---|
| Federal tax | = | $41.98 |
| F.I.C.A. | = | 6.58 |
| Union dues | = | 4.87 |
| Credit union | = | 25.00 |
| Health plan | = | 16.50 |

What is his take-home pay?

To get the answer to Exercise 1, add the two decimals 3.855 and .065. How do you do it? Reason this way: "The *decimal* 3.855 is the *mixed number* $3\frac{855}{1,000}$. The *decimal* .065 is the *fraction* $\frac{65}{1,000}$. Therefore, I can find the sum of the decimals by *adding* the two numbers."

$$3\frac{855}{1,000}$$
$$+ \quad \frac{65}{1,000}$$
$$\text{Sum} = \quad 3\frac{920}{1,000}$$

Write the sum as the decimal: 3.920.

If you add them like whole numbers and keep the decimal points lined up, you get the same answer.

Sum:  3.855
    + .065    Keep decimal points lined up.
    ‾‾‾‾‾
     3.920

*Answer:* The diameter of the rebored cylinder is 3.920 inches.

To get the answer to Exercise 2, write the decimals in a column. Keep the decimal points directly under each other. Add as in whole numbers.

| | | |
|---|---|---|
| Sunday: | 3.57 | inches |
| Monday: | 2.78 | inches |
| Tuesday: | 2.89 | inches |
| Wednesday | 3.86 | inches |
| Thursday: | 1.98 | inches |
| Friday: | 4.19 | inches |
| Saturday: | 3.88 | inches |
| Total | 23.15 | inches |

*Answer:* Total rainfall for the week is 23.15 inches.

Place the decimal point in the answer under the decimal point in the addends. If the addends are in hundredths, the answer is in hundredths. If the addends are in thousandths, the answer is in thousandths, and so on.

**Example 5:** Find the sum of the decimals:

$$3.987 \quad 1.75 \quad 6.8 \quad 5.009$$

*Solution:* Arrange the decimals with the decimal points in a straight line:

3.987
1.75
6.8
5.009
‾‾‾‾‾

To reduce the possibility of error, you may add zeros to the decimals which have fewer decimal places. You add zeroes to make the addends have the same number of decimal places. When this is done, the problem looks like this:

$$
\begin{array}{l}
3.987 \\
1.750 \longleftarrow \text{(one zero added)} \\
6.800 \longleftarrow \text{(two zeros added)} \\
\underline{5.009} \\
\end{array}
$$

*Answer:*             17.546

When you add zeros to the decimals with fewer decimal places, it is the same as writing all fractions with a common denominator. Since all decimal fractions have denominators of 10; 100; 1,000; 10,000; etc., you can write all fractions with a common denominator. Write tenths as hundredths by adding one zero on the right; as thousandths by adding two zeros on the right; as ten-thousandths by adding three zeros on the right; and so on.

**Example 6:** Find the sum of .5, .78, and .964.

*Solution:* Add two zeros to .5 to get .500, one zero to .78 to get .780. The sum is:

$$
\begin{array}{l}
.500 \\
.780 \\
\underline{.964} \\
\end{array}
$$

*Answer:*             2.244

NOTE: These added zeros do not change the final sum.

Because you can add and subtract decimal fractions like ordinary whole numbers, they are easier to work with than common fractions.

Now, to find the answer to Exercise 3, add up all the deductions listed on Mr. Buchanan's paycheck:

| | |
|---|---|
| Federal tax | $41.98 |
| F.I.C.A. | $6.58 |
| Union dues | $4.87 |
| Credit union | $25.00 |
| Health plan | $16.50 |
| Total deductions | $94.93 |

Subtract the sum from the gross pay to get his net pay (take-home pay):

Take-home pay $= \$216.45 - \$94.93$

Subtract by writing the two numbers in a vertical column. Be sure to keep decimal points lined up.

$216.45    gross
— $94.93    deductions
*Answer:*        $121.52    take-home pay

Subtract decimals the same way that you subtract whole numbers. Put the decimal point in the answer directly below the decimal point of the subtrahend and minuend.

In many problems you need to add or subtract decimals. Pay attention to the information in the problems. Draw a neat sketch when necessary. Be careful. Be accurate. Be neat. Study examples 7, 8, and 9.

**Example 7:** A piece of Nominal 2-inch Double Extra Strong wrought iron pipe has an outside diameter (OD) of 2.375 inches and a wall thickness of 0.436 inches. What is the inside diameter (ID)?

*Solution:* Draw a sketch of the pipe and label the known parts (Fig. 9-5).

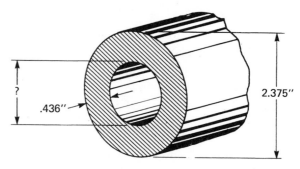

**Figure 9-5**

NOTE: This problem is often done incorrectly. You do NOT find the inside diameter by subtracting *one* wall thickness from the outside diameter. Remember—there are *two* walls. Add 0.436 to itself to get the total wall thickness. Find the inside diameter (ID) by subtraction:

$$ID = 2.375 - (.436 + .436)$$
$$ID = 2.375 - .872$$

*Answer:*        $ID = 1.503$ inches

Many manufacturing firms are involved in the production of a modern automobile. When the various parts come together for final assembly, there must be assurance they will fit. This assurance of fit is accomplished by assigning a *tolerance* or limit of accuracy to each part. *Tolerance is the amount by which the final measurement may differ from the ideal or blue-print value.* For instance, a machined part has a blueprint length of 2.550 inches, with a tolerance of $\pm.001$ inch. This means the finished part can be as long as 2.551 = 2.550 + .001 inches, or as short as 2.549 = 2.550 − .001 inches and still be accepted for assembly.

**Example 8:** The diameter of a piston is 3.750 inches with a toler-ance of $\pm.004$ inches. See Fig. 9-6. This means that the actual diameter of the piston may be .004 inch greater or smaller than 3.750 inches. Find the greatest and smallest diameter the piston can have and still be within the tolerance.

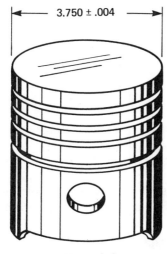

3.750 ± .004

**Figure 9-6**

*Solution:* Draw a sketch of the piston and label it with the given information.

*Answer:* The greatest diameter is the sum:

$$\begin{array}{r} 3.750 \\ +\ \ .004 \\ \hline 3.754 \end{array}$$

The smallest possible diameter is the difference:

$$\begin{array}{r} 3.750 \\ -\ \ .004 \\ \hline 3.746 \end{array}$$

**Example 9:** Use the tolerances shown in Fig. 9-7. Find the MAXIMUM width, height, and length. Find the

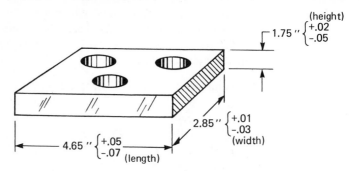

**Figure 9-7**

MINIMUM width, height, and length of the support plate.

*Solution:* Add and subtract tolerances.

$$\text{Maximum}\begin{cases} \text{height} = 1.75 + .02 = 1.77 \text{ inches} \\ \text{width} \ = 2.85 + .01 = 2.86 \text{ inches} \\ \text{length} = 4.65 + .05 = 4.70 \text{ inches} \end{cases}$$

$$\text{Minimum}\begin{cases} \text{height} = 1.75 - .05 = 1.70 \text{ inches} \\ \text{width} \ = 2.85 - .03 = 2.82 \text{ inches} \\ \text{length} = 4.65 - .07 = 4.58 \text{ inches} \end{cases}$$

*Answer:* A finished part with *any* of the dimensions in the right hand column is within the tolerance.

## PROBLEMS: Sec. 2
## Adding and Subtracting Decimals

**Problems 1 through 7.** Add the decimal numbers given in problems 1 through 7.

| 1. | (a) .6 | (b) .8 | (c) .4 | (d) .2 | (e) .8 |
|---|---|---|---|---|---|
| | .5 | .3 | .3 | .3 | .5 |
| | .7 | .1 | .5 | .9 | .9 |
| | | | | .4 | .2 |
| | | | | | .6 |

| 2. | (a) .06 | (b) .07 | (c) .01 | (d) .04 | (e) .03 |
|---|---|---|---|---|---|
| | .04 | .09 | .01 | .03 | .08 |
| | .08 | .05 | .07 | .09 | .06 |
| | | | .08 | .06 | .05 |
| | | | | | .07 |

**3.** (a) .12     (b) .32     (c) .89     (d) .31     (e) .25
      .34           .96           .90           .75           .95
      .54           .54           .06           .87           .55
                                                    .98           .45
                                                                        .05

**4.** (a) .11     (b) .09     (c) .87     (d) .03     (e) .11
      .01           .90           .78           .30           .28
      .10           .99           .88           .89           .45
                                     .08           .76           .35
                                                                        .66

**5.** (a) 2.34     (b) 3.45     (c) 2.67     (d) 1.80     (e) 2.78
      3.65         6.78         4.65         2.65         8.97
      1.93         3.98         3.33         3.76         4.78
                                         5.87         4.64
                                                           3.54

**6.** (a) 2.398     (b) 3.008     (c) 5.867     (d) 1.109     (e) 11.908
      2.345        5.056        3.335        4.908        13.855
      3.098        4.704        4.556        3.006        10.007
                                         4.665        23.605
                                                          32.040

**7.** (a) 234.98     (b) 453.67     (c) 23.70     (d) 345.987     (e)    3.81
      345.78       34.91        60.00        256.007        25.75
      121.43         4.83        45.50        468.884       143.69
                                               233.332      5346.38

**Problems 8 through 11.** Add the decimal numbers.

     **Example:**    .6
                 15.78
                  3.1

     *Solution:* Add zeros and write all decimals with two places to the right
             of the decimal point.

                .60
             15.78      Add a zero in hundredth's place
              3.10

     *Answer:*      19.48

**8.** (a) 1.7     (b) .9     (c) .7     (d) 1.6     (e) .96
      2.56        5.36       7.0         .4        6.0
      3.1                                   3.87        .7

**9.** (a) 6.965     (b) 3.7     (c) 7.075     (d) 8.6     (e) 1.38
       .54        .665         .96         .86        2.1
       .6         4.32         .3          .086       6.235
                              4.0         86.0        .17

**10.** (a) 1.567          (b) 6.45          (c) .81          (d) 1.1          (e) .61
         8.34               4.7               .9               .10               6.0
         .008               2.786            9.1               101.01          .3
         5.0                3.9               .756              .001             .306

**11.** (a) .4 + 8.6 + 34.987 + 2.8 = _____
     (b) 1.1 + 10.87 + .3 + .554 = _____
     (c) 234.7 + 115.19 + .34 + 17 = _____
     (d) 3.9 + .86 + .113 + 1.1 = _____
     (e) 45.80 + 65.08 + 80.113 + .3 = _____

**Problems 12 through 15.** Add the quantities.

**12.** (a) .11 cm.          (b) 3.6 feet          (c) 5.89 m.          (d) 5.19 liter
         .56 cm.               7.9 feet              3.7  m.               4.67 liter
         1.89 cm.              4.5 feet              4.0  m.               2.77 liter
         3.77 cm.              3.8 feet              5.18 m.               5.88 liter

**13.** (a) 45.008 gm.          (b) .1178 in.          (c) 45.8 yd.          (d) 41.8 km.
         46.987 gm.               .5567 in.               4.7 yd.               34.8 km.
         36.445 gm.               .8896 in.               16.86 yd.             26.9 km.
         81.665 gm.               .2346 in.               4.0 yd.               88.7 km.

**14.** (a) $4.78          (b) $78.11          (c) $ 0.87          (d) $456.56
         $5.98               $ 0.89               $ 2.65               $346.89
         $6.56               $ 1.45               $15.54               $265.77
                                                                        $540.05

**15.** (a) 1.1 in. + 2.88 in. + 45.87 in. + .4 in. = _____
     (b) .876 cm. + 1.67 cm. + 23.0 cm. + 56.991 cm. = _____
     (c) 1.18 m + 2.5 m. + 34.18 m. + 45.0 m. = _____
     (d) $3.78 + $1.12 + $25.54 + $0.03 = _____
     (e) 16.9 gm. + 13.5 gm. + 89.67 gm. + .345 gm. = _____

**16.** A mechanic uses four shims with thicknesses of:
     0.0625 inch     0.0500 inch     0.1375 inch     0.0025 inch
     What is the total thickness of the four shims?

**17.** Find the diameter of the pipe shown in Fig. 9-8. (Diameter is the distance across the outer circle.)

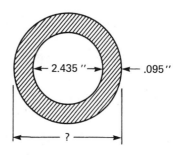

*Figure 9-8*

**18.** Find the dimension marked D in Figs. 9-9 and 9-10.

(a)

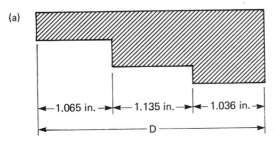

**Figure 9-9**

(b)

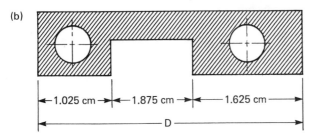

**Figure 9-10**

**19.** Johansson Gauge Blocks are used to check a dimension. Find the dimension.
.1008 + .1003 + .139 + .132 + .550 + .600 = _____ inches

**20.** The sales for five days at the Broadway Garage are:

| | | | |
|---|---|---|---|
| Monday | $156.89 | Tuesday | $225.76 |
| Wednesday | $198.54 | Thursday | $202.08 |
| Friday | $305.75 | | |

Find the total sales.

**21.** Interstate Highway 5 has 797.0 miles in California, 308.4 miles in Oregon, and 276.9 miles in Washington. How long is Interstate 5?

**22.** Interstate Highway 70 goes from Utah to Maryland. Distances through the various states are:

228.1 miles in Utah
449.6 miles in Colorado
424.2 miles in Kansas
251.6 miles in Missouri
160.3 miles in Illinois
156.3 miles in Indiana
231.8 miles in Ohio
14.4 miles in West Virginia

169.3 miles in Pennsylvania
52.3 miles in Maryland

Find the total length of Interstate 70.

23. Figure 9-11 shows a stepped pulley cone. Find the dimension labeled L.

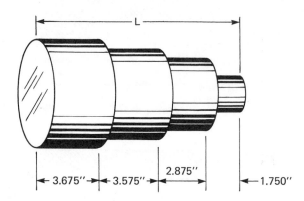

**Figure 9-11**

**Problems 24 through 31.** Subtract the decimals.

24. (a) 4.56    (b) 6.79    (c) .85    (d) 23.75    (e) 18.67
       3.45        3.55        .54       11.25         1.45

25. (a) 9.7     (b) 6.85    (c) 16.08  (d) 4.89     (e) 1.66
       8.7         3.76       11.10       1.04          .87

26. (a) 17.56   (b) 34.8    (c) 45.75  (d) 15.9     (e) 54.77
        4.2         2.75        5.2        2.           .11

27. (a) 5.345   (b) 13.778  (c) 167.180 (d) .0975   (e) 5.0954
       2.446       9.658       45.397      .0046        3.8876

28. (a) 6.0000  (b) 15.0000 (c) 3.0000 (d) 9.0000   (e) 10.000
       3.9854      11.0107     2.7064     3.3995        9.898

29. (a) 7.0     (b) 15.01   (c) 24.35  (d) 35.0     (e) 4.8
       4.657       4.3576      6.5687      .067         .865

30. (a) 4.135   (b) 14.466  (c) 38.9   (d) 11.0     (e) 19.0
       3.001       2.000       5.01        .545         3.006

31. (a) 8.0     (b) 16.0    (c) 35.0   (d) 4.0      (e) 56.0
       4.775       2.406       2.775      2.06         45.001

32. From .056 subtract 0.003.

33. Subtract 6.89 from 34.77.

34. Find the difference between 23.5 and 15.86.

35. From 78.94 take 45.87.

36. Decrease 3.998 by .664.

**37.** How much is left when 66.887 is taken from 87.992?

**38.** What is the excess of 5.75 over 4.36?

**39.** What remains after 34.87 is diminished by 17.96?

**40.** How much is 6.75 less 4.36?

**41.** From the sum of 5.75 and 2.45 subtract the sum of 1.54 and 3.77.

**42.** A "ten-thousandths" micrometer shows that a piece of machined steel is 1.7765 inches thick. When Johansson Gauge Blocks are used, the piece of metal measures 1.7758 inches thick. What is the difference in measurements?

**43.** The outside diameter of a piece of standard $\frac{5}{8}$ copper water tube is 0.750 inch. The wall thickness is 0.042 inch. What is the inside diameter?

**44.** A tapered pin has a large-end diameter of 2.450 inches. The small end has a diameter of 1.9875 inches. What is the difference in diameters?

**45.** Find the dimension marked L in Fig. 9-12.

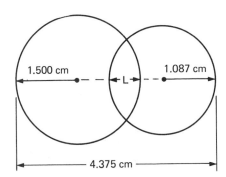

**Figure 9-12**

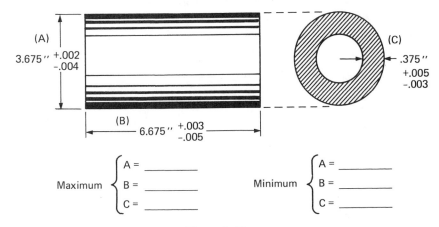

**Figure 9-13**

**46.** Find the maximum and minimum values for all dimensions in Fig. 9-13.

## SEC. 3   MULTIPLYING DECIMALS

Multiply decimal numbers in the same way that you multiply whole numbers, and then locate the decimal point in the answer. Here is how you determine the location of the decimal point in the answer.

*The number of decimal places in the answer is equal to the **sum** of the decimal places in the factors.*

**Example 10:** Multiply 3 by .6.

*Solution:* $3 \times .6 = \dfrac{3}{1} \times \dfrac{6}{10} = \dfrac{3 \times 6}{10} = \dfrac{18}{10} = 1.8$

*Answer:* 1.8

**Example 11:** Multiply .4 by .8.

*Solution:* $.4 \times .8 = \dfrac{4}{10} \times \dfrac{8}{10} = \dfrac{4 \times 8}{10 \times 10} = \dfrac{32}{100} = .32$

*Answer:* .32

**Example 12:** Multiply .06 by 1.25.

*Solution:* $.06 \times 1.25 = \dfrac{6}{100} \times \dfrac{125}{100} = \dfrac{750}{10,000} = .0750$

*Answer:* .0750

From these results notice that: *The number of decimal places in the answer is equal to the SUM of the decimal places in the factors.* For instance: to multiply .54 by 1.3 multiply $54 \times 13 = 702$. You see that .54 has 2 decimal places, 1.3 has 1 decimal place, and the answer has $2 + 1 = 3$ places. The product of $.54 \times 1.3 = .702$ (3 decimal places). Here is the

**Rule for Determining the Location of the Decimal Point in Multiplication**

*The number of decimal places in the product is equal to the SUM of the number of decimal places in the numbers being multiplied.*

**Example 13:** Multiply the numbers $.45 \times .35$.

*Solution:* Write in vertical form.

$$
\begin{array}{rl}
.45 & \text{(2 decimal places)} \\
\times \quad .35 & \text{(2 decimal places)} \\
\hline
225 & \\
135 \quad\;\; & \\
\hline
\end{array}
$$

*Answer:*        .1575        ($2 + 2 = 4$ decimal places)

**Example 14:** Multiply 23.56 × 3.65.

*Solution:*     2 3.5 6   (2 decimal places)
          3.6 5   (2 decimal places)
          ───────
          1 1 7 8 0
          1 4 1 3 6
          7 0 6 8
          ───────

*Answer:*   8 5.9 9 4 0   (2 + 2 = 4 decimal places)

In these problems, do the arithmetic, then find the number of places in the answer. (Answers follow problem 4.)

### *Practice Problems*

**1.** Add.

| (a) | 23.098 | (b) 3.0098 | (c) 13.0 |
|---|---|---|---|
| | 4.98 | 4.9975 | 0.6 |
| | 3.007 | 1.0001 | 6.906 |
| | 2.0 | 2.94 | 0.0001 |
| | 2,467.3 | | |

**2.** Subtract.
 (a) $45.906 - 23.89 = $ _____
 (b)   6.90   (c)   0.987
     $-4.758$      $-0.9865$

**3.** Multiply.
 (a) $23.065 \times 13.9 = $ ____   (b) $1.007 \times 2.08 \times 3.19 = $ ____

**4.** Do the indicated operations. (Multiply BEFORE adding and subtracting.)
 (a) $(4.85 \times 23.009) + 16.975 = $ _____
 (b) $(15.750 \times 10.95) - (4.57 \times 17.75) = $ _____

*Answer:*

    **1.** (a) 2,500.385   (b) 11.9474   (c) 20.5061
    **2.** (a) 21.016   (b) 2.142   (c) .0005
    **3.** (a) 320.6035   (b) 6.6816264
    **4.** (a) 128.56865   (b) 91.3450

# PROBLEMS : Sec. 3  Multiplying Decimals

**Problems 1 through 8.** Multiply.

1. (a) 8      (b) 7      (c) .7      (d) .3      (e) .8
      .6          .9          3          6          8
      —           —           —          —          —

2. (a) .25     (b) 45     (c) 32     (d) 28     (e) 46
      5          .7          .6          .9          .3
      —           —           —          —          —

3. (a) 26      (b) 45     (c) .78    (d) 85     (e) .89
      .17         .36         34          .45        17
      —           —           —          —          —

4. (a) .9      (b) .6     (c) .54    (d) .65    (e) .49
      .5          .8          4          .7          .5
      —           —           —          —          —

5. (a) .89     (b) .77    (c) 6.8    (d) 99     (e) 3.67
      .45         1.6         4.3         .80         .88
      —           —           —          —          —

6. (a) .675    (b) .775   (c) 56.9   (d) 198    (e) 19.6
      38          5.6         3.5         .76         3.6
      —           —           —          —          —

7. (a) .5675   (b) 67.85  (c) 23.66  (d) .6875  (e) 40.01
      23          .874        1.55        .802        1.45
      —           —           —          —          —

8. (a) .006    (b) .0005  (c) 20.001 (d) 35.375 (e) 124.875
      .005        20          4.005       .4067       50.750
      —           —           —          —          —

**Problems 9 through 12.** Multiply. After multiplying, round off answer to nearest hundredth.

9. (a) $4.56   (b) $78.95  (c) $25.55  (d) $145.95 (e) $.75
      8          .06         1.25         .55         .08
      —           —           —          —          —

10. (a) $875    (b) $755.10  (c) $1,895.00  (d) $156.98
       .667        .055          .75            .055
       —           —             —             —

    (e) $.075
        .65
        —

11. (a) 24.98 cm.   (b) 4.78 m.   (c) 75.9 in.   (d) .075 ft.
       .11             .85            .075           .5
       —               —              —              —

    (e) 465.37 liter
        .875
        —

12. (a) $56.85 × .75      (b) 45.8 cm × .095      (c) 1.08 amp × .75

13. An apprentice in a machine shop earns $3.78 an hour. What does he earn in a 40-hour week?

14. An assembly worker earns $3.08 an hour for the first 40 hours of a work week. He earns time-and-a-half (1.5 times the rate of $3.08) for overtime. What are his earnings for a work week of 44.5 hours?

15. Type M 3-inch copper water tube weighs 2.68 pounds per foot. How much does 25.8 feet weigh?

16. A sheet of 14 Stub's gauge brass weighs 3.552 pounds per square foot. What is the weight of 9.75 square feet?

17. A shop drawing shows a part to be 1.75 inches long. If the drawing is enlarged 1.5 times, how long will the part be on the drawing?

18. Standard annealed copper wire of various gauges is used in electrical work. If 11-gauge copper wire has a resistance of 1.284 ohms per 1,000 feet, what is the resistance in 2,500 feet? (HINT: Multiply $1.284 \times 2.5$.)

19. A piece of $\frac{3}{4}$-inch diameter round steel bar weighs 0.1253 pounds per inch. What is the weight of 34.57 inches of the bar? Round off answer to nearest hundredth.

20. A cubic foot of water weighs 62.5 pounds. What is the weight of 345.7 cubic feet?

21. A certain brand of diesel fuel costs $0.42 per gallon. What is the cost of 22.5 gallons?

22. One gallon of water weighs 8.33 pounds. What is the weight of 197.65 gallons? Round off answer to two places.

23. A journeyman plumber makes $8.97 an hour. His helper makes $4.73 an hour. Both men earn time-and-a-half for overtime over 40 hours. What are their combined earnings for working 48.6 hours each? (When money is involved, always round off answer to two places.)

24. A conveyor belt moves at a speed of 16.75 feet per *second*. How far will a bucket on the belt travel in 2.75 *minutes*? (Change minutes to seconds by multiplying 2.75 by 60, then multiply by 16.75.) Round off answer to two places.

25. A time study shows that it takes 1.5 minutes for a factory worker to deburr a certain part. How long will it take to deburr 45 parts?

26. In one week a salesman for the Mid-Nite Auto Supply sold 18 batteries at $24.50 each; 13 sets of spark plugs at $8.95 a set; 6 tires at $36.50 each, and other miscellaneous items amounting to $196.50. What were the total sales for the week?

27. Jerry Bilt Construction Company needs 1,800 feet of steel reinforcing rod for a construction job. Bungling Buford is given the job of finding the cost of the rod. The rod weighs 1.75 pounds per foot and costs $19.55 per hundred pounds. B. B. estimates the cost will be $605.82. Do you agree? (Note: $19.55 per hundred pounds is the same as $.1955 per one pound.)

28. A cutting tool on a lathe moves forward 0.125 inch for each revolution of the tailstock. How far will the tool move if the tailstock makes 25.5 revolutions?

**29.** Complete this labor cost report:

**LABOR COST—CROSS GARAGE**

| | | |
|---|---|---|
| (a) Auto mechanics | 88 hr. @ $4.78 per hr. | _____ |
| (b) Diesel mechanics | 145 hr. @ $5.86 per hr. | _____ |
| (c) Stock room clerk | 40 hr. @ $2.34 per hr. | _____ |
| (d) Parts salesman | 44 hr. @ $4.95 per hr. | _____ |
| (e) Shop helper | 40 hr. @ $2.05 per hr. | _____ |

Totals   (f) _____ hr.          (g) $_____

**30.** Find the total cost on the invoice:

SOLD TO:  J & R Tool Company
TERMS:   On Account
ITEMS:   (a) 6 lengths of bar stock @ $7.85 per length _____
         (b) 5 gross No. 6 fillister head machine
             screws @ $14.75 per gross                _____
         (c) 3 doz. No. 12 square head half dog
             point set screws  @ $3.64 per doz.       _____
         (d) 175 medium lock washers, nominal
             size $\frac{9}{16}$ at $.02 per washer   _____
         (e) 50 No. 1212 ASA Woodruff Keys
             @ $.17 per key                           _____

(f) Sub Total  _____
(g) Sales tax = .06 × Sub total = _____

(h) Grand Total = _____

## SEC. 4   DIVIDING DECIMALS

In many problems you will use one or more of the operations of addition, subtraction, multiplication, or division of decimals. You must be able to recognize which operations to do, and be able to do all of them.

**Example 15:** An addition problem. Find the length of the bolt shown in Fig. 9-14.

*Solution:* Length = sum of values.

*Answer:* L = .37 + 1.65 + .17 + 1.06 = 3.25

**Example 16:** A subtraction problem. Find the dimension marked X in Fig. 9-15.

*Solution:* X = difference between 3.750 and 1.375.

*Answer:* X = 3.750 − 1.375 = 2.375 cm.

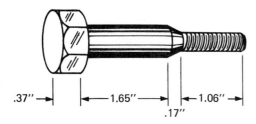

**Figure 9-14**

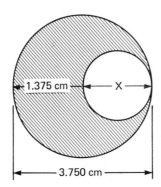

**Figure 9-15**

**Example 17:** A multiplication problem. A single sheet of aluminum is 0.070 inch thick. How high is a stack of 75 sheets?

*Answer:* $75 \times 0.070 = 5.25$ inches

**Example 18:** A division problem. A piece of sheet metal is 4 inches wide. How many .5-inch wide strips can be cut from the sheet?

*Solution* (a): To find the number of strips, divide 4 by .5. You may write this division in three equivalent forms:

(a) $4 \div .5$    (b) $\dfrac{4}{.5}$    (c) $.5\overline{)4.0}$    (NOTE: $4 = 4.0$)

In form (a), write the decimal .5 as a fraction $\dfrac{5}{10}$. Divide 4 by $\dfrac{5}{10}$ by multiplying 4 by the reciprocal of $\dfrac{5}{10}$:

*Answer* (a):    $4 \div \dfrac{5}{10} = 4 \times \dfrac{10}{5} = \dfrac{40}{5} = 8$ strips

*Solution* (b): In form (b), again write the decimal .5 as a fraction $\frac{5}{10}$. This time, multiply both numerator (4) and denominator $\left(\frac{5}{10}\right)$ by 10. This gives:

*Answer* (b):          $\dfrac{4}{\frac{5}{10}} = \dfrac{4 \times 10}{\frac{5}{\cancel{10}} \times \cancel{10}} = \dfrac{40}{5} = 8$ strips

*Solution* (c): In form (c), change the divisor .5 to a whole number by moving the decimal point *one* place to the right. In order to have a correct quotient, move the decimal point in the dividend (4.0) one place to the right. This process of changing the location of decimal point in the divisor and dividend in a division problem is demonstrated in the following: (Use a caret ( $_\wedge$ ) to indicate the new position of the decimal point in both divisor and dividend.)

$$.5_\wedge\overline{)4\ 0_\wedge 0} \qquad \text{(add one zero to right of caret)}$$

Instead of using a caret, you may use a small arrow ($\smile$) to show the movement of the decimal point to its new position.

$$.5\,\overline{)4.\,0\,0}$$

The decimal point in the quotient is placed directly above the *new* location of the decimal point in the dividend. After relocating the decimal point, carry out ordinary long division to as many places as the problem requires.

$$\begin{array}{r} 8.0 \\ 5\overline{)40.0} \\ 40.0 \\ \hline 0 \end{array}$$

*Answer* (c):   8 strips

**Example 19:** Divide 1.68 by .8.

*Solution:* Make the divisor (.8) a whole number. Move the decimal point *one* place to the right in BOTH divisor and dividend and divide.

$$\begin{array}{r} 2.\,1 \\ .8\,\overline{)1.\,6\,8} \\ 1\ 6 \\ \hline 0\ 8 \\ 8 \\ \hline 0 \end{array}$$

*Answer:* $1.68 \div .8 = 2.1$

After you relocate the decimal point, do the division as if *all numbers were whole numbers*. Put the decimal point in the quotient directly above the new location of the decimal point in the dividend. *Make no further adjustment of the decimal point.*

**Example 20:** Divide .7 by .008.

*Solution:* Add three zeros to the right of .7—two for relocation of decimal point, and one more for one decimal place in the answer.

$$
\begin{array}{r}
8\,7.5 \\
.008\,\overline{)\,.7\,0\,0\,0} \\
\underline{6\,4} \\
6\,0 \\
\underline{5\,6} \\
4\,0 \\
\underline{4\,0} \\
0
\end{array}
$$

*Answer:* $.7 \div .008 = 87.5$

You cannot expect all answers to come out even. When they do not, carry out division to three or four decimal places and round off the answer.

**Example 21:** Divide 17.2 by 1.4.

*Solution:*

$$
\begin{array}{r}
1\,2.2\,8\,6 \quad \text{rounded off}\\
1.4\,\overline{)\,1\,7.2\,0\,0\,0} \\
\underline{1\,4} \\
3\,2 \\
\underline{2\,8} \\
4\,0 \\
\underline{2\,8} \\
1\,2\,0 \\
\underline{1\,1\,2} \\
8\,0 \\
\underline{7\,0} \\
1\,0
\end{array}
$$

(Remainder is more than half of 14—see NOTE.)

*Answer:* 12.286

NOTE: Round up the quotient 12.285 to 12.286 because the remainder (10) is more than half of the divisor (14).

In general, if the remainder (after the last division) is more than half of the divisor, increase the last number in the quotient by one. If the remainder is less than half the divisor, do not increase the quotient. In most cases the nature of the problem indicates the number of decimal places in the answer.

**Example 22:** A stationary engine is used to pump water for irrigation. At full load the engine uses 4.6 gallons of fuel per hour. Find, to the nearest tenth of an hour, the number of hours the engine can operate on 55.8 gallons.

*Solution:* To get the number of hours divide:

Hours = total fuel ÷ fuel consumption per hour

Carry out the division to the hundredth's place and round off to tenths.

$$
\begin{array}{r}
1\,2.\,1\,3 \\
4.6_\wedge\overline{)5\;5.\,8_\wedge0\;0} \\
4\,6 \\
\overline{\phantom{0}9\,8} \\
9\,2 \\
\overline{\phantom{00}6\,0} \\
4\,6 \\
\overline{\phantom{00}1\,4\,0} \\
1\,3\,8 \\
\overline{\phantom{000}2}
\end{array}
$$

Round off the quotient 12.13, to the nearest tenth, to get 12.1.

*Answer:* The engine can operate 12.1 hours.

## PROBLEMS: Sec. 4  Dividing Decimals

Follow these steps to divide decimals:

Step 1.   Make the *divisor* (the number you are dividing **by**) a whole number by moving the decimal point to the right.

Step 2.   Move the decimal point in the *dividend* (the number you are dividing **into**) the same number of places to the right as it was moved in Step 1.

Step 3.   Place a decimal point directly above the new location of the decimal point obtained in Step 2.

Step 4.   Divide as with whole numbers and carry out the answer to as many places as necessary. Usually, you will not need to carry out answers to more than four places.

**Example:** Divide 45.8 by .17 and round off to the nearest hundredth.

*Solution:* Write          .17)45.8

Step 1 and 2:
(Result of Step 3)
17)4580.00

Step 4:
$$\begin{array}{r} 269.41 \\ 17\overline{)4580.00} \\ \underline{34} \\ 118 \\ \underline{102} \\ 160 \\ \underline{153} \\ 70 \\ \underline{68} \\ 20 \\ \underline{17} \\ 3 \end{array}$$   Divide as in whole numbers.

(STOP HERE)

*Answer:* 45.8 ÷ 0.17 = 269.41 (to nearest hundredth)

**Problems 1 through 4.** Divide.

1.  (a) 3)4.2     (b) 5)7.5     (c) 4)1.6     (d) 7)0.63     (e) 2)22.42

2.  (a) 3)9.69    (b) 7)92.96   (c) 6)72.06   (d) 4)91.24    (e) 5)77.45

3.  (a) 9)0.54    (b) 8).968    (c) 4)1.28    (d) 7)6.65     (e) 3)22.5

4.  (a) 5)20.0    (b) 7)63.0    (c) 9)10.80   (d) 7).063     (e) 2)200.04

**Problems 5 and 6.** Find the quotients correct to the nearest tenth.

5.  (a) 3)7       (b) 6)13      (c) 9)67      (d) 7)33       (e) 11)14

6.  (a) 17)365    (b) 8)987     (c) 21)357    (d) 54)895     (e) 23)666

**Problems 7 and 8.** Find quotients correct to the nearest hundredth.

7.  (a) 8)45      (b) 7)47      (c) 12)35     (d) 15)98      (e) 34)78

8.  (a) 14)12     (b) 45)32     (c) 57)29     (d) 125)88     (e) 78)22

**Problems 9 and 10.** Find the quotients correct to the nearest thousandth.

9.  (a) 8)1.000      (b) 4)1.000    (c) 5)1.000    (d) 4)3.000       (e) 8)7.000

10. (a) 16)11.000    (b) 8)5.000    (c) 2)1.000    (d) 32)11.000     (e) 8)3.000

**Problems 11 and 12.** Find each quotient correct to the nearest ten-thousandth.

11. (a) 16)7.0000    (b) 32)21.0000    (c) 16)5.0000    (d) 32)27.0000
    (e) 32)9.0000

12. (a) 64)3.0000    (b) 64)27.0000    (c) 64)13.0000
    (d) 32)3.0000    (e) 64)63.0000

**Problems 13 through 19.** Do the divisions.

13. (a) $.5\overline{)17.5}$ (b) $.2\overline{)22.4}$ (c) $.6\overline{)47.4}$ (d) $.8\overline{)10.72}$
(e) $.9\overline{)1.98}$

14. (a) $.3\overline{).84}$ (b) $.5\overline{).115}$ (c) $.7\overline{)87.92}$ (d) $.6\overline{)24.24}$
(e) $.2\overline{).046}$

15. (a) $.8\overline{).0016}$ (b) $.9\overline{).00108}$ (c) $.4\overline{).0008}$ (d) $.7\overline{).00035}$
(e) $.3\overline{)3.009}$

16. (a) $1.8\overline{)43.74}$ (b) $4.5\overline{)3.56}$ (c) $2.8\overline{).579}$ (d) $.24\overline{)9.84}$
(e) $1.5\overline{)35.79}$

17. (a) $.15\overline{)30.15}$ (b) $.64\overline{)34.608}$ (c) $.46\overline{)4.009}$ (d) $.25\overline{)709.0}$
(e) $1.2\overline{)24.024}$

18. (a) $.082\overline{)54.789}$ (b) $.021\overline{)346.06}$ (c) $.320\overline{)45.789}$
(d) $.046\overline{)574.302}$

19. (a) $8.7\overline{)825.63}$ (b) $.539\overline{)1875.72}$ (c) $.36\overline{)132.156}$ (d) $.0192\overline{)1.20}$

**Problems 20 and 21.** Divide and find answer to nearest tenth.

20. (a) $1.4\overline{)3.89}$ (b) $2.3\overline{).4}$ (c) $.15\overline{).467}$ (d) $3.14\overline{)877.67}$

21. (a) $6.4\overline{)16}$ (b) $.9\overline{)23}$ (c) $23.8\overline{)55}$ (d) $1.25\overline{)628}$

**Problems 22 and 23.** Divide and find answer to nearest hundredth.

22. (a) $.08\overline{)9}$ (b) $.57\overline{)6.4}$ (c) $1.1\overline{)1.86}$ (d) $24.5\overline{)17.85}$

23. (a) $27.3\overline{).895}$ (b) $.24\overline{)56.7}$ (c) $2.29\overline{)1.0109}$ (d) $76.4\overline{)56.981}$

**Problems 24 and 25.** Divide and find answer to nearest thousandth.

24. (a) $.54\overline{)78}$ (b) $.89\overline{)6.75}$ (c) $4.05\overline{)1.67}$ (d) $2.34\overline{).0162}$

25. (a) $27.89\overline{)4.8762}$ (b) $1.728\overline{).5280}$ (c) $76.5\overline{)98.7679}$ (d) $1.33\overline{)1.5688}$

**Problems 26 and 27.** Divide and find answer to nearest ten-thousandth.

26. (a) $5\overline{)17.78}$ (b) $.6\overline{)3.9}$ (c) $1.2\overline{).034}$ (d) $.9\overline{).0036}$

27. (a) $28.1\overline{)0.188}$ (b) $4.5\overline{)177.5}$ (c) $3.14\overline{).00628}$ (d) $278.9\overline{)45.897}$

28. One inch is equal to 2.54 centimeters. How many inches are in 75 centimeters? Write the answer to nearest hundredth.

29. One thousand feet of 18 gauge B & S annealed copper wire weighs 4.917 pounds. Find, to four-place accuracy, the weight of 1 foot of the wire.

30. Five hundred feet of 4 gauge B & S annealed copper wire has a resistance of 249.3 ohms.
(a) Find the resistance in ohms of 1 foot of this wire.
(b) What length of this wire will have a resistance of 1 ohm?

31. Forty complete turns of the thimble of a micrometer advances the spindle 1.000 inch. How far does one turn of the thimble advance the spindle?

32. Number 8 gauge B & S sheet steel is 0.1285 inch thick. Find the nearest whole number of sheets in a stack 1 foot high.

33. What is the speed of a car that travels a distance of 45.6 miles in 1.25 hours?

**34.** A piece of drill rod 34.51 inches long is cut into 7 pieces of equal length. What is the length of 1 piece? (Disregard waste in cutting.)

**35.** A $\frac{1}{2}$-inch carrage bolt weighs 0.1875 ounces. How many bolts are in 15 pounds?

**36.** A lab technician has 34.75 grams of iron filings to give to 26 students in a physics laboratory. How much does each student get? Give answer to nearest hundredth of a gram.

**37.** Bungling Buford just got back from a trip of 486.7 miles. He used 59.8 gallons of gas on the trip. B.B. says he got better than eight-and-a-half miles per gallon of gas. (a) Do you believe him? (b) Find his gas mileage to the nearest tenth.

**38.** An assembly worker worked 38.25 hours and got a check for $114. What is his hourly wage?

**39.** A gross (144) of brass fittings costs $51.84. What is the cost of one fitting?

**40.** The *pitch* **P** of a thread on a bolt, screw, etc., is found by dividing the number 1 by the number of threads in 1 inch.

$$\text{Pitch} = \frac{1}{\text{number of threads in 1 inch}}$$

Find the pitch to the nearest thousandth if:
(a) there are 8 threads per inch. (b) there are 12 threads per inch. (c) there are 16 threads per inch.

**41.** During a heavy downpour, 5.45 inches of rain fell in 4.75 hours. What was the rate in inches per hour?

## SEC. 5   DECIMAL EQUIVALENTS OF COMMON FRACTIONS

Machinists, mechanics, laboratory technicians, and others who use instruments of high precision use decimals in their work. Carpenters, cabinet makers, pipe fitters, and sheet metal workers usually use common fractions in their work. Some trades, such as tool and die makers, pattern makers, engineering aides, auto and diesel mechanics, etc., use both common fractions as well as decimals. These tradesmen must be able to convert from common fractions to their decimal equivalents and vice versa. Tables of decimal equivalents of common fractions are found in almost every shop. Table 2—Decimal Equivalents of Common Fractions—is on Appendix page 470. In this section, you will learn to use this table. If you use it correctly, this table is a valuable, time-saving device.

You can easily find the decimal equivalents of some common fractions. For instance, you change the fractions $\frac{1}{2}, \frac{1}{4}, \frac{3}{8}$ to decimal fractions by writing them with denominators of 10; 100; 1,000; and so on. To change the common fraction $\frac{1}{2}$ to a decimal fraction, multiply both numerator and denominator by 5, you get $\frac{5}{10}$; which is equal to the decimal .5. There-

fore, .5 is the *decimal equivalent* of the common fraction $\frac{1}{2}$.

Change the common fraction $\frac{1}{4}$ to a decimal fraction by multiplying

both numerator and denominator by 25. Choose the number 25 because $4 \times 25 = 100$, and you want a denominator that is a multiple of 10. The decimal equivalent of the fraction $\frac{1}{4} = .25$ because:

$$\frac{1}{4} = \frac{25 \times 1}{25 \times 4} = \frac{25}{100} = .25$$

To change the fraction $\frac{3}{8}$ to a decimal, you must do a bit more work.

What number do you multiply the denominator 8 by to get a multiple of 10? The smallest multiple of 10 that can be divided by 8 is 1,000. Multiply numerator and denominator by 125 to get the decimal fraction:

$$\frac{3 \times 125}{8 \times 125} = \frac{375}{1,000} = 0.375$$

The decimal equivalent of the fraction $\frac{3}{8}$ is .375. In this process you

have to do some guess work. If you think there is a better way to get the same answer—you are right! Here is what you do to change a common fraction to a decimal:

### Rule for Changing Common Fractions to Decimals

*Find the decimal equivalent of any fraction by **dividing** the denominator into numerator. Carry out the division as far as necessary.*

**Example 23:** Find the decimal equivalent of the fraction $\frac{3}{8}$.

*Solution:* Write $8\overline{)3}$

Put a decimal point after the 3 and add 3 zeros. Divide as in decimal division until the answer comes out even, or until you have as many places as you need.

$$
\begin{array}{r}
.375 \\
8\overline{)3.000} \\
2\,4\phantom{00} \\
\hline
60\phantom{0} \\
56\phantom{0} \\
\hline
40 \\
40 \\
\hline
0
\end{array}
$$   (comes out even)

*Answer:* $\dfrac{3}{8} = .375$

When the division does not come out even, round off in either the third (thousandth's) or fourth (ten-thousandth's) decimal place. You will rarely need to use decimals with more than four decimal places.

**Example 24:** Find, by division, the decimal equivalent of the fraction $\dfrac{11}{16}$ to (a) nearest thousandth (three places); and (b) nearest ten-thousandth (four places).

*Solution* (a):

$$
\begin{array}{r}
.687 \\
16)\overline{11.000} \\
9\,6 \\
\overline{\phantom{0}1\,40} \\
1\,28 \\
\overline{\phantom{00}120} \\
112 \\
\overline{\phantom{000}8}
\end{array}
$$

(Add 3 zeros. Round off in third place.)

(Remainder is half of divisor, and quotient is rounded up.)

*Answer* (a): $\dfrac{11}{16} = .688$ (to three decimal places).

*Solution* (b): Carry out the division in (a) one more step to see that:

$$\frac{11}{16} = .6875 \text{ with remainder of zero.}$$

*Answer* (b): $\dfrac{11}{16} = .6875$ (to four decimal places).

The precision of the answer depends on the problem. You will find that four-place accuracy (nearest ten-thousandth) is sufficient for most problems.

In Table 2—Decimal Equivalents of Common Fractions—on page 470, all decimals are exact. You get the decimal equivalent of each of the fractions by carrying out the division until a remainder of zero results. For instance:

$$\frac{41}{64} = .640625 \text{ (exactly)}$$

The decimal entries with fewer than six places are also exact, and six-place accuracy is obtained by adding on zeros to the decimal. For example, in the table read: $\dfrac{3}{4} = .75$. If you add on 4 zeros to get .750000, you have the decimal equivalent of $\dfrac{3}{4}$ *correct to the nearest millionth.*

Sometimes you will want to find the approximate common fraction equivalent of a decimal.

**Example 25:** An auto mechanic is installing an air conditioner in a car. He finds that a piece of tubing with a diameter of 0.525 inch must go through the firewall (Fig. 9-16). What drill size, to the nearest 64th of an inch, does he use for a close fit?

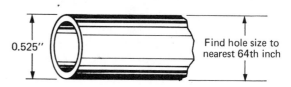

0.525″                                          Find hole size to
                                                nearest 64th inch

**Figure 9-16**

*Solution:* Let $X$ = numerator in 64ths.
Find the value of $X$ in the problem:

$$.525 = \frac{X}{64}$$

To find the value of $X$, multiply both sides of the equation by 64 and round off to the nearest whole number. Round off the answer because the decimal .525 *is not exactly* equivalent to one of the common fractions with denominator of 64.

$$X = (64)(.525)$$

$$X = 33.600$$

Round off to nearest whole number.

$$X = 34$$

*Answer:* He chooses the drill marked $\frac{17}{32}$ which is equivalent to the fraction $\frac{34}{64}$.

Table 2—Decimal Equivalents of Common Fractions—page 470, is also used to solve problems such as Example 25. In the table the decimal .525 is between the two decimals:

| | | |
|---|---|---|
| smaller | .515625 | = (33/64) |
| diameter of tubing | .525 | = _____ |
| larger | .53125 | = (17/32) |

Since the mechanic wants the hole to be slightly larger than .525, he chooses the larger value $\frac{17}{32}$ for the drill size.

From these examples, you can see that Table 2—Decimal Equivalents of Common Fractions—is a very handy, time-saving device. Study the table and become familiar with its use.

## PROBLEMS: Sec. 5 Decimal Equivalents of Common Fractions

1. Express each of the fractions as a decimal.

   **Example:** $\frac{7}{10} = .7$

   (a) $\frac{2}{10}$　　(b) $\frac{5}{10}$　　(c) $\frac{9}{10}$　　(d) $\frac{1}{10}$　　(e) $\frac{8}{10}$

2. Express each of the fractions as a decimal.

   **Example:** $\frac{17}{100} = .17$

   (a) $\frac{15}{100}$　　(b) $\frac{35}{100}$　　(c) $\frac{75}{100}$　　(d) $\frac{49}{100}$　　(e) $\frac{50}{100}$

3. Express each of the fractions as a decimal.

   **Example:** $\frac{16}{25} = \frac{4 \times 16}{4 \times 25} = \frac{64}{100} = .64$

   (a) $\frac{19}{20}$　　(b) $\frac{45}{50}$　　(c) $\frac{7}{25}$　　(d) $\frac{4}{5}$　　(e) $\frac{3}{4}$

4. Express each of the common fractions as a decimal.

   (a) $\frac{8}{10}$　　(b) $\frac{8}{100}$　　(c) $\frac{8}{20}$　　(d) $\frac{8}{25}$　　(e) $\frac{8}{5}$

5. Express each of the mixed numbers as a decimal.

   **Example:** $1\frac{1}{5} = 1\frac{2}{10} = 1.2$

   (a) $2\frac{4}{5}$　　(b) $5\frac{6}{10}$　　(c) $5\frac{7}{25}$　　(d) $6\frac{9}{100}$

6. Express each of the *improper* fractions as a decimal.

   **Example:** (i) $\frac{175}{100} = 1.75$　　(ii) $\frac{36}{25} = \frac{144}{100} = 1.44$

   (a) $\frac{150}{100}$　　(b) $\frac{16}{10}$　　(c) $\frac{45}{20}$　　(d) $\frac{98}{10}$　　(e) $\frac{17}{5}$

7. Find the decimal equivalent of each of the fractions. Divide denominator into numerator and carry out to four places. (Round off in third place.)

   **Example:** Find the decimal equivalent of $\frac{1}{6}$.

$$\begin{array}{r} .167 \\ 6\overline{)1.0000} \\ 6 \\ \hline 40 \\ 36 \\ \hline 40 \\ 36 \\ \hline 4 \end{array}$$

*Solution:*

*Answer:* $\dfrac{1}{6} = .167$ (rounded off in third place)

(a) $\dfrac{1}{8}$     (b) $\dfrac{3}{8}$     (c) $\dfrac{1}{16}$     (d) $\dfrac{7}{16}$     (e) $\dfrac{7}{8}$

8.  Find decimal equivalent of each fraction.

(a) $\dfrac{9}{64}$     (b) $\dfrac{17}{32}$     (c) $\dfrac{3}{4}$     (d) $\dfrac{15}{16}$     (e) $\dfrac{5}{8}$

9.  Find the decimal equivalent for the mixed numbers.

**Example:** $1\dfrac{2}{3} = \dfrac{5}{3} = 1.667$ (nearest thousandth)

(a) $1\dfrac{5}{8}$     (b) $4\dfrac{7}{32}$     (c) $8\dfrac{11}{16}$     (d) $9\dfrac{11}{64}$     (e) $10\dfrac{3}{16}$

10.  Write each of the fractions as a decimal correct to the nearest thousandth.

**Example:** Find the decimal equivalent of $\dfrac{12}{35}$.

$$\begin{array}{r} .343 \quad \text{(nearest thousandth)} \\ 35\overline{)12.000} \\ 10\ 5 \\ \hline 1\ 50 \\ 1\ 40 \\ \hline 100 \end{array}$$

*Solution:* Divide:

*Answer:* $\dfrac{12}{35} = .343$

(a) $\dfrac{15}{27}$     (b) $\dfrac{34}{85}$     (c) $\dfrac{27}{63}$     (d) $\dfrac{5}{21}$     (e) $\dfrac{7}{32}$

11.  Write each fraction as a decimal.

**Example:** $\dfrac{15}{1000} = .015,\ \dfrac{13}{10,000} = .0013$

(a) $\dfrac{86}{1000}$     (b) $\dfrac{563}{1000}$     (c) $\dfrac{13}{10,000}$     (d) $\dfrac{598}{1000}$     (e) $\dfrac{165}{10,000}$

(f) $\dfrac{1987}{10,000}$     (g) $\dfrac{946}{10,000}$     (h) $\dfrac{19}{10,000}$     (i) $\dfrac{2}{10,000}$

12.  Express each of the fraction measurements as a decimal measurement. Carry out the division until remainder is zero.

(a) $\dfrac{5}{8}$ inch     (b) $4\dfrac{3}{4}$ feet     (c) $\dfrac{45}{64}$ inch     (d) $\dfrac{7}{8}$ mile

(e) $2\frac{5}{32}$ yard       (f) $7\frac{11}{16}$ sq. in.       (g) $9\frac{3}{4}$ sq. mi.

13. Compare each fraction and decimal. Which is larger?

   **Example:** Which is larger, .380 or $\frac{3}{8}$ ?

   *Solution:* $\frac{3}{8}$ = .375 is less than .380

   *Answer:* .380 is the larger number.

   (a) $\frac{3}{4}$ or .78       (b) $\frac{1}{3}$ or .35       (c) .08 or $\frac{1}{16}$

   (d) $\frac{4}{9}$ or .49       (e) .8129 or $\frac{13}{16}$       (f) $\frac{7}{8}$ or .78

14. Arrange each set of fractions and decimals from *smallest* to *largest*.

   **Example:** Arrange $\frac{1}{2}$; .55; and $\frac{5}{8}$ from smallest to largest.

   *Solution:* Write each fraction as a decimal:

   $$\frac{1}{2} = .500$$

   $$\frac{5}{8} = .625$$

   *Answer:* Correct arrangement is: $.500 < .55 < .625$

   $$\left(\frac{1}{2}\right) < .55 < \left(\frac{5}{8}\right)$$

   (a) $\frac{1}{5}$; .35; $\frac{3}{9}$              (b) .533; .523; $\frac{4}{7}$

   (c) .755; $\frac{3}{4}$; $\frac{11}{16}$              (d) $\frac{5}{32}$; .125; $\frac{9}{64}$

15. For each of the numbers find the nearest 64th. See page 274 for an additional example.

   **Example:** Find the nearest 64th for the decimal .333.

   *Solution:* Multiply .333 by 64 and round off to nearest *whole* number.
   $64 \times .333 = 21.312 = 21$ (nearest whole number.)

   *Answer:* $.333 = \frac{21}{64}$ to nearest 64th.

   (a) $\frac{5}{7}$       (b) .4       (c) $\frac{4}{11}$       (d) $\frac{7}{12}$       (e) .24       (f) 1.6

   (g) $5\frac{8}{15}$       (h) $\frac{11}{14}$       (i) 5.666

16. Find nearest 32nd for each number.

   **Example:** Find the nearest 32nd for the fraction $\frac{4}{5}$.

   *Solution:* Write $\frac{4}{5}$ as a decimal (.800). Multiply by 32.

   $$32 \times .800 = 25.6$$

   Round off to nearest whole number (26).

*Answer:* $\dfrac{4}{5} = \dfrac{26}{32}$ to nearest 32nd.

(a) $\dfrac{2}{3}$     (b) $\dfrac{6}{7}$     (c) .555     (d) $\dfrac{13}{17}$     (e) $2\dfrac{3}{5}$     (f) $18\dfrac{3}{10}$

**17.** In Fig. 9-17 replace each dimension with its decimal equivalent, correct to the nearest thousandth.

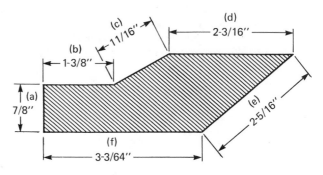

**Figure 9-17**

**18.** In Fig. 9-18 replace each fraction with an equivalent decimal, correct to nearest hundredth.

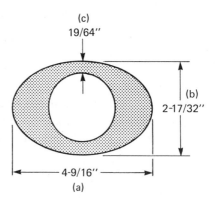

**Figure 9-18**

**19.** Write as a decimal, correct to the nearest hundredth, the fractional part of a week that is one day.

**20.** Write as a decimal, correct to the nearest hundredth, the fractional part of a year that is one month.

**21.** Write as a decimal, correct to the nearest hundredth, the fractional part of a foot that is one inch.

22. Write as a decimal, correct to the nearest hundredth, the fractional part of a day that is one hour.

23. Mr. James is a fireman. He works three days out of every seven. Write as a decimal, correct to the nearest hundredth, the fractional part of a week that he works.

24. R. B. Lloyd is a laboratory technician. He saves one hour's pay for each six hours worked. Write as a decimal, correct to the nearest hundredth, the fractional part of the wage that is saved.

25. On the last examination Keith Banners missed two problems out of a possible 15. Write as a decimal, correct to nearest hundredth, the *correct* fractional part of the test problems.

26. During a season the Honeydew Hornets won 0.6 of the games they played. The Weott Wasps won 5 games out of 8. Which team had a better standing?

27. To find a baseball player's batting average (expressed as a decimal), compare the number of hits with the number of times at bat. Express the resulting fraction as a decimal, correct to the nearest thousandth. For instance, Swingin' Willie made 7 hits in 16 times at bat. Thus, he made a hit $\frac{7}{16}$ of the times at bat. When this is expressed as a decimal it is:

$$\frac{7}{16} = .4375 = .438 \quad \text{(nearest thousandth)}$$

Therefore, Willie's batting average is .438. Find the batting average expressed as a decimal, correct to the nearest thousandth, for the following players:

| Player | At bat | Hits | Batting average |
|--------|--------|------|-----------------|
| Bill | 6 | 2 | _____ |
| Homer | 29 | 11 | _____ |
| Jake | 25 | 7 | _____ |
| LeRoy | 19 | 8 | _____ |
| Noel | 26 | 11 | _____ |
| "Fingers" | 15 | 4 | _____ |
| Carl | 21 | 2 | _____ |
| Jim | 24 | 11 | _____ |
| Charlie | 13 | 4 | _____ |

28. Refer to Problem 27. Who has the best batting average?

29. Refer to Problem 27. Who has the worst batting average?

30. At the end of a baseball season the National League (West Division) standing was as shown in the following table. Find the fraction *won* as a decimal to nearest thousandth.

|  | Team | Games won | Games lost | Fraction won (nearest .001) |
|---|---|---|---|---|
| (Example) | Cincinnati | 92 | 58 | $.613 = \dfrac{92}{150}$ |
|  | (total games played = 92 + 58 = 150) | | | |
|  | Houston | 83 | 66 | _____ |
|  | Los Angeles | 82 | 69 | _____ |
|  | Atlanta | 70 | 80 | _____ |
|  | San Francisco | 65 | 86 | _____ |
|  | San Diego | 57 | 92 | _____ |

# SEC. 6 U.S. CUSTOMARY AND METRIC CONVERSION

The technician, auto mechanic, laboratory assistant, and tradesman in today's modern technology must be familiar with the basic systems of measurement. There are two systems of measurement commonly used in almost every trade. The *U. S. Customary System* uses the inch-pound-second as basic units of measurement. The *Metric* or *International Standard (SI) System* uses the centimeter-gram-second as basic units of measurement.

Certainly you are familiar with the U. S. Customary System. After all, it is the one you grew up with and have used since childhood. But the importance of the metric system cannot be minimized. Its influence on American science and technology became so important that in 1964 the U. S. National Bureau of Standards issued the following bulletin:

> *Henceforth it shall be the policy of the National Bureau of Standards to use the units of the International System (SI), as adopted by the 11th General Conference on Weights and Measures (Oct. 1960), except when the use of these units would obviously impair communication or reduce the usefulness of a report....*

With this bulletin, the U. S. National Bureau of Standards officially went to the metric system as the standard system of measure. Although it will be years before the metric system is in common, everyday use, you can get a headstart by learning a few simple conversion factors.

You know that there are:

12 inches in 1 foot

3 feet in 1 yard

60 seconds in 1 minute

and so on for many relations among units in the customary system. You may also know that there are:

10 millimeters in 1 centimeter

100 centimeters in 1 meter

1,000 meters in 1 kilometer

The units of time are the same in both systems.

You will learn to change measures within each system. You will also learn to convert from the metric system to equivalent measures in the U. S. Customary System and vice versa.

Look at two problems to see why you need to know how to convert from one system to another.

*Problem 1.* Many cars on the road today have some parts made in Europe (metric measure) and the rest made in the U. S. (customary measure). An auto mechanic is doing a tune-up on one of these cars. He needs a 9 millimeter metric socket. His tools are American made and are in the U. S. customary measure. His sockets are $\frac{1}{4}$ inch, $\frac{3}{8}$ inch, $\frac{11}{16}$ inch, and so on.

Is there a socket in his tool box that he can use, or must he buy a new socket to do the tune-up? Continue reading to learn how to solve this problem.

*Problem 2.* A laboratory technician needs a piece of glass tubing 14 inches long. His only measuring instrument is a meter stick 100 centimeters in length. How does he convert centimeters to inches to get the correct length of tubing?

You can solve these problems, and others like them, by using a few conversion factors and common sense. First you need to be able to change measures within each system. This means that you must be able to change inches to feet, feet to yards, pounds to ounces, and so on.

## The U. S. Customary System

Memorize these basic facts of the customary system.

### Conversion Factors—U. S. Customary System

*Linear Measure*
12 inches (in.) = 1 foot (ft.)
3 feet (ft.) = 1 yard (yd.)
5,280 feet (ft.) = 1 mile (mi.)

*Weight Measure*
16 ounces (oz.) = 1 pound (lb.) avoirdupois
2,000 pounds (lb.) = 1 ton (t.)

*Volume Measure* (*Liquid*)
2 pints (pt.) = 1 quart (qt.)
4 quarts (qt.) = 1 gallon (gal.)

NOTE: For additional conversion factors, use Table 1—Conversion Factors–U. S. Customary System—page 465, in the Appendix at the back of the book.

Here are some examples of the use of these conversion factors.

**Example 26:** Change 6 feet to inches.

*Solution:* Since there are 12 inches in 1 foot, the *ratio* of $\frac{12 \text{ in.}}{1 \text{ ft.}}$ is
equal to the whole number 1. This is because 12 inches
and 1 foot represent the same length.
To change 6 feet to inches, multiply 6 by the whole num-
ber 1 which is written as the fraction $\frac{12 \text{ in.}}{1 \text{ ft.}}$.

*Answer:* $\frac{6 \text{ ft.} \times 12 \text{ inches}}{1 \text{ ft.}} = (6 \times 12) \text{ inches} = 72 \text{ inches}$

**Units of measure are divided just as if they were ordinary numbers.
This is true of conversion within any system of measure.**

**Example 27:** Change 8 yards to inches.

*Solution:* First change yards to feet by multiplying by the fraction
$\frac{3 \text{ ft.}}{1 \text{ yd.}} = 1$ (3 ft. = 1 yd.)

$$8 \text{ yd.} \times \frac{3 \text{ ft.}}{1 \text{ yd.}} = (8 \times 3) \text{ ft.} = 24 \text{ ft.}$$

Change 24 ft. to inches by multiplying by the fraction
$\frac{12 \text{ inches}}{1 \text{ ft.}} = 1$ (12 in. = 1 ft.)

$$24 \text{ ft.} \times \frac{12 \text{ inches}}{1 \text{ ft.}} = (24 \times 12) \text{ inches} = 288 \text{ inches}$$

*Answer:* 8 yd. = 288 in.

NOTE: You may do this as a combined problem:

$$8 \text{ yd.} = 8 \text{ yd.} \times \frac{3 \text{ ft.}}{1 \text{ yd.}} \times \frac{12 \text{ in.}}{1 \text{ ft.}} = (8 \times 3 \times 12) \text{ in.}$$

8 yd. = 288 inches

**Example 28:** Change 27 inches to feet.

*Solution:* Multiply 27 inches by the fraction $\frac{1 \text{ ft.}}{12 \text{ in.}} = 1$.

*Answer:* $27 \text{ in.} = 27 \text{ in.} \times \frac{1 \text{ ft.}}{12 \text{ in.}} = \frac{27}{12} \text{ ft.}$

NOTE: Write the result either as a mixed number $\left(2\frac{1}{4} \text{ ft.}\right)$ or as a
decimal (2.25 ft.) as the problem requires.

**Example 29:** How many yards are there in 4 ft. 9 in. ? Write answer
in decimal form.

*Solution:* $$4 \text{ ft.} = 4 \cancel{\text{ ft.}} \times \frac{12 \text{ in.}}{1 \cancel{\text{ ft.}}} = 48 \text{ in.}$$

$$4 \text{ ft. } 9 \text{ in.} = (48 + 9) \text{ in.}$$

$$= 57 \text{ in.}$$

Change to yards by multiplying by the fraction $\frac{1 \text{ yd.}}{36 \text{ in.}} = 1$.

$$57 \text{ in.} = 57 \cancel{\text{ in.}} \times \frac{1 \text{ yd.}}{36 \cancel{\text{ in.}}}$$

$$= \frac{57}{36} \text{ yd.}$$

*Answer:* 57 in. = 1.6 yd. (to nearest tenth)

### Rule for Conversion of Units

*To convert **old** units to **new** units, multiply the **old** units by the **conversion** fraction:*

$$\frac{\text{New Units}}{\text{Old Units}} \quad \text{(which is equal to 1)}$$

*to get **new** units.*

$$\text{Old Units} = \cancel{\text{Old Units}} \times \frac{\text{New Units}}{\cancel{\text{Old Units}}} = \text{New Units}$$

(Divide out old units.)

**Example 30:** Change 15 miles per hour to miles per minute.

*Solution:* Old Units = Miles per hour
New Units = Miles per minute
Conversion fraction:

$$\frac{1 \text{ hour}}{60 \text{ min.}} = 1$$

$$\frac{15 \text{ miles}}{1 \text{ hour}} = \frac{15 \text{ miles}}{1 \cancel{\text{ hour}}} \times \frac{1 \cancel{\text{ hour}}}{60 \text{ min.}} = \frac{15 \text{ miles}}{60 \text{ min.}} = \frac{1 \text{ mile}}{4 \text{ min.}}$$

*Answer:* 15 miles per hour = $\frac{1}{4}$ miles per minute.

## The Metric System (SI)

The major advantage of the metric system is the fact that it is a decimalized system. Changes within the system are made by multiplying or dividing by multiples of 10. Study the following table.

**Conversion Factors—International Standard (SI)**

*Linear Measure*
10 millimeters (mm.) = 1 centimeter (cm.)
10 centimeters (cm.) = 1 decimeter (dm.)
10 decimeters (dm.) = 1 meter (m.)
10 meters (m.) = 1 decameter (dam.)
10 decameters (dam.) = 1 hectometer (hm.)
10 hectometers (hm.) = 1 kilometer (km.)

*Weight Measure*
10 milligrams (mg.) = 1 centigram (cg.)
1,000 milligrams (mg.) = 1 gram (gm.)
1,000 grams (gm.) = 1 kilogram (kg.)
1,000 kilogram (kg.) = 1 metric ton (t.)

*Volume Measure*
10 milliliters (ml.) = 1 centiliter (cl.)
1,000 milliliters (ml.) = 1 liter (l.)

From this list, three of the most common conversion factors are:

10 millimeters = 1 centimer

100 centimers  = 1 meter

1,000 meters      = 1 kilometer

When changing units of the metric system, follow the ***Rule for Conversion of Units*** (page 283).

**Example 31:** Change 2.6 kilogram to grams.

*Solution:* Multiply 2.6 by the *conversion fraction* $\dfrac{1,000 \text{ gram}}{1 \text{ kilogram}} = 1$

$$2.6 \text{ kilogram} = 2.6 \,\cancel{\text{kilogram}} \times \frac{1,000 \text{ gram}}{1 \,\cancel{\text{kilogram}}}$$

$$= (2.6 \times 1,000) \text{ gram}$$

*Answer:* 2.6 kilogram = 2,600 gram

**Example 32:** Change 285 millimeters to meters.

*Solution:* Multiply 285 millimeters by the *conversion fraction* $\dfrac{1 \text{ centimeter}}{10 \text{ millimeters}}$ to change millimeters to centimeters, and then multiply it by the *conversion fraction* $\dfrac{1 \text{ meter}}{100 \text{ centimeters}}$ to change centimeters to meters.

$$285 \text{ mm.} = 285 \,\cancel{\text{mm.}} \times \frac{1 \,\cancel{\text{cm.}}}{10 \,\cancel{\text{mm.}}} \times \frac{1 \text{ m.}}{100 \,\cancel{\text{cm.}}}$$

$$= \frac{285 \text{ m.}}{1,000}$$

*Answer:*          285 mm. = 0.285 meter

Here are some of the more useful conversions between the *Metric System* and the *U. S. Customary System*.

**Conversions Between Metric and U. S. Customary Systems**

*Linear Measure*
1 centimeter = 0.3937 inch; 1 inch = 2.54 cm. (exact)
1 meter = 39.37 inches = 1.094 yards
1 meter = 3.2808 feet; 1 foot = 0.3048 m. (exact)
1 kilometer = 0.6214 mile; 1 mile = 1.609 km.

*Weight Measure*
1 gram = 0.035 ounce (avoir.) 1 ounce (avoir.) = 28.350 grams
1 kilogram = 2.205 pounds; 1 pound = .4536 kilogram

*Capacity Measure*
1 liter = 1.057 (U.S.) liquid quart; .908 dry quart
1 dry quart = 1.101 liter
1 liquid quart = 0.946 liter

NOTE: Additional conversion factors are listed in Table 1—
Conversion Factors: U. S. Customary–International
Standard—page 465, in the Appendix at the back of
the book.

Study these examples which show how various conversion factors are used.

**Example 33:** Solution to Problem 1, page 281. Solve the mechanic's problem by converting 9 mm. to inches.

*Solution:* Change millimeters to centimeters, and then to inches.

$$9 \text{ mm.} = 9 \text{ mm.} \times \frac{1 \text{ cm.}}{10 \text{ mm.}} = .9 \text{ cm.}$$

$$.9 \text{ cm.} = .9 \text{ cm.} \times \frac{.3937 \text{ inch}}{1 \text{ cm.}}$$

$$= (.9 \times .3937) \text{ inch}$$

$$= .35433 \text{ inch}$$

Use Table 2—Decimal Equivalents of Common Fractions —(page 470) to find:

$$.35433 \text{ inch} = \frac{11}{32} \text{ (to the nearest 64th of an inch)}$$

*Answer:* The mechanic can use his $\frac{11}{32}$ socket for the tune-up.

**Example 34:** Solution to Problem 2, page 281. Solve the lab technician's problem. The 14 inches is equal to how many centimeters?

*Solution:* Multiply 14 inches by the conversion fraction $\dfrac{2.54 \text{ cm.}}{1 \text{ inch}}$

$$14 \text{ inches} = 14 \cancel{\text{ inches }} \times \frac{2.54 \text{ cm.}}{1 \cancel{\text{ inch}}}$$

$$= (14 \times 2.54) \text{ cm.}$$

*Answer:*        $14 \text{ inches} = 35.56 \text{ cm.}$

**Example 35:** Change 65 miles to kilometers.

*Solution:* From the table of equivalents (Linear Measure page 285), convert miles to kilometers by multiplying by the conversion fraction $\dfrac{1.609 \text{ km.}}{1 \text{ mile}}$

$$65 \text{ miles} = 65 \cancel{\text{ miles }} \times \frac{1.609 \text{ km.}}{1 \cancel{\text{ mile}}}$$

*Answer:*        $65 \text{ miles} = 104.585 \text{ kilometers}$

From this result, you see that a speed of 65 miles per hour is equivalent to 104.585 kilometers per hour.

Memorize a few basic conversion facts. Learn how the various table are used. Then you will be able to solve difficult problems in measure conversion.

## PROBLEMS: Sec. 6 U.S. Customary and Metric Conversion

When conversion values are not given, you will find it helpful to use Table A-1 U. S. Customary-Metric on page 465 in the Appendix.

**Problems 1 through 21.** Change each value to the indicated equivalent. Find all answers correct to the nearest hundredth.

1.  Change to centimeters. (1 inch = 2.54 centimeters)
    (a) 10 in.     (b) 25 in.     (c) 45 in.     (d) 24.5 in.     (e) 89 in.
    (f) 75 in.

2.  Change to meters. (1 foot = .305 meter)
    (a) 15 ft.     (b) 25 ft.     (c) 70 ft.     (d) 150 ft.     (e) 26.9 ft.
    (f) 256 ft.

3.  Change to meters. (1 yard = .91 meter)
    (a) 2 yd.     (b) 10 yd.     (c) 35 yd.     (d) 17.5 yd.     (e) 450 yd.
    (f) 1,760 yd.

4.  Change to millimeters. (1 inch = 25.4 millimeters)
    (a) 35 in.     (b) 7 in.     (c) 3 in.     (d) 144 in.     (e) 2 ft. 4 in.
    (f) 4 ft.

5.  Change to kilometers. (1 mile = 1.61 kilometers)
    (a) 5 mi.     (b) 10 mi.     (c) 15 mi.     (d) 100 mi.     (e) 45.5 mi.
    (f) 6,987 mi.

6. Change to meters. (1 inch = .0254 meter)
   (a) 10 in.     (b) 45 in.     (c) 144 in.     (d) 12 in.     (e) 36 in.

7. Change to inches. (1 centimeter = .39 inch)
   (a) 10 cm.     (b) 25 cm.     (e) 50 cm.     (d) 100 cm.     (e) 450 cm.
   (f) 2.54 cm.

8. Change to feet. (1 meter = 3.28 feet)
   (a) 10 m.     (b) 15 m.     (c) 50 m.     (d) 100 m.     (e) 34.5 m.
   (f) .305 m.

9. Change to miles. (1 kilometer = .62 mile)
   (a) 10 km.     (b) 50 km.     (c) 75 km.     (d) 16.1 km.     (e) 165 km.
   (f) 0.5 km.

10. Change to inches. (1 millimeter = 0.04 inch)
    (a) 10 mm.     (b) 11 mm.     (c) 15 mm.     (d) 25 mm.
    (e) 45.5 mm.     (f) 195 mm.

11. Change to centimeters.
    (a) 10 in.     (b) 2 ft. 6 in.     (c) 5 yd.     (d) 36 in.     (e) 250 in.

12. Change to inches.
    (a) 15 cm.     (b) 24.5 mm.     (c) 500 cm.     (d) 1.56 m.
    (e) 250 mm.

13. Change to meters.
    (a) 24.5 in.     (b) 450 ft.     (c) 3 yd.     (d) 1 mi.     (e) 4.4 mi.

14. Change to millimeters.
    (a) 4 in.     (b) 25 in.     (c) 1 ft.     (d) 1 yd.     (e) 2 ft.     (f) 6 in.
    (g) 3 ft.

15. Change to feet.
    (a) 25 m.     (b) 45 cm.     (c) 189.5 cm.     (d) 4 km.     (e) 39.37 cm.
    (f) .15 m.

16. Change to inches.
    (a) 2.5 cm.     (b) 15 mm.     (c) 1.55 m.     (d) 22.6 mm.
    (e) 18.7 m.

17. Change to centimeters.
    (a) 15.3 in.     (b) 1 ft. 5 in.     (c) 3 yd. 1 ft. 4 in.     (d) 56 ft.

18. Change to grams. (1 ounce = 28.35 grams)
    (a) 5 oz.     (b) 10 oz.     (c) 35 oz.     (d) 16 oz.     (e) 32 oz.
    (f) 58.5 oz.

19. Change to kilograms. (1 pound = .45 kilogram)
    (a) 5 lb.     (b) 10 lb.     (c) 50 lb.     (d) 270 lb.     (e) 2,000 lb.

20. Change to ounces. (1 gram = .035 ounce)
    (a) 10 gm.     (b) 25 gm.     (c) 100 gm.     (d) 15.7 gm.     (e) 250 gm.

21. Change to pounds. (1 kilogram = 2.2 pounds)
    (a) 5 kg.     (b) 10 kg.     (c) 100 kg.     (d) 26.9 kg.     (e) 1,000 kg.

**Problems 22 and 23.** Change each metric value to its equivalent in inches
to the nearest 64th inch.

   **Example:** 15 mm. = how many 64th inch? (approximately)

*Solution:* Convert 15 mm. to inches by multiplying by the conversion fraction $\dfrac{.03937 \text{ in.}}{1 \text{ mm.}}$

$$15 \text{ mm.} = 15 \text{ mm.} \cdot \frac{(.03937) \text{ in.}}{1 \text{ mm.}} = .59055 \text{ inch}$$

To find the nearest 64th, multiply .59055 by 64 to get 37.795. Then round off to 38 which is the nearest whole number.

*Answer:* $15 \text{ mm.} = \dfrac{38}{64} = \dfrac{19}{32} \text{ in.}$

NOTE: You may use the Table of Decimal Equivalents of Common Fractions also.

**22.** (a) 5 mm.      (b) 10 mm.      (c) 25 mm.      (d) 9 mm.      (e) 11 mm.
(f) 27 mm.

**23.** (a) 5 cm.      (b) 8 cm.      (c) 12 cm.      (d) 25 cm.      (e) 2.54 cm.
(f) 11 cm.

**Problems 24 and 25.** Change each fraction (U. S. Customary) to metric to nearest millimeter.

**Example:** Change $\dfrac{3}{8}$ inch to millimeters.

*Solution:* Write $\dfrac{3}{8}$ as a decimal.

$$\frac{3}{8} = .375$$

Convert to millimeters by multiplying.

$$.375 \text{ in.} = .375 \text{ in.} \cdot \frac{(25.4) \text{ mm.}}{1 \text{ in.}} = 9.525 \text{ mm.}$$

*Answer:* $\dfrac{3}{8}$ in. = 10 mm. (to nearest millimeter)

**24.** (a) $\dfrac{1}{16}$ in.      (b) $\dfrac{3}{32}$ in.      (c) $\dfrac{1}{4}$ in.      (d) $\dfrac{9}{32}$ in.      (e) $\dfrac{7}{16}$ in.

**25.** (a) $1\dfrac{1}{8}$ in.      (b) $\dfrac{15}{16}$ in.      (c) $\dfrac{3}{4}$ in.      (d) $\dfrac{7}{8}$ in.      (e) $\dfrac{43}{64}$ in.

**26.** Figure 9-19 is dimensioned to the nearest .01 inch. Re-dimension the figure in the metric system to the nearest .01 centimeter.

**Example:** Dimension A = 6.00 inch.

Change to centimeters by multiplying by 2.54 since 1 inch = 2.54 cm.

$$6 \text{ inches} = 6(2.54) = 15.24 \text{ centimeters}$$

| Dimension | Inches | Centimeters |
|-----------|--------|-------------|
| A | 6.00 | 15.24 |
| B | 1.45 | _____ |
| C | 1.50 | _____ |
| D | 2.55 | _____ |
| E | 0.75 | _____ |
| F | 3.10 | _____ |

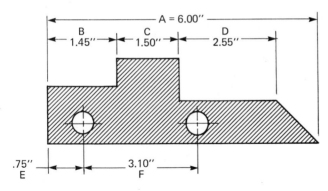

**Figure 9-19**

A liter is a measure of volume. From the Conversion Table 1, page 285, you find that:

$$1 \text{ liter} = 1.057 \text{ quarts (liquid)}$$

$$1 \text{ quart} = .946 \text{ liter (liquid)}$$

**Problems 27 through 31.** Use these values and convert customary to metric and metric to customary.

27. (a) 3 qt. = _____ liters      (b) 15 qt. = _____ liters
    (c) 24.5 qt. = _____ liters

28. (a) 1 pint = $\frac{1}{2}$ qt. = _____ liter    (b) 3 qt. 1 pt. = _____ liters

29. (a) 1 gallon (4 qt.) = _____ liters    (b) 1 gal. 3 qt. 1 pt. = _____ liters

30. (a) 3 liters = _____ qt.      (b) 5 liters = _____ qt.
    (c) 14.5 liters = _____ qt.

31. (a) 2 liters = _____ pt.      (b) 13.7 liters = _____ gal.
    (c) 50 liters = _____ gal.

32. Home movie film is 8-mm. wide. What is this film size in inches?

33. A Leica camera uses 35-mm. film. Write this dimension in inches.

34. Many professional photographers use 16-mm. movie film. What is the width of this film in inches?

35. The gas tanks on a pick-up camper will hold 44.7 gallons. Express the capacity of the fuel tanks in liters.

36. Recently a skin diver made a free dive to a depth of 145 feet. Express this depth in meters.

37. Gordon Bush is a big game hunter. Last year he bought a 7-mm. Magnum rifle for moose hunting. Express the dimension of the bore (7-mm.) in inches, to the nearest hundredth.

38. In the 1972 Olympics, Dave Wottle of the United States ran the 800-meter race in the winning time of 1 minute, 45.9 seconds. If Dave were to run a half-mile (880 yards) at the same rate, would it take him a longer time or a shorter time than it did for the 800 meters?

**39.** In the 1972 Olympics, Lasse Viren of Finland ran the 10,000-meter run in a winning time of 27 minutes, 38.4 seconds. How many miles did Lasse run? Find the answer to the nearest hundredth.

**40.** Convert the *mile per gallon* values to *kilometers per liter*.

**Example:** Convert $\dfrac{14 \text{ miles}}{1 \text{ gallon}}$ to $\dfrac{\text{kilometers}}{\text{liter}}$

*Solution:* $\dfrac{14 \text{ miles}}{1 \text{ gallon}} = \dfrac{14 \text{ miles}}{1 \text{ gallon}} \times \dfrac{1.16 \text{ km.}}{1 \text{ mile}} \times \dfrac{.264 \text{ gallon}}{1 \text{ liter}}$

$= \dfrac{(14 \times 1.61 \times .264) \text{ km.}}{1 \text{ liter}}$

$= 5.951 \text{ km./liter}$

*Answer:* 14 miles/gallon = 6.0 km./liter (nearest tenth)

(a) 15 mi./gal.    (b) 16.6 mi./gal.    (c) 19 mi./gal.    (d) 27.8 mi./gal.

**41.** Convert the *kilometers per liter* values to *miles per gallon*.

**Example:** Convert 7 km./liter to miles/gal.

*Solution:* $\dfrac{7 \text{ km.}}{1 \text{ liter}} = \dfrac{7 \text{ km.}}{1 \text{ liter}} \times \dfrac{1 \text{ mile}}{1.61 \text{ km.}} \times \dfrac{1 \text{ liter}}{.264 \text{ gal.}}$

$= \dfrac{7 \text{ mile}}{(1.61 \times .264) \text{ gal.}} = \dfrac{16.5 \text{ mile}}{1 \text{ gal.}}$

*Answer:* 7 km./liter = 16.5 miles/gallon

(a) 5 km./liter    (b) 6.5 km./liter    (c) 10 km./liter    (d) 8.2 km./liter

**42.** Cubic inches and cubic centimeters are measures of volume. Using the conversion tables you find that:

1 cubic inch = 16.387 cubic centimeters

1 cubic centimeter = 0.061 cubic inches

Use these values to make the following conversions.
(a) 15 cubic inches = _____ cubic centimeters.
(b) 25.8 cubic inches = _____ cubic centimeters.
(c) 150 cubic centimeters = _____ cubic inches.
(d) 750 cubic centimeters = _____ cubic inches.

**43.** The Ford Courier truck has an engine with a displacement of 1,800 cc. (1,800 cubic centimeters). What is the displacement in cubic inches?

**44.** A Jaguar 4.3 has an engine displacement of 4.3 liter = 4,300 cc. What is the displacement in cubic inches? (1 liter = 1,000 cubic centimeters)

**45.** A Chevrolet 327 engine has an engine displacement of 327 cubic inches. What is this displacement in:
(a) cubic centimeters?                    (b) liters?

**46.** The Pantera is a sports car powered by a 351 cubic inch Ford engine. What is the engine displacement in liters?

**47.** The Chrysler 150 limited production outboard racing engine has 4 cylinders and a total engine displacement of 96.55 cubic inches. Find the displacement of *one* cylinder in cubic centimeters.

**48.** The engine in Problem 47 has a bore and stroke of $3\frac{5}{16}$ inches $\times$ 2.8 inches. Express the bore and stroke in centimeters.

**49.** The speed of sound at sea level is about 1,100 feet per second. Express this in meters per second.

**50.** Bungling Buford measured the length and width of his front yard. As usual, B.B. goofed and used a meter stick thinking it was a yard stick. If his measurements are: length = 18.5 yards, and width = 16.5 yards, what are the actual values? (1 yard = .9144 meter)

**51.** There are many deep trenches in both the Pacific and Atlantic Oceans. In Table 9-1 the depths of some of the more famous trenches are given. Find the given depth in feet, meters, and fathoms. (1 fathom = 6 feet) (Give answer to nearest whole number.)

**TABLE 9-1**  FAMOUS OCEAN TRENCHES

*The Pacific*

| Name | Depth | | |
|---|---|---|---|
| | Feet | Meters | Fathoms |
| (a) Mariana | 36,198 | | |
| (b) Tonga | | 10,882 | |
| (c) New Britain | | | 4,998 |
| (d) Japan | 27,591 | | |
| (e) Aleutian | | 8,100 | |

*The Atlantic*

| Name | Depth | | |
|---|---|---|---|
| | Feet | Meters | Fathoms |
| (f) Puerto Rico | | | 4,589 |
| (g) Cayman | | 7,535 | |
| (h) Brazil Basin | 20,076 | | |

## Practice Test for Chapter 9

**1.** Write $\frac{45}{1,000}$ as a decimal.      *Answer:*_____

**2.** Write the numeral for sixty-five ten-thousandths.
          *Answer:*_____

**3.** Write in words 5.078.    *Answer:*_____

4. Write the numeral for the number with 5 in the ten's place, 0 in the units place, 7 in the tenth's place, 6 in the hundredth's place, and 5 in the thousandth's place. *Answer:*_____

5. Which is larger 0.8 or .08? *Answer:*_____

6. Compare the decimals. Arrange in order from smallest to largest: .45, .54, .5, .05. *Answer:*_____, _____, _____, _____.

7. Round off .678 to the nearest tenth. *Answer:*_____

8. Round off .8975 to three decimal places. *Answer:*_____

9. Find the sum. 4.78 + 3.89 + 4.85 = _____

10. Find the sum. .0045 + 3.884 + 1.01 = _____

11. Subtract. 5.674 − 3.775 = _____

12. Multiply. 4.87 × 3.66 = _____ to nearest hundredth.

13. Divide. 5.75)9.775 *Answer:*_____

14. Find the total cost of 4 fan belts at $3.75 each, 7 pressure caps at $1.95 each, 12 oil filters at $4.50 each, and 3 tune-up kits at $12.65 each. Total Cost = _____

15. Find the decimal equivalent of $\frac{15}{16}$. *Answer:* _____ (four places)

16. Find the nearest 64th inch larger than .775 inch. *Answer:*_____

17. Convert 5.75 inches to centimeters. (1 in. = 2.54 cm.) *Answer:*_____

18. Convert 54.7 cm. to inches. *Answer:*_____

19. Number 30 gauge annealed copper wire has a resistance of 115.4 ohms per 1,000 feet. What is the resistance in 1 meter? *Answer:*_____

20. Convert 150 kilometers per hour to miles per hour. *Answer:*_____

21. A foreign import sports car gets 12 kilometers per liter of fuel. An American compact gets 28 miles per gallon. Which car is the most economical to operate? *Answer:*_____

22. How many grams are in 2 pounds 12 ounces? *Answer:*_____

23. One gallon of diesel fuel cost 45.9 cents. What is the cost of one liter? *Answer:*_____

24. A pressure of 1 pound per square inch is equivalent to .07 kilograms per

square centimeter. A SCUBA tank is filled to a pressure of 2,250 pounds per square inch. What is the pressure in kilograms per square centimeter?

*Answer:*_____

**25.** Find the dimension labeled X in Fig. 9-20 and express the answer in centimeters.

*Answer:* _____

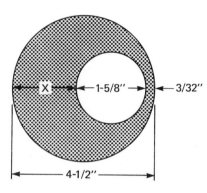

**Figure 9-20**

# 10

# Percentage

## SEC. 1   THE MEANING OF PERCENT

In Chapters 7, 8, and 9 you learned to use both common and decimal fractions to solve many practical problems. Now you will learn to use these ideas to solve problems in percent. What does *percent* mean? **Percent** means **parts per hundred**.

Here are some of the kinds of problems you will be able to solve after studying this chapter. (These problems are solved in Section 2 of this chapter.)

*Problem 1.* Your employer has just given you a 5.5 *percent* wage increase. Before the raise, you made $3.87 per hour. What is your new wage rate?

*Problem 2.* The Mid-Nite Auto Supply gives a 15 *percent* discount to mechanics who buy automotive supplies and parts from them. An auto mechanic bought a radiator that had a retail (list) price of $85.00. After deducting 15 *percent*, how much did the mechanic pay?

*Problem 3.* A resistor in an electrical circuit is rated at 500 ohms $\pm 10$ *percent*. Find the high and low values for the resistor.

*Problem 4.* On Monday a quality control inspector rejected 23 castings out of a production run of 147. On Tuesday he rejected 27 castings out of a production run of 162. On which day did the foundry have a lower *percent* of rejection?

*Problem 5.* Using a certain formula, a technician finds the horsepower of an engine is 175 hp. When the engine is tested on a Prony Brake, it delivers 160 hp. What *percent* of the formula value is the actual test value?

*Problem 6.* A parts salesman is paid a base salary of $120.00 per week plus a 12 *percent* commission on all sales he makes. In one week he sold $1,150.00 worth of auto parts. What were his earnings for the week?

Each of these problems involves the idea of percent, which means *hundredths*. There are three ways to show hundredths: *fractions, decimals,* and *percent*. Use the fraction $\frac{15}{100}$ to look at these three ideas.

*Fractions:* Recall from the study of common fractions that the fraction $\frac{15}{100}$ is read "fifteen hundredths." This means *fifteen parts out of one hundred.*

*Decimals:* This common fraction $\frac{15}{100}$ can also be written as the decimal .15. The decimal .15 means *fifteen parts out of 100.* Both of these expressions $\left(\frac{15}{100} \text{ and } .15\right)$ are ways of showing a certain number of hundredths.

*Percent:* The third way of showing hundredths is called *percent*. When the percent symbol % is used with a number—for example, 15%—it means the same as .15 or $\frac{15}{100}$.

Look at Fig. 10-1. Count the number of shaded squares and compare this number with the total number of squares (100). You will find the shaded area is *fifteen hundredths* of the total. The shaded area is

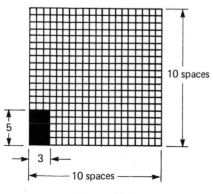

*Figure 10-1*

written as:

$$\frac{15}{100} \qquad .15 \qquad \text{or} \qquad 15\%$$

*Percent means parts per hundred.* For instance, each of the following fractions is expressed in three ways: (a) as a common fraction (b) as a decimal, (c) as a percent.

Twenty-six hundredths         (a) $\frac{26}{100}$   (b)  .26   (c) 26%

Seventy-five parts per hundred   (a) $\frac{75}{100}$   (b)  .75   (c) 75%

One hundred fifty hundredths    (a) $\frac{150}{100}$   (b) 1.50   (c) 150%

### Practice Problems

1. Write the fraction $\frac{45}{100}$ as a decimal. Write it as a percent.

*Answer:* As a decimal: .45. As a percent: 45%.

2. Write the decimal .85 as a fraction. Write it as a percent.

*Answer:* As a fraction: $\frac{85}{100}$. As a percent: 85%.

3. Write the percent 67% as a fraction. Write it as a decimal.

*Answer:* As a fraction: $\frac{67}{100}$. As a decimal: .67.

From the results of the practice problems it is possible to state these rules:

### To Change Fraction to Percent

*To write a fraction with denominator 100 as a percent, omit the denominator and write the percent symbol after the numerator.*

$$\frac{45}{100} = 45\%$$

### To Change Decimal to Percent

*To write decimal as a percent, move the decimal point TWO places to the RIGHT and attach the percent symbol.*

.45 = 45%   (Move decimal point *two* places to the **right**.)

### To Change Percent to Decimal

*To write a percent as a decimal, omit the percent symbol and move the decimal point TWO places to the LEFT.*

$3\% = .03$ (Move decimal point *two* places to the **left**.)

### To Change Percent to Fraction

*To write a percent as a common fraction, omit the percent symbol and write the number as the numerator and 100 as the denominator.*

$$12\% = \frac{12}{100}$$

Examples 4, 5, and 6 show the use of these rules.

**Example 4** (a): Express the fraction $\frac{1}{8}$ as a percent.

*Solution* (a): Express it as a decimal:

$$\frac{1}{8} = .125$$

Change it to a percent by moving the decimal point TWO places to the right and attaching the percent symbol.

*Answer* (a): $\frac{1}{8} = .125 = 12.5\%$

**Example 4** (b): Express $1\frac{11}{32}$ as a percent.

*Solution* (b): From the table of decimal equivalents:

$$\frac{11}{32} = .34375$$

Therefore, $1\frac{11}{32} = 1.34375$

Express the decimal equivalent as a percent by moving the decimal point TWO places to the right and attaching the percent symbol.

*Answer* (b): $1\frac{11}{32} = 1.34375 = 134.375\%$

**Example 5** (a): Express the decimal 1.4 as a percent.

*Solution* (a): Add one zero in order to have TWO places to the right of the decimal point. Move decimal point two places to the right, and attach the percent symbol.

*Answer* (a): $1.4 = 1.40 = 140\%$

**Example 5** (b): Write the decimal .0035 as a percent.

*Solution* (b): Move decimal point TWO places to the right and attach the percent symbol.

*Answer* (b): $.0035 = 0.35\%$

**Example 6** (a): Write the percent $17.5\%$ as a decimal.

*Solution* (a): Omit the percent symbol and move the decimal point TWO places to the left.

*Answer* (a): $17.5\% = .175$

**Example 6** (b): Express the percent $45\%$ as a common fraction in simplest form.

*Solution* (b): Write it as a fraction with a denominator of 100, using 45 as the numerator:

$$45\% = \frac{45}{100}$$

Reduce the fraction.

$$\frac{45}{100} = \frac{45 \div 5}{100 \div 5} = \frac{9}{20} \quad \text{(Divide numerator and denominator by 5.)}$$

*Answer* (b): $45\% = \frac{9}{20}$

Study Table 10-1. It will help you understand the use of percent.

**TABLE 10-1** Equivalents

| Percent | Decimal | Common fraction |
|---------|---------|-----------------|
| 5% | .05 | $\frac{1}{20}$ |
| 6.25% | .0625 | $\frac{1}{16}$ |
| 8.3% | .083 | $\frac{1}{12}$ |
| 10% | .10 or .1 | $\frac{1}{10}$ |
| 12.5% | .125 | $\frac{1}{8}$ |
| 16.7% | .167 | $\frac{1}{6}$ |
| 20% | .20 or .2 | $\frac{1}{5}$ |
| 25% | .25 | $\frac{1}{4}$ |
| 30% | .30 or .3 | $\frac{3}{10}$ |
| 33.3% | .333 | $\frac{1}{3}$ |
| 37.5% | .375 | $\frac{3}{8}$ |
| 40% | .40 | $\frac{2}{5}$ |

**TABLE 10-1** (continued)

| Percent | Decimal | Common fraction |
|---------|---------|-----------------|
| 50% | .50 | $\frac{1}{2}$ |
| 60% | .60 | $\frac{3}{5}$ |
| 62.5% | .625 | $\frac{5}{8}$ |
| 66.7% | .667 | $\frac{2}{3}$ |
| 70% | .70 | $\frac{7}{10}$ |
| 75% | .75 | $\frac{3}{4}$ |
| 80% | .80 | $\frac{4}{5}$ |
| 83.3% | .833 | $\frac{5}{6}$ |
| 87.5% | .875 | $\frac{7}{8}$ |
| 90% | .90 | $\frac{9}{10}$ |
| 100% | 1.00 or 1 | 1 |

Each percent equivalent is computed to the nearest tenth of a percent. For example, the percent equivalent of the fraction $\frac{2}{3}$ is 66.7%. This means that the *decimal equivalent* is first determined to the nearest thousandth, and the percent is then correct to the nearest tenth of a percent.

# PROBLEMS: Sec. 1
## The Meaning of Percent

1. Fill in the blanks in Table 10-2. In each row write the fraction as (1) a common fraction with denominator 100, (2) a decimal, and (3) as a percent.

**TABLE 10-2**

| | Fraction | Decimal | Percent |
|---|----------|---------|---------|
| (a) | $\frac{25}{100}$ | _____ | _____ |
| (b) | _____ | .65 | |
| (c) | _____ | _____ | 75% |
| (d) | $\frac{58}{100}$ | _____ | _____ |
| (e) | _____ | .80 | |
| (f) | _____ | _____ | 1.5% |
| (g) | $1\frac{15}{100}$ | _____ | _____ |
| (h) | _____ | 1.45 | _____ |
| (i) | _____ | _____ | 125% |

2. Express each of these numbers as a percent.
   (a) Twenty-five hundredths
   (b) 16 hundredths
   (c) Eighty-seven hundredths
   (d) 250 hundredths
   (e) 1.5 hundredths
   (f) $\frac{1}{2}$ hundredths

3. Express each of these numbers as a percent.
   (a) 55 out of a hundred
   (b) 78 out of a hundred
   (c) 23 out of 100
   (d) 156 out of 100
   (e) $\frac{1}{2}$ out of 100
   (f) $3\frac{1}{4}$ out of 100

4. Express each of these percents as a common fraction in simplest terms.

   **Example:** What is 75% as a fraction?

   *Solution:* $75\% = .75 = \frac{75}{100} = \frac{75 \div 25}{100 \div 25} = \frac{3}{4}$

   *Answer:* $\frac{3}{4}$

   (a) 50%   (b) 60%   (c) 80%   (d) 25%   (e) 12.5%
   (f) 37.5%   (g) 62.5%   (h) 6.25%   (i) $33\frac{1}{3}\%$   (j) $66\frac{2}{3}\%$
   (k) $\frac{2}{5}\%$

5. Express each common fraction as a percent, to the nearest tenth of a percent.

   **Example:** Express $\frac{1}{5}$ as a percent.

   *Solution:* $\frac{1}{5} = .20 = 20\%$

   *Answer:* 20%

   (a) $\frac{1}{4}$   (b) $\frac{3}{4}$   (c) $\frac{1}{8}$   (d) $\frac{3}{8}$   (e) $\frac{5}{8}$   (f) $\frac{4}{5}$   (g) $\frac{1}{16}$
   (h) $\frac{3}{16}$   (i) $\frac{7}{16}$   (j) $\frac{15}{32}$   (k) $\frac{1}{3}$   (l) $\frac{2}{3}$   (m) $\frac{1}{64}$   (n) $\frac{25}{64}$

6. Express each number as a percent, to the nearest tenth of a percent.

   **Example:** Express $1\frac{1}{9}$ as a percent.

   *Solution:* $1\frac{1}{9} = 1.111 = 111.1\%$, to nearest tenth of a percent.

   *Answer:* 111.1%

(a) $3\frac{1}{4}$    (b) $4\frac{1}{2}$    (c) $1\frac{2}{5}$    (d) $5\frac{3}{4}$    (e) $7\frac{1}{8}$    (f) $1\frac{5}{16}$

(g) $-2\frac{3}{5}$    (h) $\frac{23}{5}$

**7.** What percent of each square is shaded in (a) Figs. 10-2 and (b) 10-3?

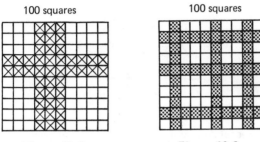

100 squares    100 squares

*Figure 10-2*         *Figure 10-3*

**8.** What percent of each square is shaded in (a) Figs. 10-4 and (b) 10-5?

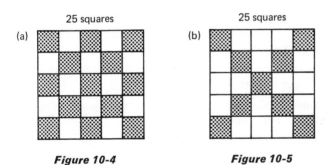

25 squares    25 squares

(a)         (b)

*Figure 10-4*         *Figure 10-5*

**9.** What percent of the circle is shaded in Fig. 10-6?

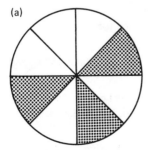

(a)

*Figure 10-6*

**10.** What percent of the triangle is shaded in Fig. 10-7?

(b)

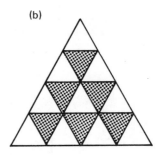

**Figure 10-7**

**11.** Eighteen students out of a class of 24 passed an examination in technical mathematics.
(a) Express as a percent the number that passed the exam.
(b) What percent of the class failed the examination?

**12.** A fuel tank with a capacity of 25 gallons has 10 gallons in it. What percent of the tank is full?

**13.** A carpenter spent 3 hours of an 8-hour day hanging doors. What percent of his work day is this?

**14.** The Piedmont Panthers played 30 baseball games and lost 9. What percent of the games did they win?

**15.** One hundred grams of oatmeal cookies has 86 grams of carbohydrates. Express this as a percent.

**16.** In each 100 units of air that you breath, about 21 parts are oxygen and 78 parts are nitrogen. Express these amounts as percentages.

**17.** A rough casting that weighs 35 pounds loses 8 pounds in a machining operation. What percent of the weight is lost?

**18.** A timekeeper finds that 15 men are absent from a work force of 165 men. What percent of the work force is absent?

**19.** When washers are stamped out on an automatic punch press, 2 square feet of sheet steel is wasted for every 12 square feet used. What is the percentage of waste?

**20.** A 520-pound bronze casting contains 182 pounds of zinc. What percent of the casting is zinc?

**21.** The U. S. Mint uses a metal called *nickel coinage* for making nickels. It is made of 75% copper and 25% nickel. Express these percentages as fractions.

**22.** A certain bronze alloy for bearings contains 82% copper, 16% tin, and 2% zinc. Express these values as fractions in lowest terms.

23. Ordinary Christmas tinsel is $\frac{3}{5}$ tin and $\frac{2}{5}$ lead. Express these values as percentages.

24. One pound of bearing babbitt is made up of .9 pound tin, .07 pound antimony, and .03 pound copper. Express these values as percentages.

25. A casting contains 34 pounds of copper and 18 pounds of zinc.
    (a) What is the total weight of the casting?
    (b) What percentage of the casting is copper?

26. A railway box car has a capacity of 98,000 pounds. If it is loaded with 56,000 pounds, what percent is this of the total capacity?

27. A certain jet engine delivers 4,400 horsepower. An auto engine delivers 160 horsepower. Express the ratio of power of the jet engine to the auto engine as a percent.

28. In one year (365 days) an average factory worker works about 235 days. (Two week vacation, holidays off, etc.) What percentage of the year does he work?

29. A painter and his helper painted a house. The painter finished half of the house while his helper finished two-fifths of it.
    (a) What fraction of the house was painted?
    (b) What percent of the job was completed?

30. A plumber ordered 450 feet of soil pipe for a job. After completing the job he had 40 feet left. What percent of the pipe was used?

31. An inspector inspected a total of 90 items during an 8-hour day. He rejected 12 because they did not meet production standards. What percent of the total was rejected?

32. Crime reports show that hardware stores, dime stores, and stores that display small items can expect to lose three items through theft for every sixty items displayed. Express this loss as a percentage.

33. An automatic electronic scanner used for inspecting production items rejected 45 and accepted 835. What percent of the total number inspected was rejected?

34. A W-200 Dodge Power Wagon weighs about 4,850 pounds. Of this weight 2,100 pounds is supported by the rear wheels and the rest is supported by the front wheels. What percent of the weight is supported by the front wheels?

35. The finished length of a precision machined part is 1.575 inches. What percent of a piece 1.850 inches long must be taken off to make the finished part?

36. When an iron casting is first poured, it is 22.00 inches. After cooling and shrinking, it is 21.84 inches. Write the shrinkage rate as a percent.

    *Hint:* Write $\dfrac{\text{loss}}{\text{original}} = \dfrac{.16 \text{ inches}}{22 \text{ inches}}$ (as a percent)

37. When a 500-ohm resistor is placed in a 120-volt circuit, 2 volts are lost. What percent of the 120 volts is lost?

**38.** A test shows that an engine delivers 180 brake horsepower to the transmission, but only 168 horsepower is delivered to the rear wheels. What percent of the original 180 horsepower is lost through friction?

**39.** A mechanical tomato picker drops or otherwise loses 45 pounds of tomatoes for every 850 pounds picked. Express the loss rate as a percent.

**40.** On his way to work one morning Sam Akers ate $\frac{1}{5}$ of his lunch. During the 10 o'clock coffee break he ate $\frac{1}{4}$ of the remainder. What percent of his lunch did he have remaining when the noon lunch whistle blew?

**41.** The ratio of antifreeze to water in an automobile radiator is 1:4. What is the percent of antifreeze in the radiator?

**42.** The recipe for a paint-brush cleaner is:

(1) Kerosene ....................2 pints
    Oleic acid ....................1 pint

(2) Aqueous ammonia (concentrated)....$\frac{1}{4}$ pint

    Denatured alcohol ...............$\frac{1}{4}$ pint

Stir (1) into (2) until uniform. Place brushes in mixture overnight. Wash thoroughly with warm water.
Find the percent of each substance in the mixture.

$\left(\text{Note:}\quad\text{The total mixture} = 3\frac{1}{2}\text{ pints.}\right)$

**43.** The total land area of California is 100,314,000 acres. The land is used as shown in this table. Fill in the blanks with the percentage value correct to one decimal place.

| | | |
|---|---|---|
| (a) Cropland in farms | 10,235,000 acres | _____ % |
| (b) Pasture and range in farms | 17,074,000 acres | _____ % |
| (c) Other rangeland | 9,226,000 acres | _____ % |
| (d) Commercial forest land | 17,317,000 acres | _____ % |
| (e) Noncommercial forest land (parks, etc.) | 25,224,000 acres | _____ % |
| (f) Other land uses (cities, towns, roads, etc.) | 21,238,000 acres | _____ % |
| Totals (Check) | _____ | _____ % |

# SEC. 2 COMPUTATIONS THAT USE PERCENT

Do you recall from Chapter 8 how to solve these problems?

(a) Find $\frac{1}{2}$ of 36.

(b) What is $\frac{4}{5}$ of 75?

$P = I/R$    Divide I by R to get P because I is over R.
$R = I/P$    Divide I by P to get R because I is over P.

When you use the value R (rate), write it as a decimal although it is usually given in percent form.

Use the I-P-R triangle to prove these answers.

1. Find I if P = $200.00 and R = 7.5%.   Answer:  I = $15.00
2. Find P if I = $25.00 and R = 5%.   Answer: P = $500.00
3. Find R if I = $16.00 and P = $800.00.   Answer: R = 2%

A business man must make a profit on what he sells in order to stay in business. When an item is sold, the retail price includes *cost, profit,* and *expenses.* That is:

*Selling Price = Cost + Profit + Expenses*

In any business—a service station, machine shop, parts house, or body and fender repair—there is a *business expense.* Business expense includes building rent, utilities, maintenance, and general operation expenses. When an item is sold, the price must include a portion of the expenses of the business. The difference between the cost and selling price is called the *margin* or *mark-up.*

*Margin = Selling Price − Cost = Profit + Expense*

The margin or mark-up includes the profit and expenses.

**Example 15:** Mr. Bush owns and operates a small cabinet shop. For each sales dollar received, Mr. Bush knows that a certain percent goes for *expense*, a percent for *cost*, and a percent for *profit*. Through years of experience, he knows that:

100% sales dollar = 45% *cost* + 40% *expense* + 15% *profit*

Mr. Bush sold a cabinet for $85.00. Find the *cost* of materials, the *expense*, the *profit*, and *the margin.*

*Solution:* From the above breakdown you see:

*Answer:*   Cost = 45% of sales dollar = 45% of $85.00 = $38.25
        Expense = 40% of sales dollar = 40% of $85.00 = $34.00
        Profit = 15% of sales dollar = 15% of $85.00 = $12.75
        Margin = $34.00 + $12.75 = $46.75

# PROBLEMS: Sec. 2
## Computations That Use Percent

**Problems 1 through 5.** Find the answers.
1.  (a) 25% of 50          (b) 30% of 60          (c) 45% of 180
    (d) 16% of 240         (e) 14% of 80          (f) 24% of 68
    (g) 35% of 75          (h) 75% of 145         (i) 80% of 120

2.  (a) 4% of 78 (b) 5% of 75 (c) 8% of 45
    (d) 2% of 98 (e) 7% of 60 (f) 9% of 126
    (g) 1% of 78 (h) 6% of 88 (i) 3% of 90

3.  (a) 11% of 15.6 (b) 15% of 1.8 (c) 4% of 46
    (d) 56% of 16.8 (e) 1.5% of 12.5 (f) 38% of 66.9
    (g) .5% of 166 (h) .2% of 89.5 (i) .8% of 178

4.  (a) 45% of 8 ft. (b) 75% of 36 in. (c) 85% of 150 cm.
    (d) 25% of 64 lb. (e) 56% of 28 oz. (f) 8% of 12 yd.
    (g) 15% of 5 gal. (h) 22% of $45 (i) 35% of 50 km.

5.  (a) 125% of $56 (b) 110% of 32 ft. (c) 145% of 18 in.
    (d) 168% of 94 lb. (e) 120% of 88 in. (f) 175% of 20 cm.
    (g) 105% of 200 yd. (h) 235% of 46 gal. (i) 500% of 72 gm.

6.  Find the answers to the nearest hundredth.
    (a) 1.5% of $19.45 (b) 5.5% of $35.40 (c) 6% of $28.95
    (d) 12% of $1.65 (e) 5.8% of $156.98 (f) 2% of $1,567.45
    (g) 50% of $66.80 (h) .5% of $4.75 (i) 1.3% of $.85

7.  Find the answers correct to the nearest cent.
    (a) 25% of $45.65 (b) 14% of $3.65 (c) 4.6% of $128
    (d) 134% of $56.25 (e) 34.5% of $89.00 (f) 1.25% of $5.78
    (g) .75% of $23.75 (h) 15.45% of $5.28 (i) .06% of $985.00

8.  Find the answers correct to the nearest hundredth.

    **Example:** Find $\frac{1}{2}$% of 48.

    *Solution:* $\frac{1}{2}$% = .5%. Therefore, $\frac{1}{2}$% of 48 = .5% of 48

    $$= .005 \times 48$$
    $$= .24$$

    *Answer:* $\frac{1}{2}$% of 48 = .24

    Change the fractions to decimals, then change percent to decimals, and do the multiplication.

    (a) $\frac{1}{2}$% of 88 (b) $\frac{3}{4}$% of 60 (c) $\frac{4}{5}$% of 50

    (d) $\frac{2}{3}$% of 45 (e) $1\frac{1}{4}$% of 45 (f) $4\frac{1}{4}$% of 82

    (g) $5\frac{1}{4}$% of 156 (h) $6\frac{1}{2}$% of 25 (i) $7\frac{3}{4}$% of 125

9.  On a plumbing job 3.5% of the pipe used is wasted. If 450 feet are used, how much is wasted?

10. Hard solder is 68% tin. How many pounds of tin are in 36 pounds of hard solder?

11. Jon got 85% correct on a test that had 20 problems. How many did he get correct?

12. A certain fabric will shrink 1.5% when washed. How much will a piece 36 inches long shrink?

**13.** A salesman earns 12% commission on all sales. What is his commission on $3,650.00?

**14.** A modern three-bedroom, two-bath home sold for $38,500. The down payment was 15%. How much was the down payment in dollars?

**15.** An inspector rejected 3.5% of a production run for an eight-hour day. If he inspected a total of 750 parts, how many did he reject?

**16.** An inspector accepted 88% of a production run for an eight-hour day. If he inspected 900 parts, how many did he accept?

**17.** George needs a score of at least 85% on the next technical mathematics test to pass the course. If the test has 40 problems, how many can he afford to miss?

**18.** An automobile engine actually burns about 88% of the fuel that it uses. If it uses 24 gallons, how many gallons does it burn?

**19.** A diesel engine burns 99.8% of the fuel it uses. If an engine uses 55 gallons of fuel, how many gallons are not burned?

**20.** A jet engine is about 78% efficient. If it uses fuel that is rated to develop 2,000 horsepower, how many horsepower does it deliver?

**21.** A long-life Never-Die battery that normally sells for $45 is advertised at 20% off. How much is the discount?

**22.** Find the amount of discount and the net price on each of the following:

|  | (a) | (b) | (c) | (d) |
|---|---|---|---|---|
| List price | $46.00 | $14.50 | $8.65 | $198.45 |
| Rate of discount | 5% | 15% | 20% | 25% |
| Discount | $_____ | $_____ | $_____ | $_____ |
| Net price | $_____ | $_____ | $_____ | $_____ |

**23.** In this problem, use the double-discount method of ten-and-five (10-and-5). Find 10% of the list price. Subtract this from the list price. Then find 5% of the remainder. Subtract this second discount to get the final net price. (NOTE. A discount of "ten-and-five" is *not* equal to a single discount of 15%.)

**Example:** Find the net price of an article that has a list price of $56.80.

*Solution:* First, find 10% of $56.80 = $5.68 (first discount).
Subtract $5.68 from $56.80 to get $51.12.
Find 5% of $51.12, which is $2.56 to nearest cent (second discount).
Subtract $2.56 from $51.12 to get the final net price.

*Answer:* $56.80 with a "ten-and-five" discount gives a net price of $48.56.

(a) List = $189.00
   First discount = $_____
   Second discount = $_____
   Net price = $_____

(b) List = $24.50
   First discount = $_____
   Second discount = $_____
   Net price = $_____

(c) List              = $5.75        (d) List              = $35.75
    First discount    = $_____           First discount    = $_____
    Second discount = $_____           Second discount = $_____
    Net price         = $_____           Net price         = $_____

(e) List              = $47.85
    First discount    = $_____
    Second discount = $_____
    Net price         = $_____

**24.** Use a 15-and-10 double-discount method to find the net price for these items:
(a) Four tires. Total list price, $192.50
(b) Rebuilt engine. List, $388.75
(c) Ring and pinion gear. List, $166.50
(d) Set of eight racing pistons. List, $96.00
(e) Four barrel carburator. List, $110.50
(f) Ignition rebuilding kit. List, $11.75
(g) Alternator. List, $89.75
(h) Rebuilt starting motor and solenoid. List, $28.45
(i) Set of wrenches. List, $88.25

**25.** Retail houses give a triple discount to some preferred customers. The Mid-Nite Auto Supply gives a triple discount of 15-10-5% to independent garage owners. Find the net price of each of the following items bought by the Westbrock Garage:
(a) One dozen fan belts @ $2.40 each.
(b) Two dozen gasket sets @ $14.50 each.
(c) Four tail pipes and mufflers.
    Total list $118.50.
(d) Six "Glass-Pak" mufflers @ $12.75 each.
(e) Fifty spark plugs. Total list $54.00.
(f) Four cases of tune-up oil at $24.00 per case.
(g) Ten sets of brake shoes at $34.50 a set.

**26.** In this problem compare the double discount method with a straight discount. Use a *double discount* of ten-and-five and find the net price on a set of shock absorbers that lists for $56.00. Compare this net price with a net price resulting from a *straight discount* of 15%.
(a) Which discount gives the lowest net price?
(b) What is the difference in the two net prices?

**27.** Compare a triple discount of 15-10-5 and a straight discount of 30% on an order of miscellaneous auto parts that lists for $140.00. Which discount gives the best net price?

**28.** Bungling Buford placed an order with Acme Steel for 7 steel plates that had a list price of $150.00. The sales clerk offered B. B. a triple discount of 15-10-5 or a straight discount of 27%. B. B. took the 27% discount. Did he do the right thing?

**29.** Find the net price of a steam boiler that lists for $685.00 with a triple discount of 40%-15%-5%.

30.  Mr. Feldman is paid a base salary of $875.00 per month. He has just learned that he is to get a 6.5% wage increase. What will be his monthly salary?

31.  The base hourly wage for a Class 1 patternmaker is $6.92. Find the new hourly wage if the patternmaker receives a 3% increase for promotion to Class 2, followed by a 1.5% differential for working the swing shift.

32.  An 18-cubic-foot freezer that regularly sells for $450.00 is on sale at a discount of 20%. If the freezer is paid for by cash, the dealer will give an additional 6% discount. What is the cash price of the freezer?

33.  A large construction firm wants to buy a new fleet of fifteen pick-up trucks. Honest Abe, Inc., will give the firm a base price of $3,250.00 per unit, followed by a double discount of 10%-and-5%. Fleet Rate Motors will give a base price of $3,400.00, followed by a discount of 18%. From which dealer should the construction firm buy the trucks?

34.  Metal will expand when subjected to an increase in temperature. A steel I-beam will expand approximately .01% of its length when exposed to the normal temperature changes during the day. How much will an I-beam 42 feet long expand during the day?

35.  A sheetmetal worker has a salary of $19,450.00 for the year. His income tax is 19.5% of his salary. How much does he pay in taxes?

36.  An electrical circuit in an automobile is fused for 18 amps. An air conditioner that is connected to the circuit uses 10 amps. What percent of the current is used by the air conditioner?

37.  In these problems, find the *original list price* when the *discount rate* and the *amount of the discount* is given.

> **Example:** If the discount rate is 12% and the discount on an item is $40.80, find the original list price.
>
> *Solution:* $12\% \times$ (original list price) = $40.80
>
> $$\text{(original price)} = \frac{\$40.80}{12\%} \quad \text{(Divide both sides by 12\%.)}$$
>
> Change 12% to the decimal .12 and carry out division
>
> *Answer:* Original price $= \dfrac{40.80}{.12} = \$340.00$

(a)  Discount rate = 15%   Discount = $45.00   Original price = _____
(b)  Discount rate = 12%   Discount = $10.80   Original price = _____
(c)  Discount rate = 8%    Discount = $16.24   Original price = _____
(d)  Discount rate = 5%    Discount = $17.50   Original price = _____
(e)  Discount rate = 7.5%  Discount = $5.50    Original price = _____
(f)  Discount rate = 16.5% Discount = $26.00   Original price = _____
(g)  Discount rate = 33.3% Discount = $75.00   Original price = _____
(h)  Discount rate = 25%   Discount = $145.20  Original price = _____

38.  Mr. Weston, a partner in the Westbrockmore Garage, bought some garage supplies at a discount of 16%. If he paid the jobber $250.00 (after discount), what was the original list price? (Note: $250.00 = (100 − 16)% = 84% of original)

**39.** When the constant-load items (radio, air conditioner, lights, etc.) in a car's electrical system are turned on, the load should equal about 85% of the charging ability of the alternator. What is the charging ability (to the nearest amp) of the alternator when the load is 43 amps?

$$\left(\text{HINT:}\quad \text{Charging ability} = \frac{43 \text{ amps}}{85\%}\right)$$

**40.** A parts salesman earns a base salary of \$150.00 per week plus a 12% commission on all sales. What are his sales for a week, if his *total* earnings are \$318.00?

**41.** Robert Paul is a machinist apprentice (2nd year) and earns \$4.08 an hour. His deductions are: 3% union dues, 1.8% F.I.C.A., 14% Federal Income Tax, 2.5% pension fund. What is Robert's take home pay for a 40-hour week?

**42.** Jim Thomsen is a student working his way through Junior College. He has a part-time job at a service station which pays him \$3.58 per hour. Last week he worked 28 hours. The owner of the station deducted 18% Federal Tax, 4.5% State Tax, and 2.3% Social Security. How much money did Jim have as take-home pay?

**43.** A contractor submitted a bid of \$5,675.00 to build a garage. His cost, after completion of the job, was \$4,445.00.
(a) What is his profit?
(b) What is the profit percent based on the cost?

$$\left(\text{HINT:}\quad \text{Express } \frac{\text{profit}}{\text{cost}} \text{ as a percent.}\right)$$

**44.** A jobber bought 250 auto floor mats from a manufacturer at \$1.32 each. He sold them to a retailer for \$1.44 each.
(a) What is his profit on one floor mat?
(b) What is his profit percent based on selling price?

$$\left(\text{HINT:}\quad \text{Find } \frac{\text{profit}}{\text{selling price}} \text{ as a percent.}\right)$$

(c) What is his total profit on the 250 mats?

**45.** An alloy called white metal is composed of 75% lead, 19% antimony, 5% tin, and 1% copper. How many pounds of each metal are in 25 pounds of white metal?

**46.** Standard type metal is 58% lead, 26% tin, 15% antimony, and 1% copper. How much of each is in a block of type metal that weighs 458 grams?

**47.** Shull Savings and Loan advertises an annual interest rate of $5\frac{1}{2}\%$ on investments held for one year. How much interest will they pay on an investment of \$3,189.00?

**48.** When investments are held for two or more years, Shull Savings and Loan guarantees to pay an annual interest rate of 6.25%. How much interest is earned on an investment of \$5,000.00 held for two years? (HINT: Find the interest for the first year, add to the principal, and compute the second year's interest based on \$5,000 + first year interest.)

49. If an investment of $3,500.00 produces interest of $252.00, what is the annual rate of interest?

50. Find the amount invested (principal) if the interest (I) is $45.00 and the rate (R) is 8.7%.

51. Find the interest rate (R) if the principal (P) is $33,575.00 and the interest (I) is $1,590.00.

52. Find the *total* interest earned on a principal of $4,800.00 invested at 5.5% (annual rate) for *three* years when the interest is added to the principal yearly.

53. The accountant for Scott Trucking has just broken down each sales dollar into cost, expense, and profit. His figures show:

$$\text{Cost} = 42\% \text{ of sales dollar}$$
$$\text{Expense} = 39\% \text{ of sales dollar}$$
$$\text{Profit} = 19\% \text{ of sales dollar}$$

Find the cost, expense, and profit of Scott Trucking for a total sales of $145,975.23 for the year.

## Practice Test for Chapter 10

1. Express nineteen twenty-fifths as a percent.    *Answer:* _____

2. Express 185 hundredths as a percent.    *Answer:* _____

3. Express 1.5% as a decimal.    *Answer:* _____

4. A fuel tank with a capacity of 86 gallons is 25% full. How many gallons of fuel are in the tank?    *Answer:* _____

5. A rough casting weighs 43 pounds. After machining, it weighs 36 pounds. What percent of the original weight remains after finishing? Express answer to nearest tenth of a percent.    *Answer:* _____

6. An auto parts jobber gives a discount of 30-25-5 percent to Mid-Nite Auto Supply. What is the discount price to Mid-Nite Auto on an item that the jobber lists at $156?    *Answer:* _____

7. An inspector rejected 60 items out of a total of 863 inspected. Express the rejection rate as a percent of the total inspected.    *Answer:* _____

8. How much interest will you pay at the end of one year on a loan of $4,568 at an annual rate of 7.75%?    *Answer:* _____

9. Sno-Grip truck tires are advertised at a reduced price of $45. If this reduced price is the result of a 20% reduction, what was the original price?    *Answer:* _____

10. A parts salesman is paid a 15% commission on all sales after the first $500. If his commission for one week is $245.00, what are his total sales?    *Answer:* _____

# 11

# Perimeters, Areas, and Volumes

## RECTANGLES, CIRCLES. AND TRIANGLES

Look at the three diagrams in Fig. 11-1. They are a *rectangle*, a *circle*, and a *triangle*.

Now look around you—wherever you are. Do you see any objects that have parts that look like these diagrams? Look at the end of your pencil. Is it round like a circle? Is there a picture on the wall? Does the frame look like a rectangle? Look at the roof on the house across the street. Does the roof-line resemble a triangle? Stop for a minute and think of as many objects as you can that look like these shapes. Here is a list of some of the objects you might think of:

Objects that use rectangles: a door, a window, this book, a piece of paper, a box of cereal, the front of a TV set, and your tool box

Objects that use circles: the steering wheel of your car, a bicycle wheel, a piston, a cylinder, the top of a soup can, the clock on the wall, and a coffee cup

Objects that use triangles: the flap on an envelope, a staircase, a sharp V-thread on a bolt or gear, and the roof of a house

Certainly, from your own experience, you can think of more objects to add to this list. As you study this chapter, you will learn a great deal

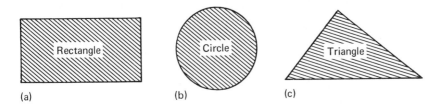

**Figure 11-1**   (*a*) *Rectangle;* (*b*) *Circle;* (*c*) *Triangle*

about these three shapes, their properties, and the various ways they are used in technology.

## SEC. 1   THE RECTANGLE

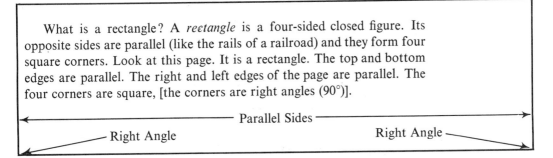

What is a rectangle? A *rectangle* is a four-sided closed figure. Its opposite sides are parallel (like the rails of a railroad) and they form four square corners. Look at this page. It is a rectangle. The top and bottom edges are parallel. The right and left edges of the page are parallel. The four corners are square, [the corners are right angles (90°)].

Parallel Sides

Right Angle                    Right Angle

## The Perimeter of a Rectangle

Every rectangle has two important properties: *perimeter* and *area*. The distance around a rectangle is called the *perimeter*. Start at the point marked *A* on the rectangle shown in Fig. 11-2, and "walk around" the

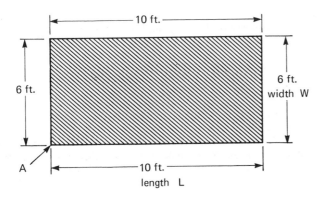

**Figure 11-2**

figure. Start at *A*, walk 10 feet, 6 feet, 10 feet, 6 feet, and return to point *A*. You have walked (10 + 6 + 10 + 6) feet = 32 feet. Therefore, the perimeter is 32 feet.

In a rectangle, one of the sides is called the *length* (*L*), and the other side, which is perpendicular to the length, is called the *width* (*W*). (It really does not matter which you call the length, just as long as you call the other one the width.) Use this rule:

### Rule for Finding the Perimeter of a Rectangle

*To find the perimeter of a rectangle, add the two lengths and the two widths.*

Use the letter P for the perimeter, and the letters L and W for length and width. The perimeter of a rectangle is:

$$P = 2L + 2W$$

**Example 1:** Find the perimeter (P) of the rectangles described in this table.

*Solution:* Use the formula $P = 2L + 2W$

*Answers:*

| Length (L) | Width (W) | Perimeter (P) |
|---|---|---|
| 2 ft. | 3 ft. | 10 ft. = 2(2) + 2(3) |
| 9 ft. | 4 in. | 26 in. |
| 1.95 cm. | 0.75 cm. | 5.40 cm. |
| 2 ft. 5 in. | 1 ft. 8 in. | 8 ft. 2 in. |

NOTE: *When adding measurements, the units must be the same.* Add feet-to-feet, yards-to-yards, and so on. Don't be guilty of adding "apples and oranges."

## The Area of a Rectangle

The second property that all rectangles have is that of *area. Area is the name of the space inside a rectangle.* It is a measure of how much the rectangle encloses.

Here is how you find the area of a rectangle:
The rectangle shown in Fig. 11-3 is

Length L = 4 inches and width W = 2 inches

Divide this rectangle into 8 smaller rectangles that are 1 inch on each side. These smaller rectangles, *with equal sides*, are called *squares.* A square, 1 inch by 1 inch, has an *area equal to 1 square inch.* The area of the bigger

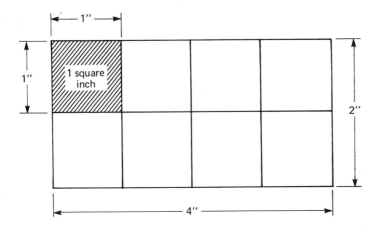

**Figure 11-3**

rectangle is equal to the number of square inches inside. The rectangle contains eight squares, 1-inch on a side. Therefore, the area of the rectangle is eight square inches (8 sq. in.). You get the number of square inches in this rectangle by multiplying the *length by the width*.

Use this rule:

### Rule for Finding the Area of a Rectangle

*To find the area (A) of a rectangle with length L and width W, multiply the length by the width.*

$$A = L \times W \text{ (square units)}$$
$$(\text{area}) = (\text{length}) \times (\text{width})$$

The units of measure for $L$ and $W$ *MUST BE THE SAME*. If $L$ and $W$ are in *inches*, then $A$ is in *square inches* (sq. in.). If $L$ and $W$ are in *feet*, then $A$ is in *square feet* (sq. ft.). If $L$ and $W$ are in *centimeters*, then $A$ is in *square centimeters* (sq. cm.), and so on.

**Example 2:** Find the area of these rectangles.

*Solution:* Use the formula $A = L \times W$ to verify the following:

| *Answers:* Length (L) | Width (W) | Area ($A = L \times W$) (*Answer*) |
|---|---|---|
| 6 in. | 5 in. | 30 sq. in. $= 6 \times 5$ |
| 5 meters | 9 meters | 45 sq. meters $= 5 \times 9$ |
| 12 in. | 12 in. | 144 sq. in. |
| 1 ft. | 1 ft. | 1 sq. ft. |

Two important results are developed in this next example.

**Example 3:** (a) Find the number of *square inches* in 1 *square foot*. See Fig. 11-4.

*Solution:* (a) Draw a square that is 1 foot by 1 foot. The area is 1 square foot. Since 1 foot equals 12 inches, each side is 12 inches long. The area, in *square inches*, is:

*Answer* (a): $A = 12 \times 12 = 144$ square inches.

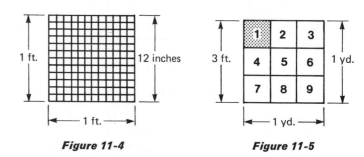

**Figure 11-4**          **Figure 11-5**

**Example 3** (b): Find the number of *square feet* in 1 *square yard*.

*Solution* (b): Draw a figure that is 1 yard by 1 yard. The area of the figure is 1 square yard (Fig. 11-5). Since 1 yard equals 3 feet, the square measures 3 feet on a side. The area, in *square feet*, is:

*Answer* (b): Area $= 3$ ft. $\times 3$ ft. $= 9$ square feet.

As Example 4 shows a figure can be divided into separate rectangles.

**Example 4:** A rancher owns the piece of land sketched in Fig. 11-6. It is made up of two rectangles, a small one and a large

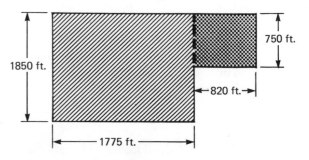

**Figure 11-6**

one. If 1 acre of land is equal to 43,560 square feet, find the number of acres of land the rancher owns.

*Solution:* Use a dotted line and divide the figure into two rectangles. The area of the large rectangle is:

$$1,850 \times 1,775 = 3,283,750 \text{ sq. ft.}$$

The area of the small rectangle is:

$$820 \times 750 = 615,000 \text{ sq. ft.}$$

Total area $= 3,283,750 + 615,000 = 3,898,750$

*Answer:* Number of acres $= \dfrac{\text{total area}}{43,560} = 89.5$ acres (nearest tenth)

# PROBLEMS: Sec. 1 Perimeters and Areas of Rectangles

1. Find the perimeter (P) and the area (A) for the rectangles whose length and width are:

|  | Length | Width | $P = 2L + 2W$<br>Perimeter (P) | $A = L \times W$<br>Area (A) |
|---|---|---|---|---|
| (a) | 2 ft. | 1 ft. | _____ | _____ |
| (b) | 4 in. | 3 in. | _____ | _____ |
| (c) | 9 cm. | 7 cm. | _____ | _____ |
| (d) | 6 yd. | 6 yd. | _____ | _____ |
| (e) | 18 m. | 13 m. | _____ | _____ |

2. How many square feet of carpet are needed for a hallway 36 feet long and 4 feet wide?

3. A window is 3 feet, 6 inches (3.50 ft.) by 4 feet, 9 inches (4.75 ft.). Find its perimeter and area. Give the answers to one decimal place.

4. Vinyl floor tiles measure 9 inches by 9 inches. How many will be needed for a bathroom floor that is a rectangle 5 feet by 6 feet?

5. A well-known brand of Indoor-Outdoor carpeting costs $4.50 a square yard. What is the cost of carpeting a kitchen that measures 12 by 15 feet?

6. A parking lot measures 186 by 56 feet.
   (a) What is the area?
   (b) What will it cost to pave the lot at $2.45 a square foot?

7. Redwood fencing costs $1.86 a foot. How much will it cost to fence in a rectangular yard 34 feet by 16 feet?

8. Allowing 35 square feet per child, how many children can be permitted to play in a playground 56 by 110 feet?

9. Find the perimeter and area of the composite rectangles shown in Figs. 11-7 and 11-8.

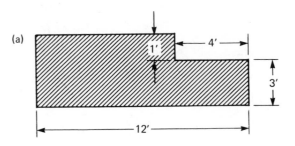

**Figure 11-7**

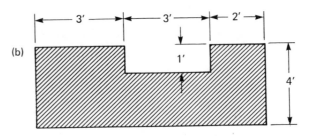

**Figure 11-8**

10. Find the perimeter and area of the composite rectangles shown in Figs. 11-9 and 11-10.

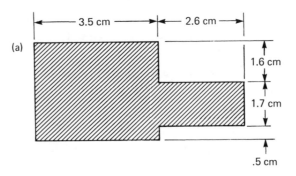

**Figure 11-9**

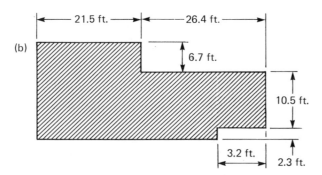

**Figure 11-10**

**Problems 11 through 17.** Figure 11-11 shows a layout for a factory, warehouse, and office building. Use the dimensions to find the answers.

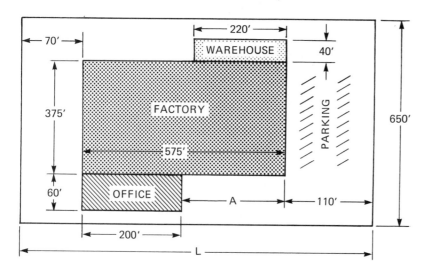

**Figure 11-11**

11. What is the length of $L$?
12. What is the value of $A$?
13. What is the area of the roof of the entire plant?
14. What is the perimeter of the entire plant?
15. What is the total ground area?
16. What percent of the total area is occupied by the *factory*?

**17.** What percent of the total area is available for parking, lawns, walkways, etc.?

**18.** Cove molding costs 12 cents per foot. What will it cost to put molding around the base of the room shown in Fig. 11-12? (Deduct 7 feet for doorways.)

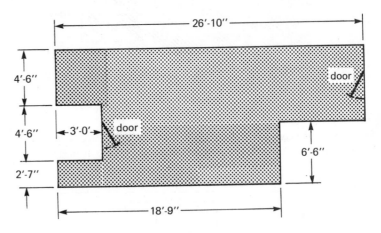

*Figure 11-12*

**19.** At 8 cents per foot, what is the cost of weatherstripping five window frames that are 34 by 56 inches, four window frames 28 by 42 inches, and six window frames 36 by 42 inches?

**20.** A small utility building is 18 feet 6 inches long, 15 feet 3 inches wide, and 10 feet 9 inches high. The walls are to be covered with All-Weather Plywood that comes in sheets 4 feet by 8 feet. Deduct 75 square feet for windows and a door. How many complete sheets are needed to cover the walls? (NOTE: A fraction of a sheet will use a full sheet of plywood.)

## SEC. 2   THE CIRCLE

An auto mechanic works with wheels, axles, pistons, and gears. A machinist works with round bars, lathe centers, pulley cones, and drills. A construction worker uses circular saws, reinforcing rods, bolts, and pipe. Each of these objects has the shape of one of our most common figures—The *circle*.

A circle is a closed curve. All points on this curve are the *same distance* from one *central point* called the *center* of the circle. The distance from the center to the curve is called the *radius* (R) of the circle. The distance around the curve (the perimeter) is called the *circumference* (C) of the circle. The distance across the circle, passing through the center, is called the *diameter*

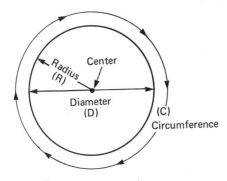

*Figure 11-13*

(D) of the circle. Look at the circle shown in Fig. 11-13 and memorize
the various parts.

The properties of the circle have fascinated people for thousands of
years. One of history's great scientists and philosophers, Euclid (200 B.C.)
studied the circle in great detail. His work in this field is a book by itself.

Two of the most important properties for you to know about the circle
are the *circumference* and *area.*

### The Circumference of a Circle

To find the approximate distance around a circle, perform this simple
experiment. Find something that is round: a tin can, the lid from a jar of
peanut butter, a roll of paper towels, or a truck tire will do. With ruler,
caliper, yardstick, or some other measuring device, measure the diameter
of the circle. (The diameter is the largest distance from one side to the
other.) Now, wrap a piece of string around your round object so that you
can *unroll* the circle and measure the circumference. Unroll the circle by
stretching out the piece of string that you wrapped around the circle.
Measure the length of the string. See Fig. 11-14.

From the result of this experiment, the *ratio of the **circumference** to
**diameter*** can be formed:

$$\frac{\text{Circumference (C)}}{\text{Diameter (D)}}$$

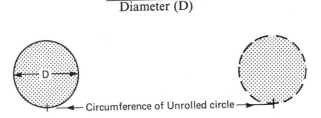

*Figure 11-14*

If you measure carefully, the ratio should be about 3.14. *This value does not depend on which circle you use.* No matter how big or small the circle, *the ratio of circumference to diameter is always the same value.* Remember this fact—*it is the most important property of a circle.*

If you measure several circles and take the average value of all the ratios, you will get a figure close to 3.1416. Mathematicians have found the value to be 3.14159266536 (to ten decimal places). This value is given the Greek letter $\pi$ (pi) pronounced "pie." Therefore:

$$\frac{\text{Circumference}}{\text{Diameter}} = 3.14159\ldots = \pi$$

This result is usually stated in the form;

$$\text{Circumference} = \pi \times \text{Diameter}$$

$$C = \pi \times D$$

*The Circumference (C) of a circle is a little more than three times the diameter (D) of the circle.* For most computations, use the value 3.1416 for $\pi$. If you want only an approximation, use 3.142 or the mixed number $3\frac{1}{7} = \left(\frac{22}{7}\right)$. Therefore,

$$C = 3.1416 \times D \text{ (for accurate results)}$$

$$C = \frac{22}{7} \times D \text{ (for good approximations)}$$

**Example 5:** Find the circumference of these circles.

*Solution:* Find the values for the circumference. First use 3.1416, then use $\frac{22}{7} = 3\frac{1}{7}$.

*Answers:*

| Given diameter | Circumference $C = \pi \times D$ | |
|---|---|---|
| | $\pi = (3.1416)$ | $\pi = \left(\frac{22}{7}\right)$ |
| *D* | *C* | *C* |
| 5 in. | 15.708 in. | $15\frac{5}{7}$ in. |
| 3 ft. | 9.4248 ft. | $9\frac{3}{7}$ ft. |
| 1.8 cm. | 5.6549 cm. | 5.655 cm. |
| 19.16 miles | 60.193 mi. | 60.271 mi. |

Study the examples in this section. You will be suprised at how quickly you can get accurate results for many problems.

Use the radius to get another formula for the circumference of a circle. The diameter is twice the radius.

$$\text{Diameter } (D) = 2 \times R \text{ (Radius)}$$

If you replace the value of $D$ by $2 \times R$ in the formula, $C = \pi \times D$, and rearrange it, you get:

$$C = 2 \times \pi \times R \text{ (Read, "Two-pi-R")}$$

**Example 6:** Find the circumference of a circle with radius $R = 8$ feet.

*Solution:* $C = 2 \times \pi \times 8$    (Use 3.142 for $\pi$.)

*Answer:* $C = 2 \times 3.142 \times 8 = 50.3$ ft.

NOTE: Tradesmen generally use the value of the diameter, not the radius. A 10-inch circle means a circle with a 10-inch *diameter*. Similarly, a piece of half-inch round bar stock refers to a *diameter* of $\frac{1}{2}$ inch, *not radius*.

In a machine shop, the term *cutting speed* means the speed, in feet per minute, with which a cutting tool such as a twist drill, lathe tool, or milling cutter engages the object to be cut. Tables of cutting speeds for various metals and cutting tools are in machine shop manuals. Cutting speed is always given in feet-per-minute (fpm).

**Example 7:** A cast-iron piston $3\frac{1}{2}$ inches in diameter is placed in a lathe and is turning at 250 revolutions per minute (rpm). Find the cutting speed of a point (A) on the surface of the piston.

*Solution:* Since point (A) is on the surface of the piston, it is on the circumference of a circle with diameter $3\frac{1}{2}$ inches (Fig. 11-15). Find the distance, in feet, traveled by point A when the piston makes one revolution.

Distance in one revolution:

$$3.1416 \times \frac{3.5 \text{ in.} \times 1 \text{ ft.}}{12 \text{ in.}} = .9163 \text{ ft. per revolution}$$

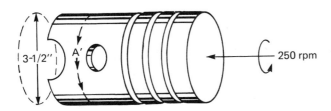

3-1/2"   A'   250 rpm

**Figure 11-15**

Now, find the distance traveled by point A in going 250 revolutions in 1 minute.

Total distance traveled in 1 minute:

250 rpm $\times$ .9163 $=$ 229.08 ft. per min.

*Answer:* The cutting speed is about 230 fpm.

You may simplify computing the cutting speed by combining the two computations into one formula. Let:

$S =$ cutting speed (in feet per minute)

$D =$ diameter of object being cut (measured in inches)

$R =$ revolutions per minute

The formula is:

$$S = \frac{\pi \times D \times R}{12} \text{ fpm}$$

When the value 3.1416 is used for $\pi$, the simplified formula is:

$$S = .2618 \, D \times R \, (\text{fpm})$$

If you need only an estimate, use either .26 $D \times R$ or $\dfrac{D \times R}{4}$ for the cutting speed.

**Example 8:** Estimate the cutting speed of a casting 8 inches in diameter, turning at 150 rpm.

*Solution:* Use the formula:

$$S = \frac{D \times R}{4}$$

$$S = \frac{8 \times 150}{4}$$

*Answer:* $\qquad S = 300$ fpm.  (estimate)

NOTE: The actual speed is slightly faster (314 fpm) since the estimate formulas give results that are less than the true value.

## The Area of a Circle

By doing a few slightly more complex experiments than that for circumference, you can find a formula for the area of a circle. When you use the radius $R$ of the circle, the area is:

$$\text{Area} = \pi \times R^2 \qquad (R^2 = R \times R)$$

Read this formula as "pi-R-square." (To *square* a number means to multiply it by itself.)

When the diameter (*D*) is used, the area is:

Area $(A) = \dfrac{\pi \times D^2}{4}$     (Read: "pi-D-square over four.")

This result, with $\dfrac{\pi}{4}$ computed to the fourth decimal place, is:

$$\text{Area } (A) = .7854D^2$$

NOTE: Memorize this last result $(A = .7854D^2)$ since the diameter is usually specified in problems, not the radius.

**Example 9:** The diameter of a circle is 8 inches. Find the area of the circle to the nearest hundredth of a square inch.

*Solution:* Use the formula $A = .7854D^2$ with $D = 8$ in., $D^2 = 8 \times 8 = 64$

*Answer:* $A = .7854 \times 64 = 50.27$ sq. in.

**Example 10:** Standard *6-inch* wrought iron pipe has an outside diameter of 6.63 inches and an inside diameter of 6.05 inches. Find the area of the ring between the two circles in Fig. 11-16.

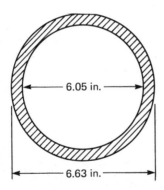

**Figure 11-16**

*Solution:* The area of the ring is the difference between the areas of the larger and smaller circles.

$$\text{Area of large circle} = .7854 \times (6.63)^2$$
$$= 34.52 \text{ sq. in.}$$
$$\text{Area of smaller circle} = .7854 \times (6.05)^2$$
$$= 28.75 \text{ sq. in.}$$
$$\text{Area of ring} = 34.52 - 28.75$$

*Answer:* Area = 5.77 sq. in.

### *Practice Problems*

Prove these answers.

**TABLE 11-1**   FINDING CIRCUMFERENCES AND AREAS OF CIRCLES

| Diameter (*D*) | $C = 3.14D$<br>Circumference (*C*) | $A = .785D^2$<br>Area (*A*) |
|---|---|---|
| 5 in. | 15.7 in. | 19.6 sq. in. |
| 3 in. | 9.42 in. | 7.06 sq. in. |
| 2.5 cm. | 7.85 cm. | 4.91 sq. cm. |
| 25 m. | 78.5 m. | 491 sq. m. |
| 4.5 yds. | 14.1 yds. | 15.9 sq. yds. |
| 14.5 mm. | 45.5 mm. | 165 sq. mm. |
| 16.4 cm. | 51.5 cm. | 211 sq. cm. |
| 240 ft. | 754 ft. | 45,216 sq. ft. |
| 82 yd. | 257 yds. | 5,278 sq. yds. |
| 1.7 mm. | 5.34 mm. | 2.27 sq. mm. |
| 3.95 m. | 12.4 m. | 12.3 sq. mm. |
| 7.85 cm. | 24.6 cm. | 48.4 sq. cm. |

**Example 11:** Some shapes are made up of both rectangles and circles. See Fig. 11-17 and find:

    (a) the distance around the metal plate (nearest tenth).

    (b) the total area of the plate (nearest tenth).

(The dotted line divides the plate into two figures—a rectangle and a half-circle or semicircle.)

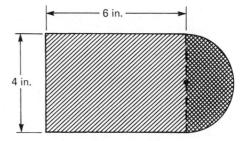

*Figure 11-17*

*Solution* (a): The distance around the figure is the sum of three sides of the rectangle and the distance around the half-circle. (Do not include the dotted line in your computations. It serves only to separate the figures.) The three sides of the rectangle add up to:

$$6 + 4 + 6 = 16 \text{ inches}$$

Find the distance around the semicircle. The width of the rectangle (4 inches) is also the diameter ($D$) of the circle. Half of the circumference of a circle with diameter 4 inches is:

$$\frac{1}{2} \times 3.1416 \times 4 = 6.2832 \text{ inches}$$

*Answer* (a): Total distance = 16 + 6.2832 = 22.2832 inches

$$= 22.3 \text{ inches (nearest tenth)}$$

*Solution* (b): Find the area of the rectangle and add the area of half a circle.

Area of rectangle = 6 × 4 = 24 square inches

Area of half-circle = $\frac{1}{2} \times .7854 \times 4^2 = 6.2832$ square inches

*Answer* (b): Total area = 24 + 6.2832 = 30.2832 square inches

$$= 30.3 \text{ square inches (nearest tenth)}$$

# PROBLEMS: Sec. 2 Circumferences and Areas of Circles

1. Find the circumference of each of these circles. Use 3.14 for $\pi$.

| Diameter ($D$) | Circumference = $\pi \times D$ |
|---|---|
| (a)  10 in. | _____ |
| (b)  21 in. | _____ |
| (c)  14 cm. | _____ |
| (d)  2.7 ft | _____ |
| (e)  14.5 m. | _____ |
| (f)  156 ft. | _____ |

2. A grinding wheel is 8 inches in diameter. How many inches will a point on the circumference of the wheel travel if the wheel makes one revolution?

3. A water tank is 12 feet in diameter. Four steel tie rods are to be wrapped around the tank. Allow 6 inches for fastening the ends of each tie rod. How many feet of tie rod are needed?

4. The equatorial radius of the earth is 6,378,388 meters.
(a) Find the equatorial diameter.
(b) If measured at the equator, how far is it around the earth?

5. Find the length of the belt connecting the pulleys shown in Fig. 11-18.

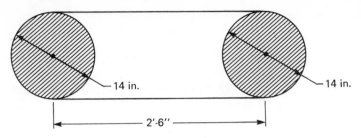

**Figure 11-18**

6. A fly is sitting on the end of the minute hand of a clock. The minute hand is 2.5 meters long. How far will the fly travel between 3:00 and 3:15 P.M.?

7. An automobile tire is 28 inches in diameter. How far will a car travel in two revolutions of the tire?

8. Find the distance around the object shown in Fig. 11-19.

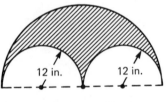

**Figure 11-19**

9. Find the perimeter of the steel plate shown in Fig. 11-20. Use $\frac{22}{7} = \pi$.

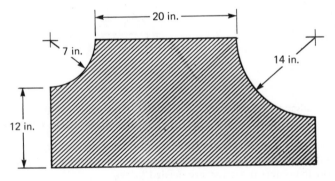

**Figure 11-20**

10. The polar radius of the earth is approximately 3,950 miles. A satellite in circular polar orbit maintains an altitude of 2,550 miles. How far does the satellite travel in making one orbit of the earth?

11. An emery wheel 6 inches in diameter is turning at 4,000 rpm. What is the rim speed in feet per minute?

12. How far will the weight W travel in four revolutions of the windlass shown in Fig. 11-21?

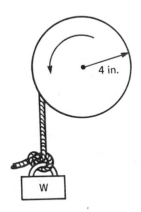

4 in.

W

**Figure 11-21**

13. Find the area of these circles. Use the formula $A = .785D^2$.

| Diameter | Area | |
|---|---|---|
| (a)  2 in. | _____ | sq. in. |
| (b)  6 cm. | _____ | sq. cm. |
| (c)  1.5 ft. | _____ | sq. ft. |
| (d) 134.5 cm. | _____ | sq. cm. |
| (e)    .00657 cm. | _____ | sq. cm. |

14. Compute the area between the circles shown in Fig. 11-22.

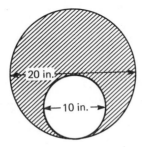

20 in.

10 in.

**Figure 11-22**

15. A circular sheet of aluminum 20 inches in diameter increases by .005 inches in diameter when the temperature is increased 180°F.

(a) What is the increase in area of the circular plate?

(b) What is the percent increase in area?

16. What is the waste in cutting the largest possible circular plate from a square piece of sheet steel 18 inches on a side?

17. What is the waste in cutting the largest circular piece of brass from a rectangular sheet 15 by 19 centimeters?

18. The side of a square is 10 inches. Find:

(a) The area of the largest inscribed circle.

(b) The waste in cutting out the circle in part (a).

19. An archway is shown in Fig. 11-23. What is the area of the archway if it is a semicircle on top of a rectangle?

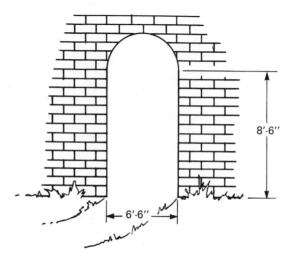

8'-6"

6'-6"

*Figure 11-23*

20. For the hydraulic press in Fig. 11-24 the ratios are in direct proportion:

$$\frac{\text{Lifting force}}{\text{Applied force}} = \frac{\text{Area of large piston}}{\text{Area of small piston}}$$

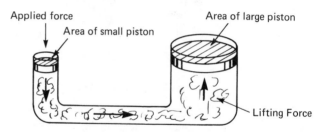

Applied force

Area of small piston

Area of large piston

Lifting Force

*Figure 11-24*

Find the lifting force when the applied force and the diameters of the large and small pistons are: (Use 3.14 for $\pi$.)

| Applied Force | Diameter of Small Piston | Diameter of Large Piston |
|---|---|---|
| (a)  25 lb. | 2 in. | 24 in. |
| (b)  150 lb. | 1 in. | 5 in. |

21.  A child rides a giant Ferris wheel that is 120 feet in diameter.
(a) How far does a child ride when the wheel makes 25 revolutions?
(b) How fast does the child travel (in feet per second) if it takes the wheel 10 seconds to make one revolution?
(c) What is the child's speed in miles per hour?
(HINT:  Use proportions and note that 88 ft./sec. = 60 mph.)

22.  Study the shape shown in Fig. 11-25. Can you find the perimeter (distance around) and the area from the single given value?

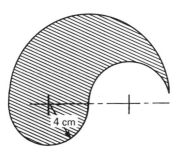

**Figure 11-25**

23.  The earth travels in an almost circular orbit around the sun. The average radius is 92,900,000 miles or 149,500,000 kilometers. Find the distance in (a) miles and (b) kilometers traveled by the earth in one year.

24.  Find the area of the steel plate in Fig. 11-26. The diameter of each of the four holes is one-half the radius of the large circle.

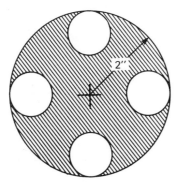

**Figure 11-26**

## SEC. 3   THE TRIANGLE

A *triangle* is a closed three-sided figure that looks like Fig. 11-27.

### The Perimeter of a Triangle

It is easy to find the distance around a triangle. The distance around the triangle is called the perimeter of the triangle.

*The perimeter of a triangle is equal to the sum of the lengths of the three sides.*

**Example 12:** Find the perimeter of the triangle shown in Fig. 11-28.

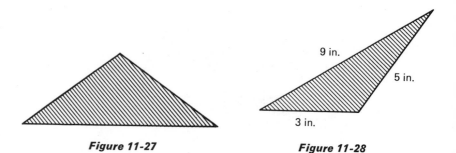

|              |              |
|:------------:|:------------:|
| *Figure 11-27* | *Figure 11-28* |

*Solution:* The perimeter (*P*) is: $P = 3 + 5 + 9 = 17$ inches.

*Answer:* $P = 17$ inches

The word *triangle* means three angles, and all triangles have three interior angles. If one of the angles of a triangle makes a square corner, then it is a 90° angle and is called a *right angle*. (There are no left angles.) A triangle with a right angle is called a *right triangle*. Figure 11-29 shows a right triangle.

An interior angle that is less than 90° is called an *acute angle*. Figure 11-30 shows an acute angle.

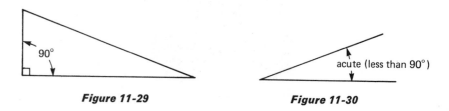

|              |              |
|:------------:|:------------:|
| *Figure 11-29* | *Figure 11-30* |

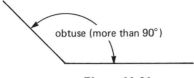

**Figure 11-31**

An interior angle that is more than 90° is called an *obtuse angle*. Figure 11-31 shows an obtuse angle.

There is a very important and useful property of the three interior angles of any triangle. This property is:

*The sum of the three interior angles is equal to 180°*

This remarkable result is proved in plane gemotery and will not be proved here.

To measure the interior angles of a triangle, you need a *protractor*. If you do not know how to use a protractor, see page 412, Chapter 13.

Angles are usually measured in *degrees* (°), *minutes* ('), and *seconds* ("). The relationship between these three measures is:

$$1° = 1 \text{ degree} = 60 \text{ minutes} = 60'$$
$$1' = 1 \text{ minute} = 60 \text{ seconds} = 60''$$
$$1° = 1 \text{ degree} = 3{,}600 \text{ seconds} = 3{,}600''$$

Read 45° 18′ 56″ "forty-five degrees, eighteen minutes, fifty-six seconds."

**Example 13:** If one of the acute angles of a *right triangle* is 56°34′, find the value of the other acute angle. See Fig. 11-32.

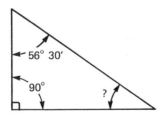

**Figure 11-32**

*Solution:* The sum of the two given angles is:

$$90° + 56°34' = 146°34'$$

Use the fact that the sum of the interior angles always

equals 180°. The value of the third interior angle is the difference:

$$180°$$
$$-146°34'$$

Borrow 1 degree (60') from 180°.
$$179°60'$$
$$-146°34'$$

*Answer:*        33°26'

Although the right triangle is most often used in technology, you should know that there are other triangles. They are: (See Fig. 11-33)

*The equilateral triangle that has all sides equal.*
*The isosceles triangle that has two sides equal.*
*The scalene triangle that has three unequal sides.*

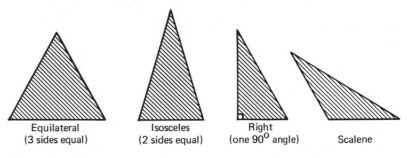

Equilateral          Isosceles          Right          Scalene
(3 sides equal)      (2 sides equal)    (one 90° angle)

*Figure 11-33*

## The Area of a Triangle

Find the area of a *right* triangle. Figure 11-34 shows that the area of *a right triangle is half of* the area of *a rectangle.* Because the area of a rectangle is the product of its length and width, the area of a right triangle is half of this value:

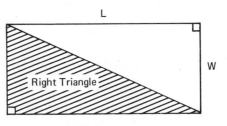

L

W

Right Triangle

*Figure 11-34*

*Area of Right Triangle* $= \dfrac{1}{2} \times$ *length* $\times$ *width*

When working with triangles, the words *base* (*b*) and *height* (*h*) are often used instead of length and width (Fig. 11-35). Using these terms, the formula for the area of a triangle is:

$$Area = \frac{1}{2} \times base \times height$$

$$Area = \frac{1}{2} \times b \times h = \frac{b \times h}{2}$$

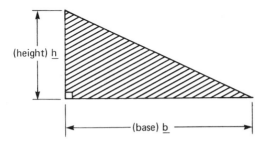

(height) h

(base) b

**Figure 11-35**

**Example 14:** Find the area of a right triangle with a base of 4 inches and a height of 7 inches.

*Solution:* $A = \dfrac{1}{2} \times 4 \times 7 = 14$ square inches

*Answer:* 14 square inches

Now find the area of a more general triangle. The shaded part of Figure 11-36 shows a triangle inside of a rectangle. The base of the triangle is the length of the rectangle. The height of the triangle (the distance from the base to the top of the triangle) is equal to the width of the rectangle. The triangle is divided into two right triangles by the (vertical) dashed line from the top to the base. The dashed line also divides the rectangle

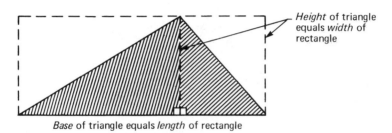

*Height* of triangle equals *width* of rectangle

*Base* of triangle equals *length* of rectangle

**Figure 11-36**

into two smaller rectangles. The area of the small right triangle is half the area of the small rectangle. The area of the large right triangle is half the area of the large rectangle. From this it is possible to state the:

### Rule for Finding the Area of a Triangle

*The area of the triangle equals half the area of the rectangle.*

Therefore, *Area of any triangle* $= \dfrac{1}{2} \times base \times height$.

The height of a triangle is the (vertical) distance from the base to the top. See Fig. 11-37.

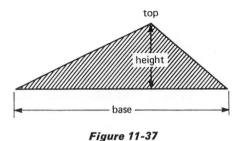

**Figure 11-37**

**Example 15:** Find the area for each of the triangles shown in Figs. 11-38 and 11-39.

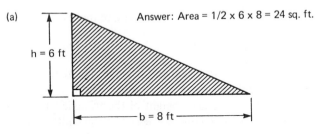

(a)    Answer: Area = 1/2 x 6 x 8 = 24 sq. ft.

h = 6 ft

b = 8 ft

**Figure 11-38**

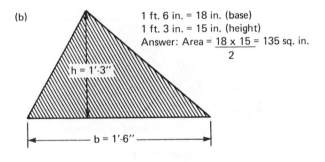

(b)

1 ft. 6 in. = 18 in. (base)
1 ft. 3 in. = 15 in. (height)
Answer: Area = $\dfrac{18 \times 15}{2}$ = 135 sq. in.

h = 1'-3"

b = 1'-6"

**Figure 11-39**

*Solution* (a): Use the formula $A = \dfrac{1}{2} \times$ base $\times$ height

*Answer* (a): Area $= \dfrac{1}{2} \times 6 \times 8 = 24$ sq. ft.

*Solution* (b): Use $A = \dfrac{1}{2} \times b \times h$

*Answer* (b): Area $= \dfrac{18 \times 15}{2} = 135$ sq. in.

When the height is not known, the area is more difficult to find. After you study the next section on finding the square root of a number, you can find the area when you know only the lengths of the sides.

# PROBLEMS: Sec. 3 Perimeters and Areas of Triangles

1. Find the perimeter of the given triangles.

| | Side one | Side two | Side three | Perimeter |
|---|---|---|---|---|
| (a) | 2 in. | 3 in. | 7 in. | _____ |
| (b) | 5 cm. | 8 cm. | 2 cm. | _____ |
| (c) | 1 ft. 6 in. | 2 ft. 7 in. | 5 ft. 8 in. | _____ |
| (d) | 1 m. 45 cm. | 3 m. 56 cm. | 7 m. 38 cm. | _____ |
| (e) | 1.09 in. | 4.88 in. | 3.08 in. | _____ |
| (f) | .96 in. | 1.85 in. | .77 in. | _____ |

2. Find the value of the missing angle. (Note: the sum of the three interior angles equals 180°.)

| | Angle A | Angle B | Angle C |
|---|---|---|---|
| (a) | 30° | 60° | _____ |
| (b) | 45° | 45° | _____ |
| (c) | 25° | 35° | _____ |
| (d) | 45° 35′ | 19° 27′ | _____ |
| (e) | 35° 15′ | 35° 15′ | _____ |
| (f) | 2° 26′ 45″ | 35° 28′ 54″ | _____ |

3. Carl measured the three angles of a triangle and got:

$$\text{Angle } A = 56° \, 28'$$
$$\text{Angle } B = 47° \, 39'$$
$$\text{Angle } C = 74° \, 53'$$

Are his measurements correct?

4. One side of an equilateral triangle measures 2 feet 8 inches. What is its perimeter?

5. Find the areas of the triangles.

| Base ($b$) | Height ($h$) | Area $= \frac{1}{2} \times b \times h$ |
|---|---|---|
| (a) 4 in. | 7 in. | _____ |
| (b) 3 cm. | 14 cm. | _____ |
| (c) 1.3 ft. | 2.6 ft. | _____ |
| (d) 1.55 in. | 4.75 in. | _____ |
| (e) 235.8 cm. | 456.9 cm. | _____ |
| (f) .07 in. | .08 in. | _____ |

**6.** A school play yard is in the shape of a right triangle with base 156 feet 8 inches (156.7 feet), and height 235 feet 9 inches (235.8 feet). What will it cost to enclose all three sides of the yard if fencing costs $6.50 per foot?

**7.** Number 4 Stubs' gauge sheetsteel weighs 9.71 pounds per square foot. Find the weight of a triangular plate with a base of 4.5 feet and a height of 3.6 feet.

**8.** A right triangle with base 4 inches and height 10 inches is cut from a brass plate that weighs 3.08 pounds per square foot. What is the weight of the triangle?

**9.** A triangular building lot has a base 178 feet and a depth of 136 feet. If the property is worth $16,500 per acre, what is the cost of the lot to the nearest dollar?

**10.** Find the perimeter and area of the piece of sheet metal shown in Fig. 11-40.

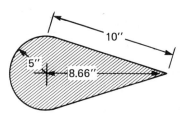

**Figure 11-40**

## SEC. 4  FINDING THE SQUARE ROOT OF A NUMBER

You have already seen that certain formulas require a number to be squared. Such a formula is:

$$A = .7854 D^2$$

in which the value $D$ is to be squared; $D^2 = D \times D$.

You will now learn how to reverse the process of squaring. This reversal is called: *Finding the square root of a number*. Finding the *square root* of a number means: *Find a number which when multiplied by itself gives the original value.*

For example, to find the *square root* of 16 means to find a number that when multiplied by itself (squared) is 16. Such a number is 4, because $4 \times 4 = 16$. Therefore, *4 is the square root of 16.*

The symbol $\sqrt{\phantom{x}}$, called a *square root sign* or *radical sign,* is used for the instruction: *Find the square root of a number.* When the symbol is placed over a number like $\sqrt{9}$, it means to find the square root of 9, which is 3.

$$\sqrt{9} = 3 \text{ because } 3^2 = 9$$

Look at Table 11-2. You probably already know the squares and square roots of many whole numbers. Study Table 11-2 and memorize any of the values you don't know.

**TABLE 11-2**

| *Square* | *Square root* |
|----------|---------------|
| $1 = 1^2$ | $1 = \sqrt{1}$ |
| $4 = 2^2$ | $2 = \sqrt{4}$ |
| $9 = 3^2$ | $3 = \sqrt{9}$ |
| $16 = 4^2$ | $4 = \sqrt{16}$ |
| $25 = 5^2$ | $5 = \sqrt{25}$ |
| $36 = 6^2$ | $6 = \sqrt{36}$ |
| $49 = 7^2$ | $7 = \sqrt{49}$ |
| $64 = 8^2$ | $8 = \sqrt{64}$ |
| $81 = 9^2$ | $9 = \sqrt{81}$ |
| $100 = 10^2$ | $10 = \sqrt{100}$ |
| $121 = 11^2$ | $11 = \sqrt{121}$ |
| $144 = 12^2$ | $12 = \sqrt{144}$ |

Use Table 3 on page 470 in the Appendix to find the approximate square roots of most whole numbers from 1 to 500. (These values are correct to the nearest .0001, and are adequate for most shop computations.)

If the square root of a number is not in the table, or when you need more accurate results, you must be able to compute the square root. In the following material you will study a method for computing square roots. The method is called *the method of averaging,* or *Newton's Method.*

### Finding the Square Root by Averaging (Newton's Method)

Sir Isaac Newton (1642–1727) discovered a method for finding the square root of a number. This method is explained in Examples 16 and 17. This method uses the fact that:

*When a number is divided by its square root, the divisor and quotient are equal.*

For instance, the square root of 64 is 8. If you divide 64 by 8, the divisor (8) and the quotient (8) are equal:

$$
\begin{array}{r}
8 \\
8\,\overline{)64} \\
64 \\
\hline
0
\end{array}
$$

Because $8 \times 8 = 64$, $8 = \sqrt{64}$

**Example 16:** Use this idea and find *equal* divisor and quotient for the square root of 14, correct to nearest hundredth.

*Solution:* You see that $\sqrt{14}$ is not a whole number because $\sqrt{9} = 3$ and $\sqrt{16} = 4$. Because 14 is between 9 and 16, the square root $\sqrt{14}$ must be between 3 and 4.

Guess the square root of 14. Try 3.5. If this is a good guess, then when you divide 14 by 3.5, the result (quotient) should be very close to the divisor 3.5.

$$
\begin{array}{r}
4.00 \\
3.5\,\overline{)14.0.00} \\
14\,0\,0 \\
\hline
0
\end{array}
$$

Evidently, 3.5 was not a good guess, since divisor and quotient are not equal. What should you try for your next guess?

For the next guess take the *average* of the divisor and quotient from the first division.

$$
\text{Second guess} = \frac{3.50 + 4.00}{2}
$$

$$
= 3.75 \text{ (average)}
$$

This second guess (3.75) is your next trial divisor.

$$
\begin{array}{r}
3.73 \\
3.75\,\overline{)14.00.00} \\
11\,25 \\
\hline
2\,75\,0 \\
2\,62\,5 \\
\hline
12\,50 \\
11\,25 \\
\hline
1\,25
\end{array}
$$

The second quotient (3.73) is close to the value of the divisor (3.75), and you are close to the value of the square root. To find your next guess, average the divisor (3.75) and quotient (3.73) to get:

$$\text{Third guess} = \frac{3.75 + 3.73}{2}$$

$$= 3.74 \text{ (average)}$$

Divide 14 by this third guess:

```
            3.74
      _____
3.74 )14.00.00
       11 22
       _____
        2 78 0
        2 61 8
        _____
          16 20
          14 96
          _____
           1 24  (remainder less than half of 374)
```

Here both divisor and quotient are 3.74. Therefore, the square root of 14 (correct to the nearest hundredth) is:

*Answer:* $\sqrt{14} = 3.74$

*Check:* $3.74 \times 3.74 = 13.9876$

$$= 13.99 \text{ (nearest 100th)}$$

Now you can see why this method of finding a square root is called the *averaging* method. First, you guess a trial divisor. Then, you get the next trial divisor by *averaging* the divisor and quotient of the preceding division, and so on.

**Example 17:** Use the method of averaging to find the square root of 150 correct to the nearest hundredth.

*Solution:* Because 150 is larger than the square of 12 ($144 = 12^2$) and less than the square of 13 ($169 = 13^2$), you decide that $\sqrt{150}$ is between 12 and 13. Try 12.5 as a first guess.

```
             1 2.0
       _____
12.5 )150.0 0
        125
        _____
         25 0
         25 0
         _____
            0
```

Because divisor and quotient are not equal, try again.

$$\text{Second guess} = \frac{12.5 + 12.0}{2}$$

$$= 12.25 \text{ (average)}$$

Divide 150.00 by this second guess:

$$
\begin{array}{r}
12.24 \\
12.25\overline{)150.00\,00} \\
\underline{122\,5} \\
27\,50 \\
\underline{24\,50} \\
3\,00\,0 \\
\underline{2\,45\,0} \\
55\,00 \\
\underline{49\,00} \\
6\,00 \text{ (remainder less than half of divisor)}
\end{array}
$$

When the divisor and quotient are averaged, the value of 12.25 is repeated. Therefore, you find that $\sqrt{150.00} = 12.25$ to nearest hundredth. (Check: $12.25^2 = 150.0625$)

*Answer:* $\sqrt{150} = 12.25$ (nearest hundredth)

Use the *method of averages* to prove these results.

    (a) $\sqrt{345} = 18.57$     Use 18.5 as first guess.
    (b) $\sqrt{729} = 27.00$     Use 26 as first guess.
    (c) $\sqrt{6.89} = 2.62$     Use 2.5 as first guess.

## PROBLEMS: Sec. 4 Square Roots

**1.** Use the averaging method of Newton to find the square roots, correct to nearest hundredth.

| Number | Square root |
|--------|-------------|
| (a)   23 | _____ |
| (b)   56 | _____ |
| (c)   178 | _____ |
| (d)  275 | _____ |
| (e)    1.6 | _____ |

**2.** Find the square root of each of these numbers by using Newton's method.

| Number | Square root |
|--------|-------------|
| (a)  45.0 | _____ |
| (b) 150.0 | _____ |
| (c)   2.6 | _____ |
| (d)   0.097 | _____ |
| (e)   0.0075 | _____ |
| (f)    0.156 | _____ |
| (g)  10.4 | _____ |

3. Find the square roots.
   (a) $\sqrt{156}$          (b) $\sqrt{2.45}$          (c) $\sqrt{.0078}$          (d) $\sqrt{.0035}$

4. Each of these computations involve finding a square root. Find:
   (a) $12.5\sqrt{89}$          (b) $\dfrac{456}{\sqrt{137}}$          (c) $\sqrt{23}\cdot\sqrt{42}$          (d) $\dfrac{\sqrt{69}}{\sqrt{27}}$

5. Which is larger (a) or (b)?
   (a) $\sqrt{28 + 57}$     or     (b) $\sqrt{28} + \sqrt{57}$
      (Do addition first)          (Do square root first)

6. Which is smaller (a) or (b)?
   (a) $\sqrt{178 - 92}$     or     (b) $\sqrt{178} - \sqrt{92}$
      (Do subtraction first)          (Do square root first)

7. Verify that: (Use table of square roots)
   (a) $\sqrt{38}\cdot\sqrt{96} = \sqrt{(38)(96)}$          (b) $\dfrac{\sqrt{78}}{\sqrt{53}} = \sqrt{\dfrac{78}{53}}$

8. Find the square roots.
   (a) $\sqrt{69.7}$          (b) $\sqrt{.0036}$          (c) $\sqrt{10.7}$
   (d) $\sqrt{.00089}$          (e) $\sqrt{27.5}$          (f) $\sqrt{175}$
   (g) $\sqrt{.0000125}$          (h) $\sqrt{12.6}$          (i) $\sqrt{75,000}$

# SEC. 5    THE PYTHAGOREAN THEOREM AND APPLICATION OF THE SQUARE ROOT

One of the oldest and most famous theorems from geometry is the Pythagorean theorem. This theorem deals with right triangles only. Figure 11-41 shows a right triangle with the sides labeled $a$ (*base*), $b$ (*height*), and the slant side $c$. The slant side of a right triangle is called the *hypotenuse*.

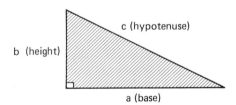

b (height)

c (hypotenuse)

a (base)

**Figure 11-41**

You can do an interesting experiment with right triangles. Draw a right triangle with base of 4 units, and height of 3 units. Draw the hypotenuse and label it $c$. Using the sides of 4 units and 3 units, construct squares with area 16 and 9 square units, respectively. See Fig. 11-42.

Construct a square of side length $c$-units-by-$c$-units, using the hypotenuse as one side of the square. See Fig. 11-43. Divide this square into smaller squares the same size as the small squares on the other two sides.

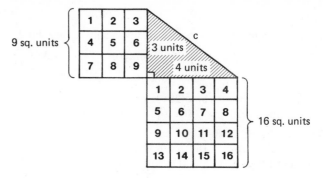

*Figure 11-42*

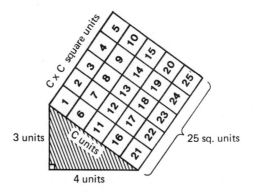

*Figure 11-43*

Count the number of small squares in this big square. If you have measured carefully, you will have EXACTLY 25 squares.

The result of this experiment is the:

### Pythagorean Theorem

*The sum of the squares of the base and height equals the square of the hypotenuse.*

Thus, $16 + 9 = 25$

This may be stated another way, using the letters $a$, $b$, and $c$ (Fig. 11-44): $a^2 + b^2 = c^2$

This result is true for right triangles ONLY. Do not apply it to other kinds of triangles. Using this result, you can find the value of any one side of a right triangle when you know the values of the other two sides.

**Example 18:** Find the length of the hypotenuse if the base is 12 and the height is 5. See Fig. 11-45.

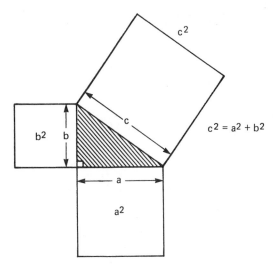

**Figure 11-44**

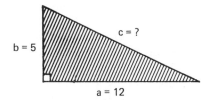

**Figure 11-45**

*Solution:* Use the Pythagorean theorem:

$$(\text{base})^2 + (\text{height})^2 = (\text{hypotenuse})^2.$$

Put in the given values:

$$(12)^2 + (5)^2 = (c)^2$$
$$144 + 25 = c^2$$
$$169 = c^2$$

You know the value of the square of the hypotenuse is 169, therefore, the hypotenuse is the square root of 169.

$$c = \sqrt{169}$$

*Answer:* $c = 13$ (from tables)

**Example 19:** Find the base ($a$) if the height ($b$) is 6 inches and the hypotenuse ($c$) is 10 inches. See Fig. 11-46.

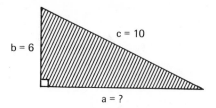

**Figure 11-46**

*Solution:* Use the Pythagorean theorem.

$$a^2 + 6^2 = 10^2$$
$$a^2 + 36 = 100$$

Decrease both sides by 36 to find value for $a^2$.

$$a^2 = 100 - 36 = 64$$

The value of $a$ is the square root of 64.

$$a = \sqrt{64}$$

*Answer:*     $a = 8$ inches

The Pythagorean theorem is useful when you are working with right triangles. Here is an example of how this theorem is used in practice.

**Example 20:** Figure 11-47 is a sketch of a roof truss. Find the length of the rafter labeled $c$: span 24 feet; pitch 1:6 (add 1 ft. for overhang)

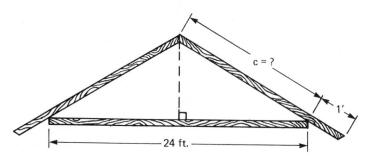

**Figure 11-47**

*Solution:* A pitch ratio of one-to-six means that the rise is one-sixth of the span: $\frac{1}{6}$ of $24 = 4$ ft. The value of 4 feet is the height of the right triangle. The base of the triangle equals $\frac{1}{2}$ of the span:

$$\text{base} = \frac{1}{2} \times 24 = 12 \text{ ft.}$$

Find the length of rafter ($c$) (not including overhang) from the Pythagorean theorem:

$$(12)^2 + (4)^2 = (c)^2$$
$$144 + 16 = c^2$$
$$160 = c^2$$

The value of $c$ is the square root of 160.

$$c = \sqrt{160}$$
$$c = 12.65 \text{ ft.}$$

Add 1 foot for overhang:

$$\text{Total length} = 13.65 \text{ ft.}$$

NOTE: Carpenters usually measure in feet and inches rather than decimal parts of a foot. Find, in inches, the equivalent of .65 feet. Do this by finding .65 of 12 inches.

$$.65 \times 12 \text{ in.} = 7.8 \text{ in.}$$
$$= 8 \text{ in. (nearest inch)}$$

*Answer:* Rafter = 13 ft. 8 in.

## Areas of More General Triangles

In Section 1 the area of a triangle is given by the formula:

Area of triangle $= \frac{1}{2} \times$ (base) $\times$ (distance to top of triangle)

See Fig. 11-48.

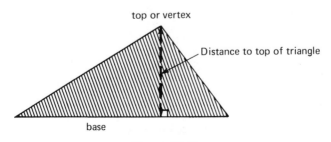

top or vertex

Distance to top of triangle

base

**Figure 11-48**

In many cases the distance from the base to top is either not available or is very difficult to find. (You can find this distance by using trigonometry. See Chapter 14.)

You need to know only the lengths of the three sides to find the area of a triangle. Heron of Alexandria, a Greek geometer and engineer (150 B.C.) developed a formula for the area of a triangle when only the lengths of the sides are known. The formula is known as:

*Heron's formula for the area of a triangle* (See Fig. 11-49):

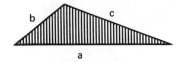

**Figure 11-49**

Area of a triangle $= \sqrt{s(s-a)(s-b)(s-c)}$

where $s = \dfrac{a+b+c}{2} = \dfrac{1}{2}$ of the perimeter

$(s-a) = \dfrac{1}{2}$ of perimeter minus the side $a$.

$(s-b) = \dfrac{1}{2}$ of perimeter minus the side $b$.

$(s-c) = \dfrac{1}{2}$ of perimeter minus the side $c$.

NOTE: *All additions, subtractions, and multiplications are done under the square root sign BEFORE the square root is found.*

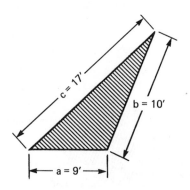

**Figure 11-50**

**Example 21:** Use Heron's formula to find the area of the triangle shown in Fig. 11-50.

*Solution:* Perimeter of triangle $= 9 + 17 + 10 = 36$ ft. One-half of perimeter $= 18$ feet $= s$

$$s - a = 18 - 9 \ = 9$$
$$s - b = 18 - 17 = 1$$
$$s - c = 18 - 10 = 8$$
$$A = \sqrt{s(s-a)(s-b)(s-c)} = \sqrt{18 \times 9 \times 1 \times 8} = \sqrt{1{,}296}$$

*Answer:* Area $= \sqrt{1{,}296} = 36$ sq. ft.

**Example 22:** An aerial photograph shows a lake in the approximate shape of a triangle. See Fig. 11-51. The lake has sides of 6.5 miles, 4.5 miles, and 9.8 miles. Find the area of the lake to the nearest square mile.

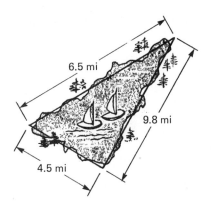

6.5 mi

9.8 mi

4.5 mi

**Figure 11-51**

*Solution:* Use Heron's formula. The three sides are:

$a = 4.5$ mi., $b = 6.5$ mi., $c = 9.8$ mi.

Perimeter $= 4.5 + 6.5 + 9.8 = 20.8$ miles

$s = \dfrac{1}{2}$ of perimeter $= 10.4$

$A = \sqrt{10.4 \times 5.9 \times 3.9 \times .6}$

$A = \sqrt{143.58}$

*Answer:*   $A = 12$ square miles (to nearest square mile)

## PROBLEMS: Sec. 5 The Pythagorean Theorem and Applications of the Square Root

1. Use the Pythagorean theorem to find the missing values.

   NOTE: $c = \sqrt{a^2 + b^2}$, $a = \sqrt{c^2 - b^2}$, $b = \sqrt{c^2 - a^2}$

   |  | Side a | Side b | Side c (*hypotenuse*) |
   |---|---|---|---|
   | (a) | 3 in. | 4 in. | _____ |
   | (b) | 5 cm. | 12 cm. | _____ |
   | (c) | 6 ft. | 11 ft. | _____ |
   | (d) | 4.5 in. | 7.5 in. | _____ |
   | (e) | 1 ft. 3 in. | 2 ft. 6 in. | _____ |
   | (f) | _____ | 6 in. | 13 in. |
   | (g) | _____ | 1.73 cm. | 2.00 cm. |
   | (h) | 4.5 ft. | _____ | 11.6 ft. |
   | (i) | .19 cm. | _____ | 1.20 cm. |

2. The transmitter tower for station KPUT is 256 feet tall. Four support cables are located 38 feet from the base. What is the length of one of the support cables if it is attached to the top of the tower?

3. A ventilating shaft is to be drilled to the tunnel shown in Fig. 11-52. What is the length of the shaft, to the nearest foot?

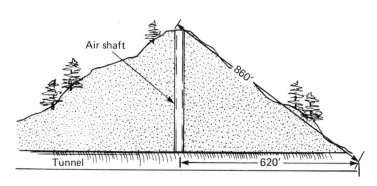

**Figure 11-52**

4. A kite flying 180 feet high is 78 feet from a boy holding the kite string at a point directly under the kite. How much kite string is out?

5. A right triangle with base 15 inches and height 20 inches is cut from a sheet of brass that weighs 2.8 pounds per square foot. What is the weight of the triangle?

6. Find the area of the support plate shown in Fig. 11-53.

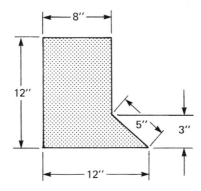

**Figure 11-53**

7. Use Heron's formula to find the areas of each of these given triangles.

| Side a | Side b | Side c | Area |
|---|---|---|---|
| (a) 20 cm. | 14 cm. | 12 cm. | _____ sq. cm. |
| (b) 10 cm. | 7 cm. | 5 cm. | _____ sq. cm. |
| (c)  6 in. | 6 in. | 6 in. | _____ sq. in. |
| (d) 19 ft. | 9 ft. | 10.5 ft. | _____ sq. ft. |
| (e)  3.25 in. | 1.25 in. | 3.00 in. | _____ sq. in. |

8. A surveyor runs a closed traverse that forms a triangle. The lengths of the sides are 225, 320, and 415 feet. What is the area of the triangular piece of ground?

9. A hole 2 inches in diameter is cut in the sheet of tin shown in Fig. 11-54. How much area remains?

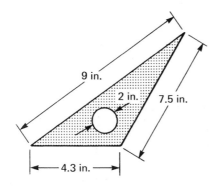

**Figure 11-54**

10. How long must a piece of steel be to make the brace shown in Fig. 11-55?

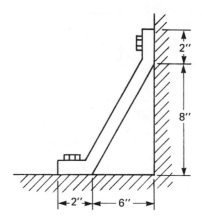

*Figure 11-55*

**11.** Find the area of the irregular shape shown in Fig. 11-56.

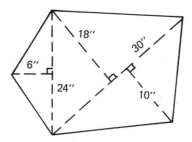

*Figure 11-56*

**12.** A ship is lost at sea. Three other ships are at corners of a triangle that encloses the lost ship, and they are separated by distances of 90, 75, and 45 miles. How many square miles must be searched in an attempt to find the missing ship?

**13.** Two ships, the *S.S. Queequeg* and the *Revenge IV*, leave port at 12:00 P.M. The *S.S. Queequeg* sails due north at 8 knots and the Revenge due east at 17 knots. How far apart are the two ships at the end of three hours?

**14.** What is the diameter of round stock if a square head bolt 1.5 centimeters on a side can be milled from it?

**15.** Test this rule-of-thumb for finding the diagonal of a square. (The diagonal is the distance between opposite corners.) "To find the approximate diagonal of a square, multiply the length of a side by 10, subtract 1% of this product, and divide the remainder by 7." Test this rule for a square of side 10 centimeters against the Pythagorean theorem.
(a) What is the difference in the two values?

(b) Divide the answer to part (a) by the result you get by using the Pythagorean theorem and change to a percent. (This is the percent error.)

Do you agree this method is both quick and accurate for many shop computations?

## SEC. 6 SURFACE AREAS AND VOLUMES

Have you ever cut a cardboard box into pieces by cutting it at each fold? Did you notice that the six pieces were just rectangles? Figure 11-57 shows a box before and after it is cut apart.

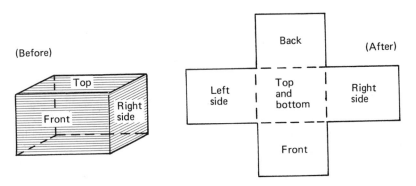

**Figure 11-57**

### Rectangular Surfaces

When you cut the box apart at the corners, you get six rectangles, (four sides and a top and bottom). If you find the area of all six rectangles, and add the areas together, you have the *surface area* of the box.

The *surface area* of an object, such as a box, a ball, or a cone, is the area of the outer surface that covers the figure. *The total surface area of an object is the sum of the areas of the parts.*

**Example 23:** Find the surface area of the box shown in Fig. 11-58.

*Solution:* When the box is cut apart, notice that certain parts match. The top and bottom match. The front and back match, and the right and left sides match. Find the surface area by adding:

Surface Area $= 2 \times$ (area of front) $+$

$2 \times$ (area of top) $+$

$2 \times$ (area of one end)

$= (12 + 12 + 8)$ square feet

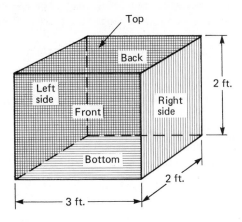

**Figure 11-58**

*Answer:* Surface Area = 32 square feet

When finding the surface area of any object, check to see if any of the parts of the surface are missing. For instance, the surface area of an *open* box should not include the top. If one end is missing, do not include its area.

**Example 24:** A plasterer always estimates the number of square feet of surface to be plastered. If four walls and the ceiling area are to be plastered, find the number of square feet of surface area. See Fig. 11-59.

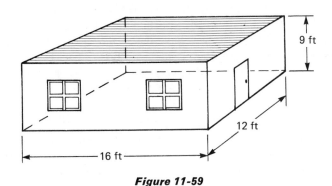

**Figure 11-59**

*Solution:* The floor is not included. The remaining surface area is the sum of the area of four walls and the area of the ceiling.

Area of four walls = $2 \times (16 \times 9) + 2 \times (12 \times 9)$
$$= 288 + 216$$
$$= 504 \text{ sq.ft.}$$

Area of ceiling $= 16 \times 12 = 192$ sq. ft.

Total surface area $= 504 + 192 = 696$ sq. ft.
(includes doors and windows)

Allow 100 sq. ft. for windows and doors.

Area to be plastered $= 696 - 100 = 596$ sq. ft.

*Answer:* About 600 square feet of area is to be plastered.

## Cylinders

Use the idea of cutting an object apart to find the surface area of a tin can. Figure 11-60 is a sketch of an ordinary tin can. When the top and bottom of the can are removed, you see that they are matching circular discs.

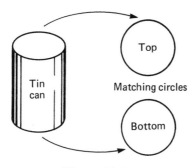

**Figure 11-60**

Now, cut down the side of the can and flatten it out. (See Fig. 11-61.) What is the shape of this part of the can after it is unrolled? You see that it is a rectangle.

The height of the rectangle is equal to the height of the can. The width

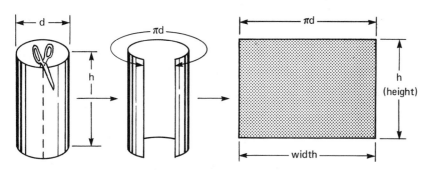

**Figure 11-61**

of the rectangle is equal to the distance around the circular top. This distance is:

Width of rectangle = circumference of top

$$c = \pi d \quad (d = \text{diameter of top})$$

The surface area of a tin can, including top and bottom, is:

Area of surface = 2 × (area of top) + (area of rectangle)

$$= 2 \times (.7854d^2) + (\pi \times d \times h)$$

**Example 25:** Find the surface area of a can with diameter 3 inches and height 6 inches.

*Solution:* Use the formula: $2 \times (.7854d^2) + (\pi \times d \times h)$

Area of top $= .7854d^2 = .7854 \times 9 = 7.069$

Area of rectangle = (width) × (height)

$$= (\pi \times d) \times (6)$$

$$= (3.142 \times 3) \times 6 = 56.556 \text{ sq. in.}$$

Surface area $= 2 \times 7.069 + 56.556$

*Answer:*  $= 70.69$ sq. in. (nearest hundredth)

Circular tubes, such as the tin can (without the ends) are called *circular cyclinders*. The surface area of a circular cylinder which is open at both ends is:

Surface area = (height) × (circumference of top)

$$= \text{(height)} \times (\pi d) \quad (d = \text{diameter of circle})$$

**Example 26:** An air duct in the shape of a circular cylinder is 17 feet 6 inches long and has a diameter of 30 inches. The duct is wrapped with insulation material that costs $1.80 a square foot. What is the cost of the insulation? (Allow 2% for waste.)

*Solution:* The amount of insulation material is equal to the number of square feet of surface area of the air duct plus 2% of surface area.

Surface area of the duct = (length) × ($\pi$ × diameter)

$$= (17.5 \text{ ft.}) \times (3.142 \times 2.5 \text{ ft.})$$

$$= 137.46 \text{ sq. ft.}$$

Surface area (to nearest square foot) = 137 sq. ft.

Allow 2% of surface area for waste of material.

Waste = 2% of 137 = 2.74 sq. ft.

$$= 3 \text{ sq. ft. (nearest sq. ft.)}$$

Total number of square feet of insulation = 140 sq. ft.

Cost at \$1.80 per sq. ft. = 140 × \$1.80 = \$252.00

*Answer:* Cost = \$252.00

## Cones

There are many geometric objects of different sizes and shapes. One that is used quite often in industrial and technical work is the *cone*. A cone is a familiar object, and it is used to hold everything from strawberry ice cream to feed grain for hogs. The conical shape is also quite common in nature. The pine cone and some sea shells are good examples. Figure 11-62 shows the cone in comparison with a circular cylinder. The sharp point of a cone is called the *vertex*.

If a cone is cut from the base to the vertex along a straight line and unrolled, the resulting surface looks like a piece of pie. Figure 11-63 shows the result of unrolling a cone. The unrolled surface is not one of the

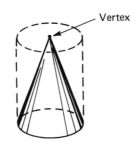

**Figure 11-62**

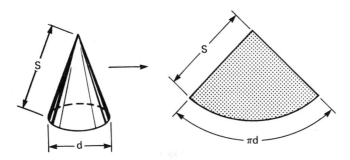

**Figure 11-63**

figures you have already studied. Scientists use two formulas for finding the surface area of a cone. The first formula is in terms of the *diameter* of the base (*d*) and the *slant* height (*S*) of the cone. (The slant height is shown in Fig. 11-63 by the letter *S*.) The formula for the surface area is:

$$\text{Area} = \frac{1}{2} \times \text{(perimeter of base)} \times \text{(slant height)}$$

$$\text{Area} = \frac{\pi d \times S}{2}$$

The second formula is in terms of the radius of the base (*r*) and the altitude (h) (distance from vertex to base). The formula for the surface area is:

$$\text{Surface area} = \pi r \sqrt{r^2 + h^2} \qquad h = \text{altitude}$$

$$r = \text{radius of base}$$

This formula uses the radius (*r*) of the base and not the diameter (*d*), and the vertical height (*h*), not slant height (*S*) (See Fig. 11-64).

The first formula is the one most often used, because both the slant height and diameter can be easily measured.

> **Example 27:** A drinking cup, in the shape of a cone, is made from paper that costs .25 cents per square inch. The cup has a diameter of 2.5 inches and a slant height of 3 inches. Find, to the nearest penny, the cost of paper for 1 cup (Fig. 11-65).

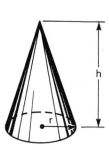
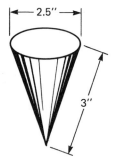

Figure 11-64              Figure 11-65

*Solution:* The cost is the product:

(.25 cents per sq. in.) × (surface area of cup)

The surface area of the cup is:

$$\text{Area} = \frac{3.142 \times 2.5 \times 3}{2}$$

$$= 11.8 \text{ sq. in. (to nearest tenth of a sq. in.)}$$

Cost = .25 × 11.8 = 2.95 cents.

*Answer:* Cost of paper for 1 cup = 3 cents (to nearest penny)

## Volume of Familiar Objects

Do you know how to find the capacity of a box, cylinder, or cone? For instance, if a cylindrical water tank is 10 feet high and has a diameter of 6 feet, can you find how many gallons of water it will hold? To answer this kind of question you must know about *volume.*

The *volume* of an object, such as a box, a tin can, or a cone, is a measure of its *capacity.* The units that express volume are *cubic units.* Thus a box has a volume of so many *cubic feet,* or a cylinder has a volume of so many *cubic inches* or *cubic centimeters.* The volume of a box-shaped object (Fig. 11-66) is:

Volume = (length) × (width) × (height)

or, because (length) × (width) = area of base, the formula is:

Volume = (area of base) × (height)

Remember that volume is written in *cubic units.*

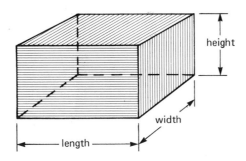

**Figure 11-66**

**Example 28:** A storage tank in the shape of a box has these dimensions: Length = 8 feet

Width = 6 feet

Height = 9 feet

What is the volume?

*Solution:* The volume is:

Volume = (8) × (6) × (9) cubic feet

= 432 cubic feet.

*Answer:* Volume = 432 cubic feet.

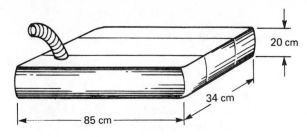

**Figure 11-67**

**Example 29:** The fuel tank on a small foreign sports car (Fig. 11-67) is a box measuring:

85 centimeters long

34 centimeters wide

20 centimeters high

Find the volume in liters. (1 liter = 1,000 cubic centimeters)

*Solution:* The volume in cubic centimeters is:

Volume = $(85) \times (34) \times (20) = 57,800$ cubic centimeters

The volume in liters is found by dividing the number of cubic centimeters by 1,000:

Volume in liters = $\dfrac{57,800 \; \text{cubic centimeters}}{1,000 \; \text{cubic centimeters}} \times 1$ liter

*Answer:*          Volume = 57.8 liters

**Example 30:** The fuel tank of the sports car in Example 11-29 is to be filled at an American service station where the unit of volume is the gallon. Since 1 liter = .2642 U. S. gallon, what is the capacity of the fuel tank in U. S. gallons?

*Solution:* Use the conversion fraction $\dfrac{.2642 \; \text{gallons}}{1 \; \text{liter}}$

57.8 liters = $57.8 \; \text{liters} \times \dfrac{.2642 \; \text{gallons}}{1 \; \text{liter}} = (57.8 \times .2642)$ gallons

*Answer:* Volume in U. S. gallons = 15.3 gallons to the nearest tenth of a gallon.

The formula for finding the volume of a circular cylinder is similar to that for a box. The formula is:

Volume of a cylinder = (area of base) × (height)

The area of the base is the area of a circle.

Volume of cylinder = $(.7854d^2) \times$ (height)

**Example 31:** Find the volume of a barrel that has a diameter of 2.5 feet and a height of 3 feet.

*Solution:* The volume is given by:

$$\text{Volume} = (.7854 \times 6.25) \times (3)$$

$$= 14.73 \text{ cubic feet (to nearest .01 cu. ft.)}$$

Since 1 cubic foot = 7.48 gallons, what is the capacity of the barrel in gallons?

Use the conversion fraction $\dfrac{7.48 \text{ gallons}}{1 \text{ cubic foot}}$

$$14.73 \text{ cubic feet} = 14.73 \; \cancel{\text{cubic feet}} \times \frac{7.48 \text{ gallons}}{1 \; \cancel{\text{cubic foot}}}$$

$$= (14.73 \times 7.48) \text{ gallons}$$

*Answer:* Volume (gallons) = 110.18 gallons (to nearest .01 gallon)

Auto and diesel mechanics must know how to find the cubic inch displacement of an engine. The cylinder dimensions in an engine are given in terms of *bore* and *stroke*. The *bore* of a cylinder is the same as the diameter.

$$\text{bore} = \text{diameter}$$

The *stroke* is the distance traveled by the piston in completing one-half revolution of the engine. The stroke is equal to the height of the cylinder.

$$\text{stroke} = \text{height}$$

**Example 32:** A late-model 8-cylinder engine has a bore of 4.00 inches and a stroke of 3.48 inches. What is the displacement in cubic inches, to the nearest cubic inch?

*Solution:* The total cubic inch displacement is:

$$8 \times (\text{volume of 1 cylinder}) = 8 \times (.7854 \times 4^2) \times (3.48)$$

$$\text{Displacement} = 349.85 \text{ (to nearest .01 cubic inch)}$$

*Answer:* Engine displacement = 350 cubic inches (to nearest cubic inch).

The formula for the volume of a cone is easy to remember. Engineers and scientists have found that:

*The volume of a cone is equal to one-third of the volume of a cylinder with the same base and height.*

Therefore, volume of cone $= \dfrac{1}{3} \times (\text{area of base}) \times (\text{height})$

$$= \frac{1}{3} \times (.7854d^2) \times (\text{height})$$

$$(d = \text{diameter of base})$$

This formula can be simplified to:

$$\text{Volume of cone} = (.2618d^2) \times (\text{height})$$

**Example 33:** Find the volume of a conical storage tank that is 5 feet high and has a base diameter of 4 feet. (See Fig. 11-68.)

*Solution:* Volume $= (.2618 \times 4^2) \times 5$

*Answer:*     $V = 20.9$ cubic feet (to nearest .1 cubic feet)

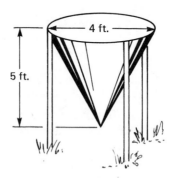

*Figure 11-68*

## PROBLEMS: Sec. 6 Surface Areas and Volumes

1. Find the surface area and volume of a box with these dimensions: width $= 2$ feet, length $= 4$ feet, height $= 3$ feet.

2. Find the surface area and volume of an *open* box (no top) with these dimensions: width $= 35$ centimeters, length $= 14$ centimeters, and height $= 25$ centimeters.

3. A bedroom 12 feet by 14 feet by 8 feet high is to be plastered on all walls and the ceiling. Find the number of square feet of area to be plastered.

4. An auxiliary fuel tank is mounted on a truck. The tank is 1.8 feet wide, 3.5 feet long, and 1.7 feet high.
   (a) What is the volume of the tank, in cubic feet?
   (b) If gasoline weighs approximately 42 pounds per cubic foot, what is the weight of fuel in a full tank?

5. Visualize a cubic yard. Fill in the blanks with the correct word or number. A cubic yard contains _____ cubic feet. It has _____ corners and _____ edges. It has _____ faces, each face is _____ feet square. Each face contains _____ square feet. You can think of

it as being made up of cubic feet arranged in _____ layers of _____ cubic feet in each layer.

6. Fill the blanks with the correct word or number. Imagine a cubic yard. If the six outside faces of a cubic yard are painted red, then cut up into one-foot cubes, then _____ cubes would be red on three sides, _____ would have paint on two sides, _____ would have paint on one side, and _____ would have no paint on any side.

7. A concrete walkway is 3 feet wide, 3 inches deep, and 100 feet long. How many cubic yards of concrete does it contain?

8. How many cubic yards of concrete are needed to surface 1 mile of roadway 30 feet wide and 6 inches deep? (1 mile = 5,280 feet)

9. The inside dimensions of a refrigerator are: 18 inches deep, 28 inches wide, and 36 inches high.
   (a) What is the volume in cubic inches?
   (b) What is the volume in cubic feet?

10. How many cubic centimeters are in 1 cubic meter?

11. A rectangular tank (box) is 24 inches long, 18 inches high, and 15 inches wide.
    (a) Find the volume in cubic inches.
    (b) What is the capacity of the tank in gallons? (1 gal. = 231 cu. in.)

12. A cylindrical air duct is 27 inches in diameter and 23 feet 6 inches long. Ignore any allowance for joints. What is the surface area?

13. A high-pressure steam pipe 4 inches in diameter and 12 feet long is to be wrapped with insulating material that cost $.09 per square inch. Allow 2% waste. How much will it cost to wrap the pipe?

14. Allowing .25 inches overlap, what are the dimensions of a label for a can 2.7 inches in diameter and 4.5 inches high.

15. An oil storage tank is 30 feet high and 45 feet in diameter.
    (a) What is the capacity in cubic feet?
    (b) What is the capacity in gallons? (1 cu. ft. = 7.5 gal.)

16. Find the cubic-inch displacement of an eight-cylinder engine with a bore of 3.80 inches and a stroke of 3.85 inches.

17. What is the displacement, in liters, for an engine with four cylinders, bore = 9.4 centimeters, stroke = 11.6 centimeters. (1 liter = 1,000 cu. cm.)

18. How many cubic inches of metal are in a piece of cast iron pipe 8 feet long, if the inside diameter is 6 inches and the wall thickness is $\frac{5}{16}$ inch?

19. A piece of brass round stock is 8 feet 9 inches long and 2 inches in diameter. Find its weight if brass weighs 520 pounds per cubic foot.

**20.** Find the volume of a conical storage tank with a height of 18 feet and a base diameter of 12 feet.

**21.** When sand is dumped from a conveyor belt, it forms a conical pile. How many cubic yards of sand are in the pile shown in Fig. 11-69?

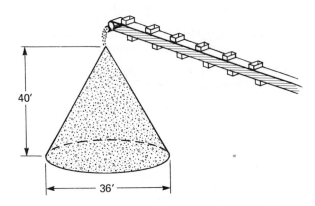

*Figure 11-69*

**22.** Farmer Jess has a conical pile of field corn stored near his barn for winter feeding of hogs. He wants to cover the pile with a canvas tarp. If the pile is 12 feet in diameter and has a slant height of 9 feet, how many square feet of canvas will he need?

**23.** A reducer for an air duct is made by cutting the top off a cone. If the reducer is to go from 20 inches to 8 inches in diameter with a slant height of 25 inches, find the amount of sheet metal needed, including waste. (See Fig. 11-70.)

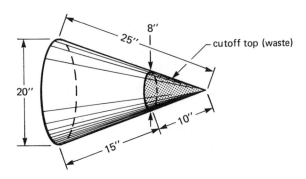

*Figure 11-70*

# PROBLEMS: Secs. 1-6 Perimeters, Areas, and Volumes

1. Find the perimeter and area of a rectangle 5 feet 6 inches long and 4 feet 9 inches wide.

2. Square floor tile 10 inches by 10 inches are to be used to cover a kitchen floor that is 12 feet by 14 feet. Allowing 2% waste, how many floor tile are needed?

3. A steel rod 22 inches long is bent into a circle with the ends just touching. What is the diameter of the circle? $\left(\text{Use } \frac{22}{7} \text{ for } \pi.\right)$

4. A spool of thread is 1.5 inches in diameter. How many revolutions will the spool turn when 2 feet of thread is unwound from the spool?

5. The outside dimensions of a picture frame are: length = 30 inches, width = 20 inches. The picture inside the frame has dimensions of 26 inches by 16 inches. What percent of the total framed area is filled by the picture?

6. The diameter of a truck tire is 42 inches. How many revolutions will the tire make in 1 mile? (1 mile = 5,280 ft.)

7. Two circles are drawn with the same center (concentric circles). If the radius of the larger circle is 15 centimeters and the radius of the smaller circle is 10 centimeters, how much area is between the circles?

8. An 18-inch water line connects two towns that are 135 miles apart. How many gallons of water are needed to fill the pipe? (1 cu. ft. = 7.5 gal.)

9. An oil well is 16,500 feet deep. The well pipe has an inside diameter of 9 inches.
   (a) How many gallons of oil are in the pipe?
   (b) If the oil weighs 55 pounds per cubic foot, how much does the oil in a full pipe weigh?

10. The diameter of one circle is 8 inches. The diameter of another circle is 16 inches.
   (a) What is the ratio of the diameters? (larger: smaller)
   (b) What is the ratio of their areas?

11. The area of a circle is 560 square inches. What is the diameter?

   Use the formula: $\qquad d = \sqrt{\dfrac{\text{Area}}{.785}}$

12. A truck bed is 7 feet wide, 5 feet deep, and 10 feet long.
   (a) What is the capacity in cubic feet?
   (b) What is the capacity in cubic yards?

13. A trench for a gas main is 2 feet wide, 4 feet deep, and 3,000 feet long. If 30% of the trench is filled by the pipe, how many cubic feet of fill is needed to cover the trench?

14. A coal-carrying railroad car is 8 feet wide, 42 feet long, and is filled to a depth of 6 feet. If coal weighs 50 pounds per cubic foot, how many tons of coal are in the car? (1 ton = 2,000 pounds)

15. A walkway 75 feet long and 4 feet wide is to be paved to a depth of 3 inches with material costing $1.50 per cubic foot. Find the cost of paving the walkway.

16. A squelch basin is made by cutting a barrel in half lengthwise which forms a trough with semicircular ends. Find the capacity of the squelch basin if the barrel is 6 feet long and 3 feet in diameter.

17. While measuring the gasoline in an underground tank, a service station attendant found that the gas was 4 feet deep in an 8-foot diameter tank that is 12 feet long. (The cylindrical tank is on its side.) How many gallons of gasoline are in the tank? (1 cu. ft. = 7.5 gal.)

18. A stalactite is a conical shaped deposit of calcium material pointing downward from the roof of a cave. A stalactite is 12 feet long and has a diameter at the base of 3 feet.
    (a) Find the volume of the stalactite in cubic inches.
    (b) If it takes 2 years to deposit 1 cubic inch of calcium, what is the approximate age of the stalactite?

19. Find the volume of the steel rivet shown in Fig. 11-71.

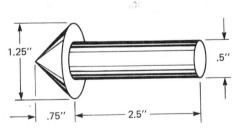

**Figure 11-71**

20. A conical church steeple is 36 feet in diameter and has a slant height of 65 feet. It takes 3.6 bundles of shingles to cover 100 square feet. How many complete bundles are needed to shingle the roof of the steeple?

## SUMMARY AND COMMENTS

In this chapter you have studied the figures most often used in technology and the various trades. Look around you to see rectangles, triangles, circles, cylinders, boxes, and cones. Know the various shapes and the important formulas that go with them.

Study the figures and formulas on the following pages. How many formulas do you remember? Memorize a few important formulas. You will find them useful—especially if you need the area or volume of a simple figure and do not have your text or handbook with you.

In some trades you will use shapes and objects not studied in this text. For these special figures, check a handbook or other source.

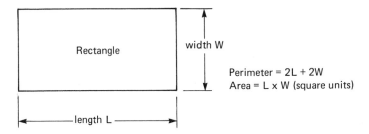

Perimeter = 2L + 2W
Area = L x W (square units)

**Figure 11-72**   *Rectangle*

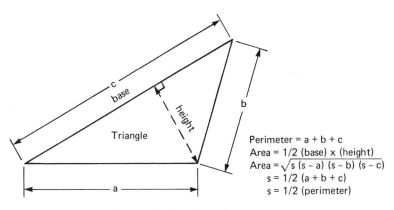

Perimeter = a + b + c
Area = 1/2 (base) x (height)
Area = $\sqrt{s\,(s-a)\,(s-b)\,(s-c)}$
s = 1/2 (a + b + c)
s = 1/2 (perimeter)

**Figure 11-73**   *Triangle*

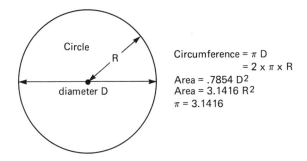

Circumference = $\pi$ D
= 2 x $\pi$ x R
Area = .7854 $D^2$
Area = 3.1416 $R^2$
$\pi$ = 3.1416

**Figure 11-74**   *Circle*

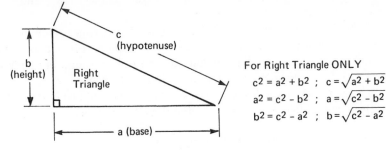

For Right Triangle ONLY

$$c^2 = a^2 + b^2 \; ; \; c = \sqrt{a^2 + b^2}$$
$$a^2 = c^2 - b^2 \; ; \; a = \sqrt{c^2 - b^2}$$
$$b^2 = c^2 - a^2 \; ; \; b = \sqrt{c^2 - a^2}$$

**Figure 11-75**   *Right triangle*

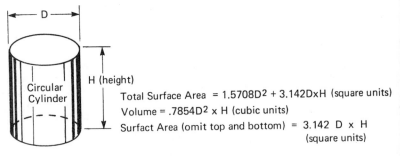

Surface Area =
2(LW + LH + WH)
(square units)

Volume = LxWxH
(cubic units)

**Figure 11-76**   *Rectangular solid*

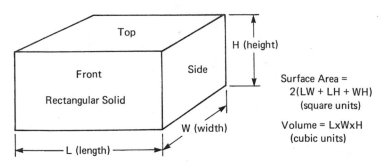

Total Surface Area = 1.5708D$^2$ + 3.142DxH (square units)

Volume = .7854D$^2$ x H (cubic units)

Surfact Area (omit top and bottom) = 3.142 D x H
(square units)

**Figure 11-77**   *Circular cylinder*

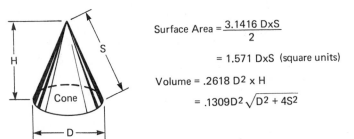

Surface Area = $\dfrac{3.1416 \, DxS}{2}$

= 1.571 DxS (square units)

Volume = .2618 D$^2$ x H

= .1309D$^2 \sqrt{D^2 + 4S^2}$

**Figure 11-78**   *Cone*

## Practice Test for Chapter 11

1.  The floor of a living room is 18 feet by 22 feet. How much will it cost to carpet the floor if the carpet costs $12.50 a square yard?    *Answer:* _____

2.  A guy wire supporting a telephone pole is connected 22 feet up the pole and anchored 9 feet from the base of the pole. What is the length of the guy wire? Give answer to nearest foot.    *Answer:* _____

3.  What is the area of a triangle of base 16 centimeters and height 15 centimeters?    *Answer:* _____

4.  A sea anchor is a canvas bag shaped like a cone. What is the volume if the diameter of the base is 18 inches and the height is 30 inches? Express answer in cubic feet.    *Answer:* _____

5.  A concrete grain silo is in the shape of a cylinder with an outer diameter of 55 feet and a wall thickness of 2 feet. Find the volume of the silo if it is 96 feet tall. Express answer in bushels. (1 cu. ft. = 0.803 bushel)
    *Answer:* _____

6.  A box is 2 inches high, 14 inches long, and 8 inches deep. When wrapped in paper, an extra 20% is needed for overlap and ends. How much paper is needed to wrap the box?    *Answer:* _____

7.  What is the cubic inch displacement of a six-cylinder engine with a bore of 3.85 inches and a stroke of 3.25 inches?    *Answer:* _____

8.  A contractor dug a basement 18 feet wide, 22 feet long, and 8 feet deep. The earth removed weighed 85 pounds per cubic foot. What was the weight of the total earth removed?    *Answer:* _____

# 12

# Essentials
# of
# Trigonometry

## SEC. 1  THE SINE AND TANGENT RATIOS

Most of the measurements made by technicians, mechanics, or other tradesmen, are *direct* measurements. They are obtained by direct reading of a scale, a ruler, micrometer, protractor, or other measuring instrument. Finding the length of a bolt, the weight of a sack of nails, or the area of a wall to be painted are all results of direct measurement.

To measure the temperature of the sun, the distance to the moon, the width of a river, or the height of a mountain, *indirect* measurement is used. The right triangle is used to get indirect measurements. This kind of measurement, which makes use of triangles, is called *trigonometry*. The word *trigonometry* means *triangle measurement*. A right triangle has one square corner that is 90°. (See Fig. 12-1.) The other two angles in a right triangle are acute and are called *complementary* angles because they always add up to 90°. The sum of the three interior angles of a triangle is always equal to 180°.

Look at the right triangle shown in Fig. 12-1. The side opposite the right angle is called the *hypotenuse*. The other two sides, which were called base and height in Chapter 11, are called *legs* of the triangle. There is an important relationship between the three sides of a right triangle. Do you recall the Pythagorean theorem? This theorem states that for every right

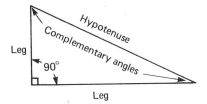

**Figure 12-1**

triangle:

$$(\text{hypotenuse})^2 = (\text{leg})^2 + (\text{leg})^2$$

*If you know any two sides of a right triangle, you can find the third side by applying the Pythagorean theorem.*

You will now learn that it is possible to find the values of the two acute interior angles of a right triangle if you know the lengths of two sides. Memorize the two names for the legs of a right triangle. This is important.

In the right triangle shown in Fig. 12-2 the three angles have been labeled A, B, and C. Angle C is the right angle. Imagine you are standing at angle A and looking across the triangle at leg **a**. When you are in this position, leg **a** is called the *opposite leg* and leg **b** is called the *adjacent leg*. If you place yourself at angle **B**, then leg **b** becomes the *opposite leg* and leg **a** the *adjacent leg*. These names of *opposite* and *adjacent* are used only with angles A and B.

The *opposite leg* (opp. leg) is the leg facing the angle you are working with.

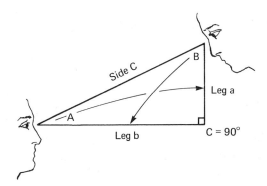

**Figure 12-2**

The *adjacent leg* (adj. leg) is the side of the right triangle—not the hypotenuse—that is part of the angle you are working with.

Study the triangle shown in Fig. 12-3. Make sure you understand the correct naming of the legs with reference to the two angles A and B.

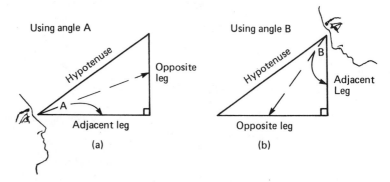

**Figure 12-3**   (*a*) *Using angle A; (b) Using angle B*

An important and valuable set of facts about certain ratios of sides of right triangles has been discovered. Two of these facts are:

*For all right triangles having the same size angle,*

$$\text{the ratio } \frac{\text{opposite leg}}{\text{hypotenuse}} \text{ is always the same.}$$

*For all right triangles having the same size angle*

$$\text{the ratio } \frac{\text{opposite leg}}{\text{adjacent leg}} \text{ is always the same.}$$

For instance, in Fig. 12-4 both triangles have angle A = 30°. The Pythagorean theorem gives the lengths of the sides. Now find the two

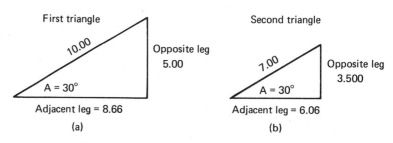

**Figure 12-4**   (*a*) *First triangle; (b) Second triangle*

ratios, using the 30° angle:

$$\begin{array}{ccc} & \text{(First triangle)} & \text{(Second triangle)} \\ \dfrac{\text{side opposite 30° angle}}{\text{side adjacent angle}} = & \dfrac{5.00}{8.66} = 0.577 & \dfrac{3.50}{6.06} = 0.577 \\[3mm] \dfrac{\text{side opposite 30° angle}}{\text{hypotenuse}} = & \dfrac{5.00}{10.00} = 0.500 & \dfrac{3.50}{7.00} = 0.500 \end{array}$$

These two ratios are so important that they have been given special names:

The ratio $\dfrac{\text{leg opposite angle}}{\text{hypotenuse}}$ is called the *sine* of the *angle*.

The ratio $\dfrac{\text{leg opposite angle}}{\text{adjacent side}}$ is called the *tangent* of the *angle*.

Thus, in Fig. 12-4,

$$\text{sine of angle} = \text{sine of } 30° = \frac{5.00}{10.00} = .500$$

$$\text{tangent of angle} = \text{tangent of } 30° = \frac{5.00}{8.66} = .577$$

For convenience, the words sine and tangent are shortened to *sin* for sine and *tan* for tangent:

$$\text{sine of } 30° = \sin 30° = .500$$

$$\text{tangent of } 30° = \tan 30° = .577$$

The ratios sine and tangent for all angles between 0° and 90° have been worked out by mathematicians once and for all. Tables of these ratios are in most handbooks. In this book Table 12-1 starts at 1° and increases by degrees up to 90°. Values for the two ratios are given to four decimal places.

SPECIAL NOTE: You can find six possible ratios by using the three sides of a right triangle. Each of these ratios has a special name. They are:

$$\text{sine } A = \frac{\text{leg opposite angle } A}{\text{hypotenuse}} \qquad \text{cosecant } A = \frac{\text{hypotenuse}}{\text{leg opposite angle } A}$$

$$\text{cosine } A = \frac{\text{leg adjacent angle } A}{\text{hypotenuse}} \qquad \text{secant } A = \frac{\text{hypotenuse}}{\text{leg adjacent angle } A}$$

$$\text{tangent } A = \frac{\text{leg opposite angle } A}{\text{adjacent leg}} \qquad \text{cotangent } A = \frac{\text{leg adjacent angle } A}{\text{opposite leg}}$$

In this course of study you need to have values for the sine and tangent ratios only. In more advanced studies, you will need all of the ratios.

Look at Fig. 12-5 and memorize the names and ratios.

$$\text{Sine of angle } A = \sin A = \frac{\text{leg opposite angle } A}{\text{hypotenuse}}$$

**TABLE 12-1**  VALUES OF TRIGONOMETRIC FUNCTIONS—SINE AND TANGENT

| Angle | Sin | Tan | Angle | Sin | Tan |
|---|---|---|---|---|---|
| 1° | .0175 | .0175 | 46° | .7193 | 1.0355 |
| 2° | .0349 | .0349 | 47° | .7314 | 1.0724 |
| 3° | .0523 | .0524 | 48° | .7431 | 1.1106 |
| 4° | .0698 | .0699 | 49° | .7547 | 1.1504 |
| 5° | .0872 | .0875 | 50° | .7660 | 1.1918 |
| 6° | .1045 | .1051 | 51° | .7771 | 1.2349 |
| 7° | .1219 | .1228 | 52° | .7880 | 1.2799 |
| 8° | .1392 | .1405 | 53° | .7986 | 1.3270 |
| 9° | .1564 | .1584 | 54° | .8090 | 1.3764 |
| 10° | .1736 | .1763 | 55° | .8192 | 1.4281 |
| 11° | .1908 | .1944 | 56° | .8290 | 1.4826 |
| 12° | .2079 | .2126 | 57° | .8387 | 1.5399 |
| 13° | .2250 | .2309 | 58° | .8480 | 1.6003 |
| 14° | .2419 | .2493 | 59° | .8572 | 1.6643 |
| 15° | .2588 | .2679 | 60° | .8660 | 1.7321 |
| 16° | .2756 | .2867 | 61° | .8746 | 1.8040 |
| 17° | .2924 | .3057 | 62° | .8829 | 1.8807 |
| 18° | .3090 | .3249 | 63° | .8910 | 1.9626 |
| 19° | .3256 | .3443 | 64° | .8988 | 2.0503 |
| 20° | .3420 | .3640 | 65° | .9063 | 2.1445 |
| 21° | .3584 | .3839 | 66° | .9135 | 2.2460 |
| 22° | .3746 | .4040 | 67° | .9205 | 2.3559 |
| 23° | .3907 | .4245 | 68° | .9272 | 2.4751 |
| 24° | .4067 | .4452 | 69° | .9336 | 2.6051 |
| 25° | .4226 | .4663 | 70° | .9397 | 2.7475 |
| 26° | .4384 | .4877 | 71° | .9455 | 2.9042 |
| 27° | .4540 | .5095 | 72° | .9511 | 3.0777 |
| 28° | .4695 | .5317 | 73° | .9563 | 3.2709 |
| 29° | .4848 | .5543 | 74° | .9613 | 3.4874 |
| 30° | .5000 | .5774 | 75° | .9659 | 3.7321 |
| 31° | .5150 | .6009 | 76° | .9703 | 4.0108 |
| 32° | .5299 | .6249 | 77° | .9744 | 4.3315 |
| 33° | .5446 | .6494 | 78° | .9781 | 4.7046 |
| 34° | .5592 | .6745 | 79° | .9816 | 5.1446 |
| 35° | .5736 | .7002 | 80° | .9848 | 5.6713 |
| 36° | .5878 | .7265 | 81° | .9877 | 6.3138 |
| 37° | .6018 | .7536 | 82° | .9903 | 7.1154 |
| 38° | .6157 | .7813 | 83° | .9925 | 8.1443 |
| 39° | .6293 | .8098 | 84° | .9945 | 9.5144 |
| 40° | .6428 | .8391 | 85° | .9962 | 11.4301 |
| 41° | .6561 | .8693 | 86° | .9976 | 14.3007 |
| 42° | .6691 | .9004 | 87° | .9986 | 19.0811 |
| 43° | .6820 | .9325 | 88° | .9994 | 28.6363 |
| 44° | .6947 | .9657 | 89° | .9998 | 57.2900 |
| 45° | .7071 | 1.0000 | 90° | 1.0000 | ∞ |

NOTE: ∞ is the symbol for infinity and, as used here, indicates no value for tan 90°.

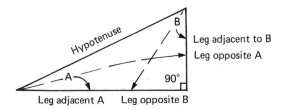

**Figure 12-5**

Tangent of angle A $=$ tan A $= \dfrac{\text{leg opposite angle A}}{\text{leg adjacent to angle A}}$

Sine of angle B $=$ sin B $= \dfrac{\text{leg opposite angle B}}{\text{hypotenuse}}$

Tangent of angle B $=$ tan B $= \dfrac{\text{leg opposite angle B}}{\text{leg adjacent to angle B}}$

**Example 1:** Find the sine and tangent values for the two acute angles A and B shown in Fig. 12-6.

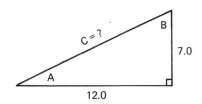

**Figure 12-6**

*Solution:* First, determine the length of the hypotenuse. Do this by using the Pythagorean theorem:

$$\text{hypotenuse} = \sqrt{(12)^2 + (7)^2}$$
$$c = \sqrt{193}$$
$$\text{hypotenuse} = c = 13.9$$

*Answer:*  $\sin A = \dfrac{7}{13.9} = .504$     $\tan A = \dfrac{7}{12} = .583$

   $\sin B = \dfrac{12}{13.9} = .863$     $\tan B = \dfrac{12}{7} = 1.714$

Examples 2 through 7, which follow, show how to find sine and tangent values from Table 12-1. They also show how to find the angle when these values are known.

**Example 2:** Find the value of sin 56°. (Use Table 12-1.)

*Solution:* Look down the column marked *angle* until you find 56°. The first column to the right gives values for the sine ratio. Read the decimal .8290.

*Answer:* sin 56° = .8290

**Example 3 :** Find tan 87°. (Use Table 12-1.)

*Solution:* Look down the *angle* column until you find 87°. The second column to the right gives values for the tangent ratio. Opposite 87° read 19.0811 in the tan column.

*Answer:* tan 87° = 19.0811

**Example 4:** Find the size of an angle whose sine is the decimal .2419. (Use Table 12-1.)

*Solution:* Look down the column headed sin until you find the decimal .2419. In the angle column to the left read the angle 14°.

*Answer:* angle A = 14° and sin A = .2419

**Example 5:** Find the angle whose sine value is .9710. (Use Table 12-1.)

*Solution:* Look in the sin column for the value .9710. You do not find .9710 exactly, but you do find .9703 and .9744. Here you see that:

.9703 corresponds to 76°

.9744 corresponds to 77°

Therefore, you know that the answer to the question is an angle between 76° and 77°. Because the table is in degrees only, determine the nearest degree value for .9710. Because .9710 is closer to .9703 than to .9744, take 76° as the answer.

*Answer:* Sin 76° = .9710 (approximately)

Although there is some error in this approximation process, it usually is not significant. For extreme accuracy use a more extensive table.

**Example 6:** Find angles A and B for the triangle shown in Fig. 12-7.

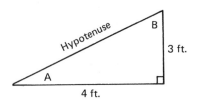

**Figure 12-7**

*Solution:* Do not use the value for the hypotenuse. Use the tangent ratio to find either angle A or angle B.

From the figure:     $\tan A = \dfrac{3}{4} = .7500.$

From Table 12-1: The angle with tangent value .7500 is 37° (approximately). Because the two angles A and B must add up to 90°, and A = 37°,

*Answer:* $B = 90° - 37° = 53°$

**Example 7:** Find the lengths of the two legs and the size of angles A and B shown in Fig. 12-8. In the figure the hypotenuse is 16.0 mm., and the value of sin A = .7310.

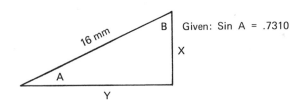

Given: Sin A = .7310

**Figure 12-8**

*Solution:* From Table 12-1 look in the sin column until you find a value nearest to .7310. The value .7314 is nearest and this corresponds to 47°. Because

angle A = 47°,     angle B = 90° − 47° = 43°.

Recall from the definition of the sine of an angle that:

$$\text{sine of angle} = \frac{\text{opposite leg}}{\text{hypotenuse}}$$

Here you know that:

$$\sin A = .7310$$

$$\text{Sin } A = \frac{\text{opposite leg}}{\text{hypotenuse}} = \frac{x \text{ (unknown)}}{16 \text{ mm.}} = .7310$$

Substitution of these values gives:

$$.7310 = \frac{x}{16 \text{ mm.}}$$

Find the value of $x$ by multiplying both sides by 16.

$$x = 16 \,(.7310) = 11.7 \text{ mm. (nearest tenth)}$$

Use the value of sin B to find the length of the other leg.

$$\sin B = \sin 43° = .6820 \quad \text{(from table)}$$

$$.6820 = \frac{y}{16 \text{ mm.}} \quad \text{(multiply both side by 16.)}$$

and

$$y = 16 \,(.6820) = 10.9 \text{ mm.} \quad \text{(nearest tenth)}$$

*Answer:* $x = 11.7$ mm., $y = 10.9$ mm.

The Pythagorean theorem and a table of values for the sin and tan ratios enable you to solve problems involving both direct and indirect measurement. In the next section you will study applications of trigonometry to problems in technology.

## PROBLEMS: Sec. 1 The Sine and Tangent Ratios

1. Use Table 12-1 to find the sine and tangent for each of these angles.

| Angle A | Sine A | Tangent A |
|---------|--------|-----------|
| (a) 20° | | |
| (b) 25° | | |
| (c) 30° | | |
| (d) 42° | | |
| (e) 45° | | |
| (f) 57° | | |
| (g) 60° | | |
| (h) 79° | | |
| (i) 89° | | |

2. Find the nearest angle for the sine or tangent values.

   (a) $\sin A = .3584$  (b) $\tan B = 6.3138$
   (c) $\tan G = .0875$  (d) $\sin D = 1.0000$
   (e) $\tan H = 5.6710$  (f) $\sin B = .9207$
   (g) $\sin A = .1875$  (h) $\tan A = 15.0976$

3. Use either sine or tangent ratios to find the size of the angle marked $x$ in Fig. 12-9(a) through (f).

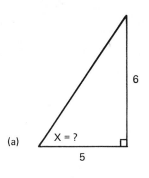

(a) X = ?
6
5

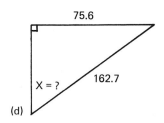

(d)
75.6
X = ?
162.7

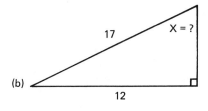

(b)
17
X = ?
12

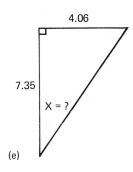

(e)
4.06
7.35
X = ?

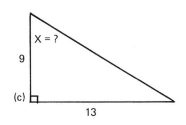

(c) X = ?
9
13

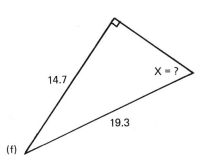

(f)
14.7
X = ?
19.3

*Figure 12-9*

387

**4.** Use the sine ratio to find the length of the side marked $x$ in Fig. 12-10(a) through (f).

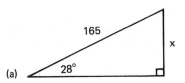

(a)

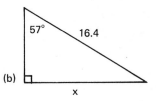

(b)

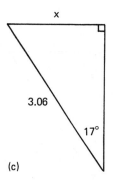

(c)

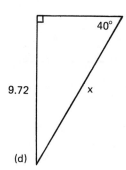

(d)

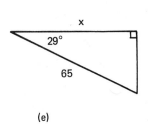

(e)

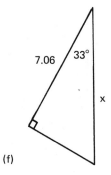

(f)

***Figure 12-10***

**5.** Use the tangent ratio to find the length of the side marked $x$ in Fig. 12-11(a) through (f).

(a)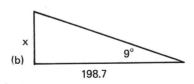

(b)

(c)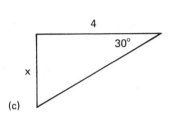

(d)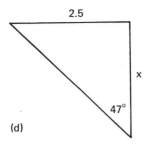

(e)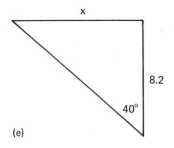

(f)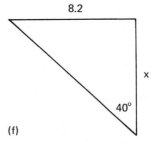

**Figure 12-11**

**6.** A right triangle has sides of 5, 12, and 13 (hypotenuse). What are the values of the interior angles?

**7.** Find each of the indicated lengths in Fig. 12-12.

A B = _____

B C = _____

C D = _____

A D = _____

D E = _____

E A = _____

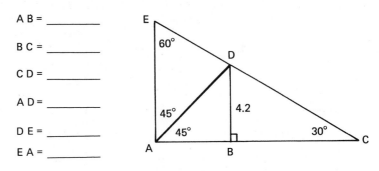

**Figure 12-12**

**8.** Use Fig. 12-13 with angles A, B, C, and sides a, b, c, and find values for the missing parts.

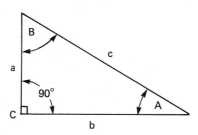

**Figure 12-13**

(a) A = 30°,   B = _____, a = 12,   b = _____, c = _____
(b) A = _____, B = 60°,   a = _____, b = 50,    c = _____
(c) A = 45°,   B = _____, a = _____, b = _____, c = 10
(d) A = _____, B = 18°,   a = 6.7,   b = _____, c = _____
(e) A = _____, B = _____, a = 140,   b = 570,   c = _____
(f) A = _____, B = _____, a = 15,    b = _____, c = 56
(g) A = 5°,    B = _____, a = 1,678, b = _____, c = _____

## SEC. 2   APPLYING TRIGONOMETRY
## TO TECHNOLOGY

The broad application of trigonometry to many different technologies is most valuable to the technician. Study the following examples.

**Example 8:** The navigator of a ship wants to determine the distance from the ship to a lighthouse sighted on shore. He measures an angle of 12° from the horizontal (sea level) to the beacon of the light house that is 570 feet above sea level. What is the distance between the ship and the lighthouse? (Fig. 12-14)

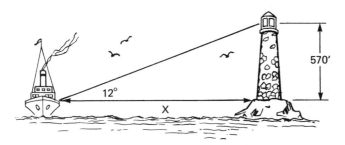

**Figure 12-14**

*Solution:* Use the tangent ratio. Call the unknown distance $x$. Use the complement angle $90° - 12° = 78°$

$$\tan 78° = \frac{\text{opposite leg}}{\text{adjacent leg}} = \frac{x}{570}$$

From Table 12-1 tan 78° = 4.7046

The equation for $x$ is: $\frac{x}{570} = 4.7046$

To solve this equation for $x$, multiply both sides by 570.

$$x = (4.7046)(570) = 2,682$$

*Answer:* $x = 2,682$ feet (nearest foot)

**Example 9:** A surveyor wanted to know the length of a lake. From a sketch (Fig. 12-15) he decide that he needed to find

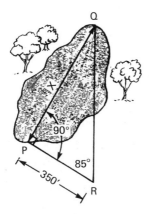

**Figure 12-15**

the distance from point P to point Q. He set up his transit at point R (350 feet from P) and sighted P, then Q. The angle PRQ is 85°. What is the distance from P to Q?

*Solution:* Call the unknown distance $x$. Use the tangent ratio:

$$\tan 85° = \frac{x}{350}$$

From Table 12-1 tan 85° = 11.4301
Solve for $x$ by multiplying both sides by 350.

$$x = 350\,(11.4301) = 4{,}001 \text{ feet}$$

*Answer:* The lake is 4,001 feet from P to Q.

**Example 10:** A Boy Scout wants to estimate the height of a tree. He finds that the length of the shadow cast by the tree is 26 paces (1 pace = 2.5 feet for the scout). Standing at the tip of the shadow, he estimates the angle from the base to the top of the tree is 20 degrees. What is the approximate height of the tree? (Draw a figure.)

*Solution:* Use the tangent ratio.

$$\tan 20° = \frac{\text{height of tree}}{\text{length of shadow}}$$

$$.3640 = \frac{x\,(\text{height of tree})}{26 \text{ paces}}$$

$$x = 26(.3640)(2.5) \text{ feet}$$

$$= 23.7 \text{ feet}$$

*Answer:* Height of tree = 24 feet (nearest foot)

**Example 11:** Use trigonometry to find the dimensions $x$ and $y$ in Fig. 12-16.

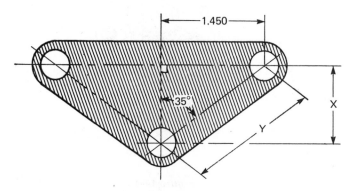

*Figure 12-16*

*Solution:* First find the value of $x$. Because the one given angle is 35°, the other acute angle must be $90° - 35° = 55°$. Use the tangent ratio.

$$\tan 55° = \frac{x}{1.450}$$

$$1.4281 = \frac{x}{1.450}$$

$$x = (1.4281)(1.450) \text{ (Multiply both sides by 1.450.)}$$

*Answer:*   $x = 2.070$ cm.

To find the value of $y$ use the Pythagorean theorem.

$$y^2 = (1.450)^2 + (2.070)^2$$
$$y^2 = 2.103 + 4.285 = 6.388$$
$$y = \sqrt{6.388} = 2.527$$

*Answer:*   $y = 2.527$ cm.

**Example 12:** A short piece of aluminum is bent into an offset bracket as shown in Fig. 12-17. Find the dimensions $x$ and $y$. Express answers in inches and fractions of inches to nearest sixteenth.

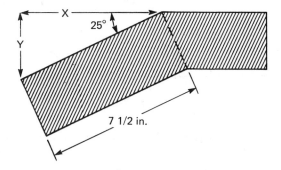

*Figure 12-17*

*Solution:* First find the value for $y$ by using the sine ratio.

$$\sin 25° = \frac{y}{7.5}$$

$$.4226 = \frac{y}{7.5}$$

$$y = 7.5(.4226) \text{ (Multiply both sides by 7.5.)}$$
$$y = 3.170 \text{ inches}$$

Change .170 inch to 16ths by multiplying by 16:

$$.170 \times 16 = 2.72 = 3 \text{ (approximately)}$$

Therefore, $y = 3\frac{3}{16}$ inches to the nearest 16th.

Use the other acute angle and the sine ratio:

$$\sin 65° = \frac{x}{7.5}$$

Do computations similar to the steps that you used to find $y$:

$$x = 6.797 \text{ inches} = 6\frac{13}{16} \text{ inches}$$

*Answer:*      $x = 6\frac{13}{16}$ inches

$$y = 3\frac{3}{16} \text{ inches}$$

## PROBLEMS: *Sec. 2 Applying Trigonometry to Technology*

1. In laying out a sprinkler system, a landscape designer made the sketch shown in Fig. 12-18. How much pipe is needed?

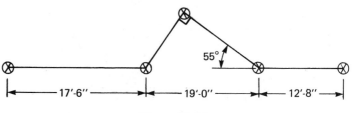

**Figure 12-18**

2. A ladder 18 feet long just reaches to the top of a wall. If the angle at the base of the ladder is 48°, what is the height of the wall?

3. Find the length of the rafter shown in Fig. 12-19.

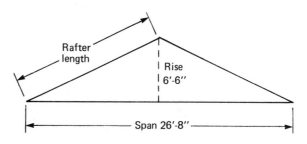

**Figure 12-19**

4. The angle of elevation to the top of a statue is 27° from a point 357 feet from the statue. What is the height of the statue?

5. The transmitter antenna for station KPUT is 675 feet tall. Four guy wires are connected to the top and are anchored 150 feet from the base of the antenna. Find the length of one of the guy wires.

6. A submarine is diving at an angle of 18°. How far will the sub travel horizontally in diving to a depth of 500 feet?

7. A man starts rowing directly across a river which is 250 feet wide. The current takes him downstream. After landing he is 45 feet below the point at which he had intended to arrive. How far did he actually travel?

8. A jet is flying at a speed of 600 miles per hour and is climbing at an angle of 16°. What is the change in altitude after 4.5 minutes?

9. A roadbed has a rise of 6 feet for every 100 feet horizontally. What is the angle of inclination of the road?

10. When an airplane is at 4,500 feet, a landmark is sighted at an angle of 28° below the horizontal. What is the distance from the landmark to a point directly under the plane?

11. From an airplane 6,500 feet above the surface of the water, a pilot sees two ships directly ahead. The angles of depression to the two ships are 14° and 25°. How far apart are the two ships?

12. A ship leaves the dock and travels 14 miles due north. It then changes its direction to due west and travels another 9 miles. How far is the ship from the dock?

13. A weather balloon is sighted at an altitude of 16,500 feet and one-half mile downwind from the point at which it was released. How far is the balloon from the release point?

14. During a kite-flying contest, Jon let out 750 feet of kite string. The angle of inclination of the kite with the ground was 78°. What was the altitude of the kite?

15. Find the dimension D in Fig. 12-20 to the nearest 64th inch.

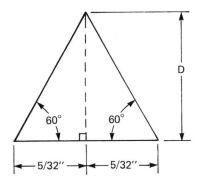

**Figure 12-20**

**16.** A road rises 500 feet vertically in a horizontal distance of 2,800 feet.

(a) Express the ratio $\dfrac{500 \text{ rise}}{2{,}800 \text{ run}}$ as a per cent.

(b) Find the angle of inclination of the roadbed.

*Hint:* Compute $\tan A = \dfrac{500}{2{,}800}$ for angle A.

**17.** A building stands on level ground. From a point P on the ground the angle of elevation to the sill of a window on the third floor, which is 32 feet above ground, is 28°. When the top of the building is viewed from point P, the angle of elevation is 78°. What is the height of the building?

**18.** In Fig. 12-21, holes are to be drilled at points A, B, C, and D. Determine the lengths of AD, DB, DC, and BC to the nearest .001 inch.

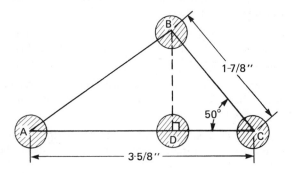

**Figure 12-21**

**19.** In Chapter 11 it was shown that the area of a triangle is:

$$\text{Area} = \frac{1}{2}(\text{base}) \times (\text{altitude})$$

When the angle C between two consecutive sides $a$ and $b$ is known, a formula for the area is:

$$\text{Area} = \frac{1}{2}(b) \times a \,(\sin C)$$

$b$ is the base and C is the angle between sides $a$ and $b$. In this formula the value $a \,(\sin C)$ is equal to the altitude of the triangle. See Fig. 12-22. Use this new formula to find the area of these triangles:

| | Side a | Side b | Angle C |
|---|---|---|---|
| (a) | 10 ft. | 5 ft. | 50° |
| (b) | 18 in. | 24 in. | 35° |
| (c) | 1 ft. 5 in. | 2 ft. 7 in. | 68° |
| (d) | 35 cm. | 46 cm. | 48° |
| (e) | 1.06 mm. | 3.67 mm. | 18° |
| (f) | .657 cm. | 3.554 cm. | 59° |
| (g) | 1.6 mi. | 5.8 mi. | 11° |

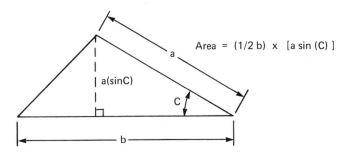

**Figure 12-22**

**20.** Find the lengths of the steel beams A, B, and C for the bridge truss layout shown in Fig. 12-23.

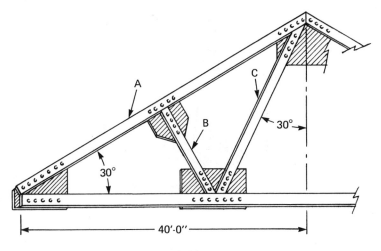

**Figure 12-23**

## Practice Test for Chapter 12

**1.** Use Table 12-1 of sine and tangent values to find sin A and tan A for the given angle A.

(a) $A = 10°$           (b) $A = 25°$

*Answer:* (a) _____

*Answer:* (b) _____

**2.** Find the nearest angle for the sine values.

(a) $\sin A = .3584$        (b) $\sin A = .5050$

*Answer:* (a) _____

*Answer:* (b) _____

3. Find the nearest angle for the tangent values.
   (a) tan A = .1763                          (b) tan A = 2.0500

   *Answer*: (a) _____

   *Answer*: (b) _____

4. One angle of a right triangle is 36° and the hypotenuse is 14 inches long.
   How long are the other two sides?          *Answer*: _____

5. Figure 12-24 is a sketch of a 600-foot radio tower. AC and AD are two anchor
   cables (in the same vertical plane) anchored at C and D on a level with the
   base of the tower. Use the angles in the figure to find:
   (a) the length of cable AC.          *Answer*: (a) _____
   (b) the length of cable AD.          *Answer*: (b) _____
   (c) the distance CD between the anchor points.

   *Answer*: (c) _____

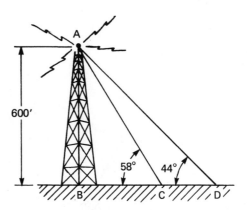

**Figure 12-24**

# 13

# Graphs

## SEC. 1   PICTURE, BAR, AND LINE GRAPHS

"One picture is worth a thousand words" is a familiar statement. The use of pictures in business reports, magazines, technical journals, and other publications to show numerical data is often superior to other ways of showing the same information. Look at the evening paper, a technical journal, or the owner's manual for an automobile. In each of these you will see pictures that show numerical facts clearly, accurately, and often colorfully.

Whenever a picture is used to show a relationship between objects and numbers, it is called a *graph*. Many types of graphs are used by business, industry, and technology. In this chapter you will study the following types of graphs:

> *Picture graphs* (pictographs)
> *Bar graphs*
> *Line graphs*
> *Circle graphs*. (Pie graphs)

Study the various types of graphs. Each one has a special way of showing information.

## The Pictograph

A *pictograph* is a graph that uses pictures to show numerical facts. Symbols or pictures which relate to the subject of the graph are used to show the amount of the quantity. In a pictograph each symbol used represents a definite number of things being shown. The graph gives the information that is true for a certain date in time. A pictograph is an easy way to show information that can be understood at a glance.

**Example 1:** The pictograph (Fig. 13-1) shows the number of new homes built during a five-year period. Use these steps to develop this pictograph:

1. Since the graph is about houses, use a small picture of a house as the symbol.
2. Let each symbol represent 100 homes. (This is the *scale* in the graph.)
3. Let less than 50 houses be represented by half of a house.
4. Give the graph a suitable title and clearly show the scale.

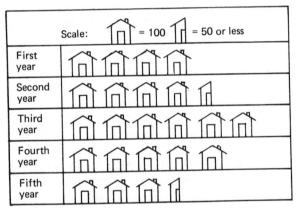

**Figure 13-1**   *New home construction (5-year period)*

The pictograph is quite versatile. Although most pictographs use a single symbol, some may use several symbols. Figure 13-2 is an example of a pictograph that uses four symbols.

**Example 2:** Each year the State of California publishes a pictograph which shows the percent of workers in four broad catagories: white-collar workers, blue-collar workers, service workers, and farm workers. One of these pictographs is shown in Fig. 13-2.

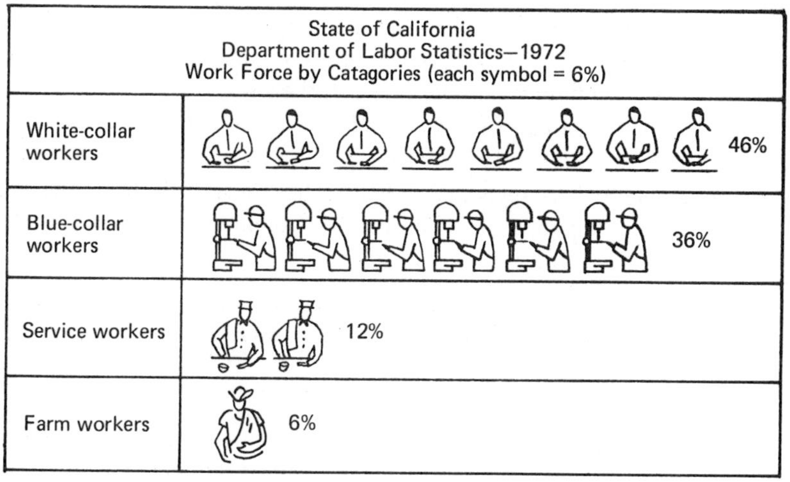

**State of California**
**Department of Labor Statistics—1972**
**Work Force by Catagories (each symbol = 6%)**

| White-collar workers | | 46% |
| Blue-collar workers | | 36% |
| Service workers | | 12% |
| Farm workers | | 6% |

*Figure 13-2*

Four different symbols are used. Each symbol represents 6 percent of the total work force in the state during that year. (In another year the same symbol may represent a different percent.) The graph has a suitable title, the scale is clearly shown, and the total work force is represented.

Here are some suggestions for making a good pictograph:

1. Get the numerical information for the graph.
2. Decide upon a symbol (or symbols) which relates to the subject.
3. Choose a scale and clearly show how much each symbol represents.
4. Give the graph a suitable title and give any dates or other necessary information.

## Bar Graphs

Another graph that is used to compare numerical quantities is the *bar graph*. The bar graph is used to compare numerical facts about different things at the same time. Or it is used to compare numerical facts about one thing at *different times*. The quantities that are compared are shown by parallel bars or heavy parallel lines. The lines may be drawn either horizontally or vertically. The length of a bar is used to show a known numerical fact. The length depends on the choice of the scale. The length of each bar in a bar graph shows the quantity it represents. The width of all bars in a bar graph are the same. The bars are usually separated from one another because it makes the graph easier to read.

To make a bar graph, use two lines called *axes*, a horizontal line and a vertical line, are used. Show the scale on each axis clearly. For accurate comparison of two numerical quantities, start each scale at zero.

The following examples show several uses of the bar graph.

**Example 3:** The registration figures for the Fall term at the five campuses of the Peralta Community College District are:

| Laney | Merritt | Alameda | North Peralta | Feather River |
|-------|---------|---------|---------------|---------------|
| 8,800 | 5,600 | 4,100 | 1,400 | 600 |

*Procedure:* To construct the bar graph, draw two lines—one line horizontal and one line vertical. Label the vertical axis with the names of the five colleges. Let each unit (square) along the horizontal axis represent 500 students—this is the scale for determining the length of each bar. Place a zero (0) at the point where the axes cross. (See Fig. 13-3.)

To determine the length of each bar, divide the number of students registered at each college by the unit of 500. Thus:

$$\text{Length of bar for Alameda} = \frac{4,100}{500} = 8.2 \text{ units}$$

$$\text{Length of bar for Feather River} = \frac{600}{500} = 1.2 \text{ units}$$

$$\text{Length of bar for Laney} = \frac{8,800}{500} = 17.6 \text{ units}$$

$$\text{Length of bar for North Peralta} = \frac{1,400}{500} = 2.8 \text{ units}$$

$$\text{Length of bar for Merritt} = \frac{5,600}{500} = 11.2 \text{ units}$$

Fractional parts of units are approximated on the graph in Fig. 13-3.

**Example 4:** This is an example of a *double* bar graph. Al and Jon are both students in a technical mathematics class.

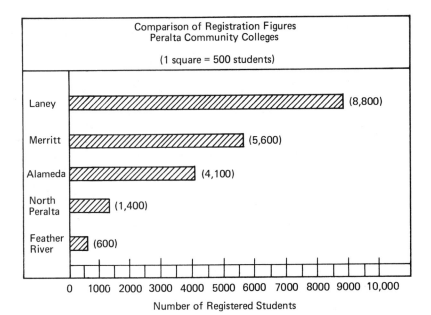

**Figure 13-3**

They decide to use a bar graph to compare their test scores. Their tests are scored on a basis of 100%. Table 13-1 gives the results of their tests.

**TABLE 13-1**  TEST RESULTS BY PERCENT CORRECT

| Test number | Al's score (%) | Jon's score (%) |
|---|---|---|
| 1 | 76 | 83 |
| 2 | 68 | 54 |
| 3 | 87 | 65 |
| 4 | 75 | 89 |
| 5 | 80 | 95 |
| 6 (Final) | 85 | 88 |

The results in Table 13-1 are shown in the double-bar graph in Fig. 13-4. Notice that the vertical axis is graduated in percent —this is the scale.

The test numbers are shown along the horizontal axis. The two bars are drawn side by side, which makes it easy to compare the results. The use of different colors or methods of shading the bars also makes the comparison easier to see.

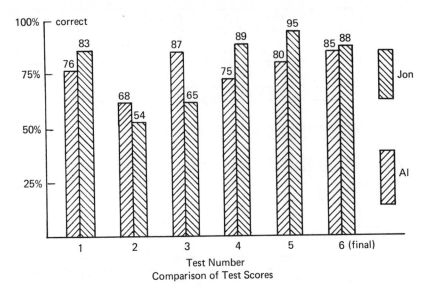

Comparison of Test Scores

*Figure 13-4*

Bar graphs may have several bars, or they may have only one bar. *Single-bar graphs* (also called 100% graphs) are useful to show the relationship of the various parts to the whole. Here are two examples of single bar graphs.

**Example 5:** For each sales dollar taken in by the J & R Tool Company, the accountant knows that:

$0.35 goes for wages
$0.47 goes for cost of material
$0.11 goes for business expense
$0.07 is profit.

Total  $1.00

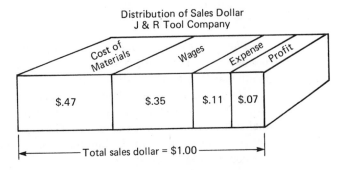

*Figure 13-5*

The single bar graph in Fig. 13-5 shows the various parts of the sales dollar. The total sales dollar is represented by $1.00. The parts are shown in proper proportion to the total.

**Example 6:**  An alloy used for setting type for a newspaper, magazine, etc., is called *type* metal. A standard formula for type metal is 78% lead, 6% tin, and 16% antimony. The single-bar graph in Fig. 13-6 shows the percent amount of each substance in 100% type metal.

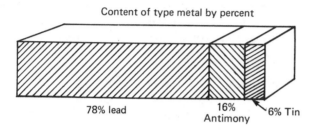

Content of type metal by percent

**Figure 13-6**

Suggestions for making bar graphs:

1. Get all of the numerical information necessary for the graph.
2. Round off numerical values so they can be easily represented.
3. Plan the bar graph carefully. Make a sketch to decide whether to use horizontal or vertical bars. (A graph should be pleasing to the eye as well as informative.)
4. Decide upon a suitable scale and clearly show it on the graph.
5. Clearly label the horizontal and vertical axes.
6. Draw the bars all the same width. Leave space between the bars.
7. Give the graph a suitable title.

## Line Graphs

A *line graph* is sometimes called a *broken-line graph*. Line graphs use one or more lines to show the changes that take place between numerical quantities. They are also used to show different values of a quantity at different times. Line graphs show an increase or decrease in a quantity by the rising or falling of a line in the line graph.

To make a line graph, use *squared* graph paper. This paper has both horizontal and vertical lines ruled at regular intervals. Some graph paper may have as few as four lines per inch and other paper may have ten or more lines per inch. The desired accuracy of the resulting graph will determine the particular graph paper to use. Start by using inexpensive graph paper with no more than four or five lines per inch.

To make a line graph draw a horizontal line and a vertical line as axes. The point where the two axes cross is called the *origin*. The vertical and horizontal scales do not have to be equal. Be sure that the graph will fit on the paper. A bit of planning will avoid loss of time and will result in a neat, accurate, and attractive graph.

To locate a point (that is, to *plot* a point) on the line graph, start at the origin and move to the right along the horizontal axis and find a number that corresponds to a known value. Start at this point on the horizontal axis and move upward (vertically) until you are opposite a known value on the vertical axis. Make a dot (·) at this point on the graph paper. Use known values from both horizontal and vertical axes and plot points of the graph in the manner described.

Finish the line graph by connecting the consecutive points with straight lines.

**Example 7:** Table 13-2 shows *average* temperatures for twelve months for a small eastern town. Make a line graph that shows the increase and decrease of average monthly temperatures.

*Procedure:* To make the line graph choose the time scale on the horizontal axis, starting with the month of January. Let each horizontal unit represent one month. Start with zero at the origin and let each vertical unit represent 10 degrees of temperature.

**TABLE 13-2**   AVERAGE TEMPERATURE FOR MONTHS IN DEGREES FAHRENHEIT

| Jan. | Feb. | March | April | May | June | July | Aug. | Sept. | Oct. | Nov. | Dec. |
|------|------|-------|-------|-----|------|------|------|-------|------|------|------|
| 32°  | 34°  | 38°   | 50°   | 59° | 65°  | 70°  | 75°  | 68°   | 54°  | 42°  | 38°  |

1. Start at the origin (January, 0°) and go vertically 32 units. This point on the temperature axis represents the *average* temperature for the month of January.
2. To plot the next point, go one unit to the right to the month of February. Go vertically 34 units. The point represents the average temperature for the month of February.
3. Continue in this manner until you have plotted all twelve points.
4. Start at the first point and connect each successive point with a straight line. The result is the line graph shown in Fig. 13-7.

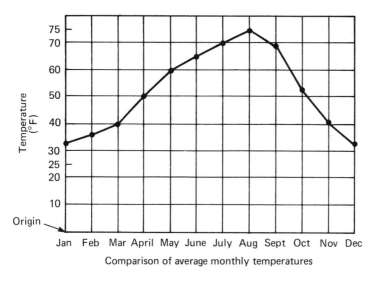

Comparison of average monthly temperatures

*Figure 13-7*

**Example 8:** This is a *double-line graph*. Two line graphs are drawn to compare two sets of numerical values. Between the ages of 10 and 16, boys and girls gain weight at different average rates. The average annual gain in weight for boys and girls between the ages of 10 and 16 is given in Table 13-3. Make a double line graph that shows a comparison of weight gained per year.

*Procedure:* To make the line graph, let the horizontal axis be the age in years, starting at 10 years and ending at 16 years. Use a scale of five units for one year. Let the vertical axis be the number of pounds of weight gained. Let one unit be 1 pound.

**TABLE 13-3**

| Age | 10 | 11 | 12 | 13 | 14 | 15 | 16 |
|-----|-----|-----|-----|-----|-----|-----|-----|
| | Average weight gained per year (in pounds) | | | | | | |
| Boys | 8 | 8.5 | 9 | 11 | 14 | 13 | 10 |
| Girls | 8 | 10.5 | 12 | 10 | 9.5 | 6.5 | 3 |

Start at the origin and mark off values at 4, 8, 12, and 16 pounds on the vertical axis. Draw the line graph for the boys first. The origin represents the age of 10 years and zero pounds. To plot the first point, go up the vertical (pounds gained) axis to 8. This point corresponds to: (boys) age = 10, pounds gained = 8.

Find the next point by going horizontally until you reach 11 (years). Go vertically to reach 8.5. The result is the point (11 years; 8.5 pounds). Continue plotting points in this manner until all of the boys' age and weight values are represented.

Start at the first point (10 years, 8 pounds) and connect successive points with straight lines. The result is the line graph for the average weight gain per year for boys between the ages of 10 and 16 years.

Now plot the points for the girls. Connect successive points with a broken or dotted line to distinguish it from the graph for the boys. The resulting graph is shown in Fig. 13-8.

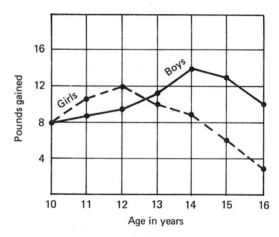

Comparison of weight gain for boys and girls

**Figure 13-8**

Here are some suggestions for making line graphs:

1. Plan your graph. Make sure it will fit on the ruled paper.
2. Clearly indicate the quantities represented on both the horizontal and vertical axis. Show the scale to be used on both axes.
3. Connect successive points with straight lines, dotted lines, or use color. Make your graph as informative and attractive as possible.
4. Give the graph a suitable title.

# PROBLEMS: Sec. 1 Picture, Bar, and Line Graphs

1. Make a horizontal bar graph that shows the comparison between the five most densely populated states in the United States in 1970:

   *Density of population by states*

   | | |
   |---|---|
   | New Jersey | 953 persons per square mile |
   | Rhode Island | 906 persons per square mile |
   | Massachusetts | 727 persons per square mile |
   | Connecticut | 624 persons per square mile |
   | Maryland | 397 persons per square mile |

2. Use the values in Table 13-4. Make a bar graph that shows the strip mining of coal in one year.

   **TABLE 13-4**

   | State | Coal produced by strip mining (millions of tons) |
   |---|---|
   | Kentucky | 56.8 |
   | Ohio | 37.6 |
   | Illinois | 29.0 |
   | Pennsylvania | 28.0 |
   | W. Virginia | 21.1 |

3. Make a bar graph that shows the comparison between the five longest underwater tunnels in North America.

   | Name | Waterway | Length |
   |---|---|---|
   | BART Trans-Bay | San Francisco Bay | 3.9 miles |
   | Brooklyn-Battery | East River | 9,117 feet |
   | Holland Tunnel | Hudson River | 8,557 feet |
   | Lincoln Tunnel | Hudson River | 8,216 feet |
   | Baltimore Harbor Tunnel | Patapsco River | 7,650 feet |

4. Table 13-5 shows the percent of the earth occupied by the seven continents. Make a bar graph that shows this data.

   **TABLE 13-5**

   | Continent | Percent of earth occupied |
   |---|---|
   | Asia | 29.5% |
   | Africa | 20.0% |
   | North America | 16.3% |
   | South America | 11.8% |
   | Antartica | 9.6% |
   | Europe | 6.5% |
   | Australia | 5.2% |

5. Make a single (100%) bar graph that shows the composition of the air. (Excluding smog, of course!)

Composition of air by gases

| Gas | Percent present |
|---|---|
| Nitrogen | 78% |
| Oxygen | 21% |
| Neon, Helium, Krypton, etc. | 1% |

6. Make a double-bar graph that compares the number of people employed and the number of unemployed people for the years shown.

| Year | Employed | Unemployed |
|---|---|---|
| 1963 | 67,800,000 | 4,100,000 |
| 1966 | 72,900,000 | 2,900,000 |
| 1969 | 77,900,000 | 2,800,000 |
| 1972 | 79,100,000 | 5,000,000 |
| 1973 | 80,500,000 | 5,100,000 |

7. Draw a broken-line graph that shows the *purchasing power of the dollar.* Use year 1967 as an index. Thus, for 1967 one dollar was worth $1.00 in purchasing power.

| Year | Purchasing power of the dollar compared to 1967 |
|---|---|
| 1950 | 1.22 |
| 1952 | 1.13 |
| 1954 | 1.14 |
| 1956 | 1.10 |
| 1958 | 1.06 |
| 1960 | 1.05 |
| 1962 | 1.06 |
| 1964 | 1.06 |
| 1966 | 1.00 |
| 1968 | .98 |
| 1970 | .91 |
| 1972 | .87 |

8. Draw a broken-line graph that shows the population growth of the state of New York over 20-year intervals from 1810 to 1970. (Values are to nearest thousand.)

| Year | Population of New York State |
|---|---|
| 1810 | 959,000 |
| 1830 | 1,918,000 |
| 1850 | 3,097,000 |
| 1870 | 4,383,000 |
| 1890 | 6,003,000 |
| 1910 | 9,114,000 |
| 1930 | 12,588,000 |
| 1950 | 14,830,000 |
| 1970 | 18,241,000 |

9. Draw a double broken-line graph that shows the years of life expectancy of smokers (40 cigarettes per day) compared to a nonsmoker. (Source: The American Cancer Society)

| At age: | Nonsmoker can expect to live: | Smoker can expect to live: |
|---|---|---|
| 25 | 49 years | 40 years |
| 30 | 44 years | 36 years |
| 35 | 40 years | 31 years |
| 40 | 35 years | 27 years |
| 45 | 30 years | 23 years |
| 50 | 26 years | 19 years |
| 55 | 21 years | 16 years |
| 60 | 18 years | 13 years |
| 65 | 14 years | 11 years |

10. Draw a broken-line graph that shows the population per square mile of land area in the United States for the dates shown.

| Year | Population per square mile of land area |
|---|---|
| 1820 | 5.5 |
| 1840 | 9.8 |
| 1860 | 10.6 |
| 1880 | 16.9 |
| 1900 | 25.6 |
| 1920 | 35.6 |
| 1940 | 44.2 |
| 1960 | 50.5 |
| 1970 | 57.5 |

11. Draw a broken-line graph that shows the increase in automobile factory sales for the years indicated. (Values to nearest 1,000)

| Year | Factory sales (Cars, Trucks, Busses) |
|---|---|
| 1900 | 4,000 |
| 1910 | 187,000 |
| 1920 | 2,228,000 |
| 1930 | 3,363,000 |
| 1940 | 4,472,000 |
| 1950 | 8,003,000 |
| 1960 | 7,869,000 |
| 1970 | 8,239,000 |

# SEC. 2 CIRCLE GRAPHS

*Circle graphs* are sometimes called *pie graphs*. They are used to show the relationship of the various parts of a quantity to the whole. Circle graphs are used in the same way that 100% bar graphs are used. Circle graphs are used to show such things as: How your tax dollar is spent, how a city uses its water supply, or how the family budget is distributed. They are also used in advertising because they are attractive and easily understood.

In general, circle graphs are not as accurate as other graphs. Use them only when necessary.

To construct a circle graph use a protractor. Figure 3-9 shows a protractor. A *protractor* is an instrument used for drawing or measuring angles. It may also be used to divide a circle into any number of parts.

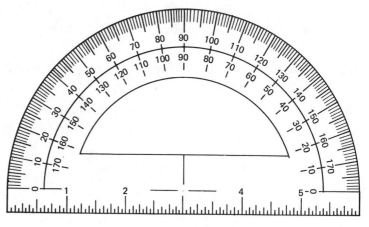

**Figure 13-9**

Look at a protractor. There are two rows of numbers going around the curved section. Start at the left and notice that the *outer row* begins with 0° and, going clockwise, ends with 180° on the right. The *inner row* of numbers starts with a 0° at the right and, going counterclockwise, ends with 180° on the left. These two rows of numbers are for drawing and measuring angles. An *angle* is the space between two lines that meet at a point. Angles are measured in degrees.

Figure 13-10 shows an angle. The point (B) where the two lines meet is called the *vertex* of the angle. The angle is usually named by the three letters A, B, and C, and is written;

angle ABC.

Sometimes the symbol ∠ is used for angle; thus,

angle ABC = ∠ABC.

The letter (B) at the vertex of the angle is written between the letters that name the lines.

Here is how you find the number of degrees in an angle ABC. (See Fig. 13-11.)

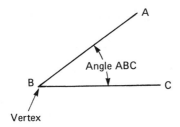

**Figure 13-10**

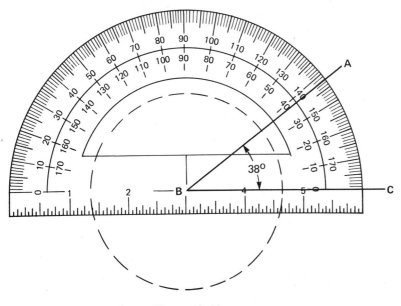

**Figure 13-11**

Step 1. *Imagine* the vertex (B) of the angle is at the center of a circle. (The size of the circle is not important.)

Step 2. Place the center (0) of the protractor at the vertex B of the angle.

Step 3. Place the protractor in such a position that the base line matches one of the sides of the angle. (Side BC in the figure.)

Step 4. The side of the angle that is along the base line of the protractor cuts across the zero degree mark (0°) on one of the two rows of numbers on the protractor. (In the figure the side BC cuts across the zero degree mark on the right side of the protractor.)

Step 5. The other side of the angle (side BA in the figure) cuts

across a degree mark on the curved portion of the protractor.

Step 6.  When the side of the angle along the base line cuts through
the zero (0°) on the right, use the *inner set* of numbers for
reading the angle. When the side of the angle along the
base line cuts through the zero on the left, use the *outer set*
of numbers for reading the size of the angle. In the figure
the inner set of numbers is used. The angle ABC is 39°.

Use the protractor to measure the angles in Figs. 13-12 to 13-15.

Circle graphs are used to show the relationship of the parts of a whole
quantity to each other *and* to the whole quantity. The entire circle is the
whole quantity. The circle is divided into parts called *sectors* of the circle.
Figure 13-16 shows a sector of a circle. Each sector of the circle is meas-

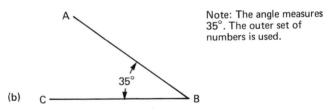

Note: The angle measures
30°. The inner set of
numbers is used.

**Figure 13-12**   *The angle measures 30°. The inner set of numbers is
used.*

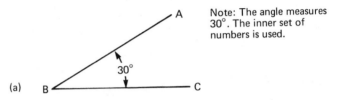

Note: The angle measures
35°. The outer set of
numbers is used.

**Figure 13-13**   *The angle measures 35°. The outer set of numbers is
used.*

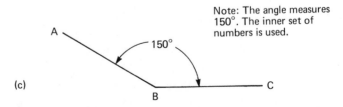

Note: The angle measures
150°. The inner set of
numbers is used.

**Figure 13-14**   *The angle measures 150°. The inner set of numbers
is used.*

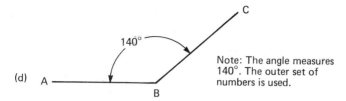

Figure 13-15 *The angle measures 140°. The outer set of numbers is used.*

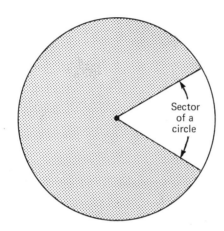

Figure 13-16

ured with a protractor. The sum of all the angles of the sectors must add up to 360°.

Find the size of a sector in a circle graph by writing the proportion:

$$\frac{\text{Sector of circle (degrees)}}{\text{Whole circle (360 degrees)}} = \frac{\text{Amount represented by sector}}{\text{Total quantity}}$$

From this direct proportion you see that:

$$\text{Sector (in degrees)} = \left(\frac{\text{Amount represented by sector}}{\text{Total Quantity}}\right) \times 360°$$

Here are some suggestions for making circle graphs:

1. Find the amount of the whole quantity and the amount of each item of the whole.
2. Find the fraction (or percent) that each item is of the whole.
3. Multiply the fraction (or percent) from Step 2 by 360°. The result is the number of degrees in the sector that represents the part of the whole. (Values are usually rounded off to nearest degree.)
4. Draw a circle and use a protractor to mark off the sectors that represent each part of the whole. (The total circle must be used.)

5. Clearly label each sector with appropriate names or numerical values.
6. Given the graph a suitable title.

Examples 9 and 10 show how circle graphs are used.

**Example 9:** The city of Redwing buys its water from a large municipal utilities district. The city used the water in the following amounts: homes, 45%; schools, hospitals, parks, etc., 21%; business and industrial, 27%; leakage, evaporation, and other losses, 7%. Make a circle graph to show this information.

*Procedure:* In this case the whole is 100%; the various parts are 45%, 21%, 27%, and 7%. Divide the circle graph into four sectors. Find the number of degrees in each sector by multiplying each percent value by 360°. (See Fig. 13-17.)

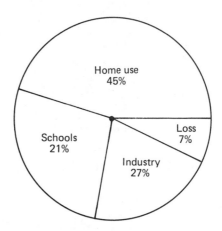

**Figure 13-17**   *How Redwing uses its water supply*

Size of sector showing
home use
$$= 45\% \times 360°$$
$$= .45 \times 360$$
$$= 162°$$

Size of sector showing
school use, etc.
$$= 21\% \times 360°$$
$$= .21 \times 360$$
$$= 76° \text{ (nearest degree)}$$

Size of sector showing
industry use
$$= 27\% \times 360°$$
$$= .27 \times 360$$
$$= 97° \text{ (nearest degree)}$$

Size of sector showing
losses                      $= 7\% \times 360°$
                            $= .07 \times 360$
                            $= 25°$ (nearest degree)
Check: $162° + 76° + 97° + 25° = 360°$. Draw a circle
and mark sectors with a proctractor. Label sectors and
give the graph a title. (See Fig. 13-17.)

**Example 10:** The board of directors of International Stump wants
to send a circle graph to its stockholders to show
the distribution of each sales dollar. The company
accountant shows:
Distribution of sales dollar (100¢)
Supplies   Labor Taxes   Business Expense   Profit
  62¢      19¢   4¢           9¢             6¢
Show this information in a circle graph.

*Procedure:* The total (whole) is equal to 100¢. Express each part of
the sales dollar as a percent of the whole:

    Supplies $= 62¢$ of sales dollar $= 62\%$
    Labor    $= 19¢$ of sales dollar $= 19\%$
    Taxes    $= \phantom{0}4¢$ of sales dollar $= \phantom{0}4\%$
    Expense $= \phantom{0}9¢$ of sales dollar $= \phantom{0}9\%$
    Profit     $= \phantom{0}6¢$ of sales dollar $= \phantom{0}6\%$

Find the number of degrees in each of the five sectors
by multiplying each percent value by 360°.

The results are shown in the circle graph Fig. 13-18.

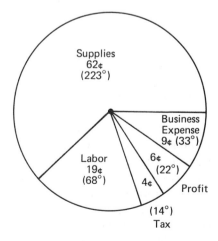

**Figure 13-18**   *Distribution of International Stump sales dollar*

## PROBLEMS:  Sec. 2  Circle Graphs

**1.** Seawater bronze is: 45% copper, 33% nickel, 16% tin, 5% zinc, and 1% bismuth. Draw a circle graph that shows the composition of Seawater bronze.

**2.** Between January 1, 1961, and October 7, 1972, the United States armed forces suffered a total of 45,882 deaths in Southeast Asia. Of this total, the Army reported 30,566 killed, the Navy 1,416, the Marine Corps 12,935, and the Air Force 965. Draw a circle graph that shows this information.

**3.** Gordon Bush makes $540 a month working at a service station. His deductions are: 18% federal income tax, 8% state income tax; 5% retirement fund; 3% FICA (Social Security). Draw a circle graph showing all deductions and take-home pay.

**4.** In a recent survey of educational background and type of employment it was found that:

   5% of the employed have grade school or less.

   27% of the employed have high school.

   48% of the employed have technical education.

   20% of the employed have 4 or more years college education.

Draw a circle graph showing this data.

**5.** For the fiscal year 1972–1973, the city of Piedmont, California, allotted a budget of $1,601,538 in the following way:

| Cost Unit | Amount | Budget (%) |
|---|---|---|
| Public Works | $326,838 | 20.4 |
| Police Department | 344,912 | 21.5 |
| Fire Department | 314,561 | 19.6 |
| Administration | 189,311 | 11.8 |
| Hospitalization Plan | 21,479 | 1.4 |
| City Employees Pension | 49,712 | 3.0 |
| Police & Fire Pension | 98,009 | 6.2 |
| Recreation Department | 115,901 | 7.2 |
| Park Department | 140,815 | 8.8 |

Make a circle graph that shows the expenditures of Piedmont.

## SEC. 3  GRAPHS IN TECHNOLOGY

Every technician, mechanic, and tradesman looks forward to an advancement in position and an increase in wages. To be eligible for advancement, he or she must understand graphs and be able to use them.

The technician who uses all available methods for problem solving will rapidly climb the ladder of advancement. As a person in a position of responsibility, you may be required to make and use many types of graphs.

In Secs. 13-1 and 13-2 you learned to construct and read four of the most common types of graphs. Now you will study the graphs that are often used in trade journals, engine specifications, building codes, structural requirements, etc. You will also learn to make and read graphs of many trade formulas.

To *graph a formula* means to show the relationship between the quantities that can change in value.

**Example 11:** Graph the formula for the circumference (*C*) of a circle with diameter *D*.

$$C = \pi D \qquad \text{(Note: Use 3.142 for } \pi \text{)}$$

*Procedure:* Follow the method for a line graph. Use ruled paper and draw horizontal and vertical axes. Label the horizontal axis the *diameter* (*D*). Label the vertical axis the *circumference* (*C*). Start both axes with the number 0. To make the graph, find the values for the circumference (*C*) for various values of the diameter (*D*). Do this by making a table of values. Use the formula to compute the value for *C*, as *D* is assigned different numbers. Plot the values for *D* and *C* using the methods for a line graph. Look at the result in Fig. 13-19. The horizontal and vertical scales in Fig. 13-19 are NOT the same. Each horizontal unit equals 1 inch. Each vertical unit equals 5 inches. When the points are connected, the graph is a *straight* line.

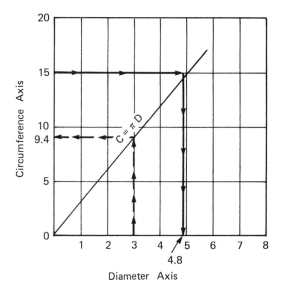

**Figure 13-19**

| Table of Values for the Formula: $C = 3.142D$ | | | | |
|---|---|---|---|---|
| $D$: 0 | 1 | 2 | 4 | 6 |
| $C$: 0 | 3.142 | 6.284 | 12.568 | 18.852 |

### Rule for Straight Line Graphs

*When one (letter) value in a formula is equal to the product of a number and another (letter) value, the graph of the formula will ALWAYS be a straight line.*

For instance, the formulas $E = 10R$, $W = 117I$, and $D = 45T$ will have a straight line graph. (You will graph each of these formulas in the exercises.) Once you have made the graph of a formula you can use it to get numerical results that are not in the original table. Look again at the graph of the formula $C = 3.142D$ shown in Fig. 13-19. Using the graph, you can get good approximations for the circumference of a circle with a diameter *any* value between 0 and 6 inches.

Getting new values between those already known is called *interpolation*.

**Example 12:** Use interpolation to find an approximate value for the circumference of a circle with diameter $D = 3$ inches. (Use Fig. 13-19.)

*Procedure:* Find the value 3 on the horizontal diameter axis. Go vertically until you reach the graph of the formula. Move horizontally to the LEFT until you reach the vertical (circumference) axis. From the graph, you see that you are approximately 9.4 units up the vertical axis.

*Answer:* The circumference of a circle with diameter 3 inches is approximately 9.4 inches.

NOTE: Actual value is 9.426 inches.

You may reverse the roles of the letters *C* and *D*. That is, you can use the graph to find the approximate diameter of a circle when the circumference is known.

**Example 13:** Use the graph in Fig. 13-19 and find the diameter of a circle with circumference equal to 15 inches.

*Procedure:* Start at the origin. Go up the vertical (circumference) axis to 15. Stay at this level and go to the RIGHT until you come to the graph. At the point where you intersect the graph, go vertically DOWN until you reach the horizontal axis. The value on the horizontal axis is approximately 4.8.

*Answer:* The diameter of a circle with circumference 15 inches is approximately 4.8 inches.

NOTE: The actual value is $D = 4.78$.

In many cases you may have to use a graph to obtain the answer to a problem. Manufacturers often use graphs to show information about their products. An example of this is the performance of grinding wheels. Machine work is not always turned to its final value on a lathe or milling machine. Instead, it is often turned or planed to dimensions that are within .010 to .001 inch of the finished dimension. Any excess is removed by a grinding process. A precision grinder can make cuts of .001 inch or less and leave a very smooth surface. For this reason, grinding is a good way to get critical dimensions. Grinding also produces a polished product.

For most precision grinding jobs, you must know the *surface* or *rim speed* of the grinding wheel. Most of this type of grinding is done at surface speeds between 5,000 and 6,000 feet per minute. But the speed of a grinding motor is usually stated in revolutions per minute (rpm), and not feet per minute (fpm). You must be able to find the number of revolutions per minute for surface speeds of 5,000 or 6,000 feet per minute for various diameters of grinding wheels.

The graph in Fig. 13-20 shows two curves. The lower curve shows the relationship between the diameter of a grinding wheel (in inches) and

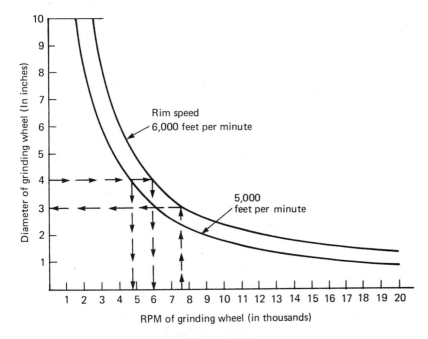

**Figure 13-20**

the number of revolutions per minute for a surface speed of 5,000 feet per minute. The outer curve shows the same relationships for a surface speed of 6,000 feet per minute.

Using the graph and interpolation, you can find the number of revolutions per minute for a particular diameter of a grinding wheel. You can also find the diameter of a grinding wheel for a specified number of rpms and surface speed. (The exact speed of the grinding wheel will depend on the particular machine used. It will also depend on the type of metal being ground, the amount of force between grinder and metal, and other factors.)

Use the graph in Fig. 13-20 to determine the diameter of a grinding wheel or the rpm for a grinder.

**Example 14:** Find the number of rpms for a grinding wheel 4 inches in diameter that is to have a surface speed of 5,000 fpm, and for 6,000 fpm.

*Procedure:* Use interpolation to determine the rpm. The vertical axis shows various diameters from 1 to 10 inches. Start at 4 inches and go horizontally until you intersect the two curves. At the point where you intersect the first (inner) curve (5,000 fpm curve), go down to the rpm axis. Estimate the value 4,900 rpm on the (horizontal) rpm axis.

*Answer:* For a surface speed of 5,000 fpm, a 4-inch grinding wheel turns about 4,800 rpm. (The actual computed value is 4,775 rpm.) For a surface speed of 6,000 fpm, the graph shows an approximate value of 5,700 rpm. (The actual value is 5,730 rpm.) Although both of these values are estimates, they are close enough for most shop jobs.

**Example 15:** A motor for a grinder shows an rpm value of 7,500. What size grinding wheel should be attached for a surface speed of 6,000 fpm?

*Procedure:* Use Fig. 13-20. The value of 7,500 rpm on the horizontal axis is between the 7 and 8. Start at this point and go vertically until you intersect the 6,000 fpm curve (the outer curve). At the point of intersection, go to the left until you reach the vertical (diameter) axis. You see the value is about 3 inches.

*Answer:* A 3-inch wheel turning at 7,500 rpm produces a surface speed of 6,000 fpm. (Actual value: 3-inch wheel turning at 7,639 rpm gives 6,000 fpm.)

From these examples you can see that graphs are valuable in solving many problems. In most cases you may use the graph to get approximate answers and avoid possible difficult computations. You may use graphs also as a means for checking computations. When you make a computation and a graph of the formula is available, use it to verify the result. In this way you can avoid major errors.

Each of the following examples shows the methods for graphing formulas. Study them because they are graphs of familiar trade formulas.

**Example 16:** In direct current circuits you can find the power ($P$) by using the formula:

$$P = RI^2$$

$R$ = resistance in ohms

$I$ = current in amperes

Obtain a graph of the power curve for a fixed resistance of 10 ohms and for current values between 0 and 15 amps. Graph the formula: $P = 10I^2$ for $I$ between 0 and 15 amps

*Procedure:* Use ruled graph paper and draw a horizontal (current) axis and a vertical (power) axis. To find the horizontal and vertical scales notice that when $I = 0$, $P = 0$; but when $I = 15$, $P = (10)(15)^2 = 10(225) = 2,250$ watts. Compress the vertical scale to fit on the paper. To find the shape of the curve, compute several values for the power. (See the table below Fig. 13-21.) Plot the points

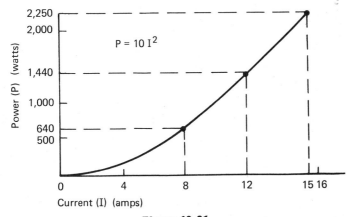

**Figure 13-21**

$$P = 10I^2$$

| $I$ (amps) | 0 | 2 | 4 | 6 | 8 | 10 | 12 | 14 | 15 |
|---|---|---|---|---|---|---|---|---|---|
| $P$ (watts) | 0 | 40 | 160 | 360 | 640 | 1000 | 1440 | 1960 | 2250 |

and draw a SMOOTH curve connecting the points. Do not connect points with straight lines—this is not a broken line graph.

The table shows values of $P$ for various values of $I$. By using this curve you can estimate the power ($P$) for different values of the current ($I$). You can also find the current needed for various power values.

# PROBLEMS: Sec. 3  Graphs in Technology

1. Use the table of values below to graph the formula $E = 10R$. The resulting graph is a straight line.

| $E = 10R$ | | | | | | |
|---|---|---|---|---|---|---|
| $R = 0$ | 1 | 2 | 4 | 6 | 8 | 10 |
| $E = 0$ | 10 | 20 | 40 | 60 | 80 | 100 |

2. Get the straight-line graph of the formula $W = 117I$ by making a table of values.

3. Get the straight-line graph of the formula $D = 45T$.

4. The cutting speed for a piece of work $\frac{1}{2}$ inch in diameter for various speeds (in rpm) is given in the following table. Graph the table values and connect all points with a *smooth* curve.

| *Cutting speeds of $\frac{1}{2}$-inch diameter tailstock* | | | | | | | | |
|---|---|---|---|---|---|---|---|---|
| *Revolutions per minute* (rpm) | | | | | | | | |
| 153 | 229 | 306 | 382 | 458 | 534 | 612 | 633 | 764 |
| *Cutting speed in feet per minute* (fpm) | | | | | | | | |
| 20 | 30 | 40 | 50 | 60 | 70 | 80 | 90 | 100 |

5. Use the graph from Problem 4. What is the approximate speed (in rpm) for a cutting speed of 75 fpm?

6. The surface speed (in fpm) of a 6-inch diameter grinding wheel turning at various speeds is given in Table 13-5. Graph the table values (straight-line graph).

   What is the approximate speed in rpm of a 6-inch grinding wheel that is to have a surface speed of 2,000 fpm?

**TABLE 13-5**

| Revolutions per minute (rpm) | Surface speed 6-in. diameter (fpm) |
|:---:|:---:|
| 0 | 0 |
| 500 | 785 |
| 1,000 | 1,571 |
| 1,500 | 2,356 |
| 2,000 | 3,142 |
| 2,500 | 3,927 |
| 3,000 | 4,712 |

7. Table 13-6 shows the electrical resistance in ohms per 1,000 feet of standard annealed copper wire of different diameters. The diameters are in mils (1 mil = .001 inch).

**TABLE 13-6**

| Diameter (mils) | Resistance in ohms per 1,000 feet of wire |
|:---:|:---:|
| 200 | 0.25 |
| 100 | 1.02 |
| 80 | 1.62 |
| 50 | 4.10 |
| 30 | 10.35 |
| 20 | 26.18 |
| 10 | 105.2 |
| 5 | 423 |

Plot the data and draw a smooth curve through the points. This is not a broken-line graph.

**TABLE 13-7**

| Altitude above sea level (in miles) | Barometric pressure (in inches of mercury) |
|:---:|:---:|
| 0 | 29.92 |
| 1 | 24.5 |
| 2 | 20.0 |
| 3 | 16.2 |
| 4 | 13.5 |
| 5 | 11.0 |
| 6 | 8.9 |
| 7 | 7.3 |
| 8 | 6.0 |
| 9 | 4.9 |
| 10 | 4.0 |

8. Approximate barometric pressures (in inches of mercury) at different altitudes above sea level are shown in Table 13-7. Plot the data and draw a smooth curve through the points. This is not a broken-line graph.
At approximately what altitude is the barometric pressure 15 inches of mercury?

9. Extend the vertical axis downward to include negative values. Draw a double-line graph that shows the comparison between the two sets of data. The freezing temperature for two of the more common antifreeze solutions for various percent compositions are given in Table 13-8.

**TABLE 13-8**

| | Freezing temperature (°F) | |
| *Percent by volume* | *Prestone* | *Denatured alcohol* |
|:---:|:---:|:---:|
| 10 | +25 | +25 |
| 20 | +16.5 | +17.5 |
| 25 | +11 | +14 |
| 30 | +5 | +5 |
| 35 | −3 | −1 |
| 40 | −12 | −11 |
| 45 | −25 | −18 |
| 50 | −38 | −25 |

10. Plot and draw a smooth curve showing the relation between an angle A and the value of sin A. Use the horizontal axis for angle measures. Plot values for every 5° of angle. Use values from the Table 12-1 on page 382.

11. Plot and draw a smooth curve showing the relation between an angle A and the value of tan A. Use the horizontal axis for angle measure. Plot values for every 5° of angle. Use values from Table 12-1 on page 382.

12. Plot two curves that compare the horsepower delivered at the rear wheels for various rpm values and two different engine setups.

| | Horsepower delivered | |
| *Engine speed* (rpm) | *Supercharged* (*dual exhaust*) | *Supercharged* (*without dual exhaust*) |
|:---:|:---:|:---:|
| 2,000 | 77 | 73 |
| 2,200 | 82 | 78 |
| 2,400 | 88 | 84 |
| 2,600 | 96 | 92 |
| 2,800 | 105 | 99 |
| 3,000 | 113 | 106 |
| 3,200 | 117 | 110 |
| 3,400 | 119 | 112 |
| 3,600 | 118 | 112 |

**13.** Plot two curves that compare the horsepower delivered at the rear wheels for various rpm values and two different engine setups.

| | Horsepower delivered | |
| --- | --- | --- |
| Engine speed (rpm) | Unsupercharged (dual exhaust) | Unsupercharged (standard exhaust) |
| 2,000 | 65 | 60 |
| 2,200 | 71 | 65 |
| 2,400 | 75 | 69 |
| 2,600 | 79 | 74 |
| 2,800 | 83 | 77 |
| 3,000 | 85 | 79 |
| 3,200 | 86 | 80 |
| 3,400 | 84 | 77 |
| 3,600 | 80 | 73 |

## Practice Test for Chapter 13

**1.** The sources of revenue for a small town are:

Property Taxes = $1,258,538   (78.6% of total)

Fees, Grants, etc. = $245,000   (15.2% of total)

Reserves = $98,000   (6.2% of total)

Draw a vertical bar graph that shows this information.

**2.** Use the information from Problem 1 and draw a pie graph (circle graph).

**3.** Table 13-9 gives corresponding temperature readings for both the Fahrenheit and Celsius temperature scales. Plot this data and graph the result. It is a straight line.

**TABLE 13-9**

| Fahrenheit temperature (°F) | Celsius temperature (°C) |
| --- | --- |
| 32° | 0° |
| 122° | 50° |
| 212° | 100° |

Normal body temperature is 98.6°F. What is the equivalent Celsius temperature?          *Answer:* _____

**4.** Table 13-10 gives the corresponding values between U. S. gallons and liters. Plot the data and draw the resulting graph.

**TABLE 13-10**

| U. S. Gallons | Liters |
|:---:|:---:|
| 1 | 3.86 |
| 2 | 7.72 |
| 3 | 11.57 |
| 4 | 15.44 |
| 5 | 19.30 |
| 10 | 38.60 |
| 15 | 57.90 |
| 20 | 77.20 |

5. Use the graph from Problem 4 to determine approximately how many U. S. gallons are equal to 50 liters.          *Answer:* _____

# 14

# Applied
# Algebra

Algebra is a highly useful mathematical tool. It is used by scientists and engineers to solve problems in electronics, bridge design, rocket engines, space hardware, and so on. To the scientist, *algebra* is important in the search for knowledge. It is used to develop new theories and is applied to existing ones. To the technician, engineering aide, engine mechanic, and tradesman, *algebra* is important in solving practical problems. Practical algebra is a *must* for the modern-day technician.

A technician who understands practical algebra is a valuable person. He or she knows that a formula is not a meaningless collection of numbers and letters. He or she also knows that the solution of a trade problem with algebra does not involve mumbo-jumbo or sleight of hand.

Many industries require their technicians to have a working knowledge of algebra before they can be promoted to positions that have more responsibility and higher earnings.

The language of algebra is made up of *signs of operation* and *symbols that represent numbers*. The signs of operation are the directions that tell you what to do. These signs are the familiar signs from arithmetic: $+, -, \cdot, \div$. You already know that each of these signs has a particular meaning in arithmetic. Their meaning is the same in algebra.

Algebra is NOT some mysterious language that is only used and understood by a select few. It uses the same rules and numbers that you studied in arithmetic. If you learned the arithmetic in the preceding chapters, you are well on your way to mastering algebra.

In arithmetic you use the Hindu-Arabic numerals 0, 1, 2, 3, . . . , 9 to represent numbers. These numerals are used to represent *specific* numbers such as 1,876; 0.0098; and 34.56. In algebra you will use letters such as $a, b, c, d, . . . , x, y, z$ to represent *general* numbers. Algebra uses both *general* and *specific* numbers.

When a number is represented by the digit symbols from arithmetic, it is called an *arithmetic number*. Some arithmetic numbers are 78,900; 56.009; 0.045; and 167. When a number is represented by a letter, like $a, c, t$, or $x$, it is called a *literal* number. Expressions in algebra that use letters are called *literal expressions*. Literal means "using letters." Examples of *literal expressions* are:

$$a + b \qquad m - n \qquad p \div q \qquad a \cdot b$$

Some expression require both arithmetic and literal numbers. These expressions are called *algebraic expressions*. Examples of algebraic expressions are:

$$3a + 5b \qquad 5p - 3q \qquad 10a \cdot b^2 \qquad 45n \div 6m$$

The *terms* of an algebraic expression are the parts that use either literal or arithmetic numbers. For instance, in the expression $3c + 8d$ the *terms* are $3c$ and $8d$.

Table 14-1 shows how to identify and correctly name the parts of an algebraic expression.

**TABLE 14-1**

| Expression | Operation | Literal numbers | Arithmetic numbers | Terms |
|---|---|---|---|---|
| $2s + 6t$ | add | $s, t$ | 2, 6 | $2s, 6t$ |
| $(4x) \cdot (3z)$ | multiply | $x, z$ | 4, 3 | $4x, 3z$ |
| $a + b - c$ | add and subtract | $a, b, c$ | none | $a, b, c$ |
| $5y - 10$ | subtract | $y$ | 5, 10 | $5y, 10$ |
| $\dfrac{25k}{6h}$ | divide | $k, h$ | 25, 6 | $25k, 6h$ |
| $(2t \div 3s) \cdot 5u$ | divide and multiply | $t, s, u$ | 2, 3, 5 | $2t, 3s, 5u$ |

Each letter in an algebraic expression stands for a number. Which number that letter stands for depends on how the expression is used in solving a problem.

In arithmetic the dot ($\cdot$) is used to show multiplication. Although the cross ($\times$) could be used, it might be confused with a literal number. To avoid any possible confusion between the $\times$ for multiplication and the $x$ used for a literal number, the dot is used ($\cdot$) to show multiplication. The product of $x$ and $y$ is written as $x \cdot y$. The dot appears above the line, not on the line as in 8.5, which shows a decimal, not a product.

To avoid even this possible error, the dot is often omitted when literal numbers are involved. The expression $xy$ without a dot means $x$ times $y$. Similarly, $ab$ means $a$ times $b$, $5s$ means 5 times $s$ and $\pi D$ means $\pi$ times $D$. If both factors in a product are arithmetic numbers, however, you *must* use a symbol for multiplication, otherwise $8 \cdot 5$ might be mistaken for 85.

Algebra is a tool—a tool of mathematics. The terms in expressions are statements of facts about a problem. These statements must be written clearly and accurately. In this chapter you will learn how to "say something in algebra." You will learn how to express yourself in the language of science and technology.

## Writing Algebraic Expressions

It is important for you to learn to write algebraic expressions for word statements. Read the six examples that follow. Try to get the answer before looking at the solution.

**Example 1(a):** Write an algebraic expression for the sum of two numbers $p$ and $q$.

*Solution (a):* Use the sign $+$ which means addition.

*Answer (a):* The expression $p + q$ shows the sum of the two numbers. The sum can also be written as $q + p$ because the addition of two numbers can be done in either order (Recall the Commutative Law of Addition, page 51).

**Example 1(b):** Write an algebraic expression for the product of the numbers 3 and $t$.

*Solution (b):* When one of the two factors is a literal number omit the dot that indicates multiplication.

*Answer (b):* The expression for this product is $3t$.

**Example 1(c):** Write an algebraic expression for the division of $5y$ by $7z$.

*Solution (c):* Use the symbol $\div$ which means division.

*Answer (c):* $5y \div 7z$. Because division can also be written as a fraction, you may write the answer $\dfrac{5y}{7z}$.

**Example 1(d):** Write an algebraic expression for the subtraction of $3x$ from $8t$.

*Solution* $(d)$: Use the minus sign $(-)$ to show subtraction.

*Answer* $(d)$: $8t - 3x$.

**Example 1(e):** Write an algebraic expression for the statement, "three less than five times $x$."

*Solution* $(e)$: The symbol for three is 3. The sign for "less" is $-$. Five times $x$ is written as $5x$.

*Answer* $(e)$: $5x - 3$.

**Example 1(f):** Write an algebraic expression for the statement, "multiply the sum of $m$ and $n$ by twice $p$."

*Solution* $(f)$: Write the sum of $m$ and $n$ as $m + n$. Write twice the number $p$ as $2p$. Write the product of $2p$ and $m + n$ as $2p(m + n)$.

*Answer* $(f)$: $2p \, (m + n)$

### TABLE 14-2

| Word or phrase in English | Stated in algebra |
|---|---|
| "a certain number" | $x$ (or some other letter) |
| "a (or the) number" | $x$ (or some other letter) |
| "an unknown quantity" | $x$ (or some other letter) |
| *(Addition)* | |
| A number *plus* 3. | |
| A number *added* to 3. | |
| A number *increased* by 3. | $x + 3$ |
| The *sum* of a number and 3. | |
| Three *more* than a certain number. | |
| *(Subtraction)* | |
| A number *minus* 3. | |
| A number *less* 3. | |
| *Subtract* 3 from a number. | $x - 3$ |
| A number *diminished* by 3. | |
| *Take* 3 from a certain number. | |
| A number *decreased* by 3. | |
| *(Multiplication)* | |
| The *product* of a number and 3. | |
| Three *times* a certain number. | $3x$ |
| A number *multiplied* by 3. | |
| Three *times as large as* a number. | |
| *(Division)* | |
| A number *divided* by 3. | |
| The *quotient* of a number and 3. | $\dfrac{x}{3}$ or $x \div 3$ |
| One-third *of a certain* number. | |
| The *ratio* of a number and 3. | |

NOTE: Parentheses ( ) are used to group the sum of *m* and *n*. If the answer is written $2pm + n$ without grouping *m* and *n* inside parentheses, it does NOT mean the same as the original statement. In the original statement the *sum* is multiplied by $2p$, not just the number *m*. No multiplication sign is needed because literal numbers are used.

Pay attention to the wording of a statement in order to make the correct change into the language of algebra. To help you make these changes, Table 14-2 shows certain words and phrases and how they are to be changed into the signs and symbols of algebra.

Here are some practice problems. Study each statement. As soon as you are *certain* of the operations and the numbers, write what you think is the correct algebraic expression. Check your results with the answers that follow these problems.

### Practice Problems

1.  Write each of the given statements in algebraic language.
    (a) The sum of $5r$ and $3t$.
    (b) The sum of $100x$, $36y$ and $67z$.
    (c) $3u$ divided by $6v$.
    (d) $5c$ subtracted from $7d$.
    (e) $10h$ diminished by $5k$.
    (f) Eighteen more than seven $q$.
    (g) Five times $z$ added to twice $x$.
    (h) $79t$ increased by $52s$.
    (i) The quotient of $3b$ and $7c$.
    (j) The product of $p$, $r$, and $t$.

2.  Each statement (a) through (h) uses the phrase "*the number*." Express each statement in algebra. Replace the phrase "*the number*" with the letter $x$.
    (a) Fives times *the number*.
    (b) 100 added to 3 times *the number*.
    (c) 6 less than 4 times *the number*.
    (d) *The number* divided by $1.7y$.
    (e) Twice *the number* decreased by $m$.
    (f) The product of $3y$ and 6 times *the number*.
    (g) The quotient of *the number* and three.
    (h) *The number* diminished by five $y$.

*Answers:* **1.** (a) $5r + 3t$;  (b) $100x + 36y + 67z$;  (c) $\dfrac{3u}{6v}$ or $3u \div 6v$;

   (d) $7d - 5c$;  (e) $10h - 5k$;  (f) $18 + 7q$;  (g) $5z + 2x$;

   (h) $79t + 52s$;  (i) $\dfrac{3b}{7c}$ or $3b \div 7c$;  (j) *prt*.

**2.** (a) $5x$;  (b) $100 + 3x$;  (c) $4x - 6$;  (d) $\dfrac{x}{1.7y}$ or $x \div 1.7y$;

   (e) $2x - m$;  (f) $(3y)\cdot(6x)$ or $18xy$;  (g) $\dfrac{x}{3}$;  (h) $x - 5y$.

Each of the signs $(+, -, \div, \cdot)$ means that an operation is to be done. When $+$ is between two numbers, you find the *sum*. When $-$ is between two numbers, you *subtract*, and so on. The equality sign $(=)$ is **not** a sign of operation. The equality sign, when it is shown between two numbers or expressions, is used to show how these *quantities are related*. The statement "$5 = t$" means that the number 5 and the number $t$ are *equal*. Similarly, the statement "$E = IR$" means that the number represented by $E$ is *equal to* the product of the two numbers $I$ and $R$. The statement, "three feet plus five feet equals eight feet" is written in arithmetic as $3$ ft. $+ 5$ ft. $= 8$ ft.

In algebra the sign $(=)$ is used to show equality between two algebraic expressions. For instance, five $x$ plus three $x$ equals eight $x$ is stated in algebra as: $5x + 3x = 8x$. In word statements, the phrase *is equal to* is usually replaced by the single word *is*.

**Is** means **is equal to,** and is expressed by the equals $(=)$ sign. For instance, the statement, "17 added to $x$ is 185" is written

$$17 + x = 185$$

The statement, "five less than a certain number is 53" is written

$$x - 5 = 53$$

## PROBLEMS:  Sec. 1  Signs, Symbols, and Expressions

1. Match the written statement on the left with the correct algebraic expression on the right. Use $x$ for "a number."
   (a) Twice a number plus five
   (b) $10v$ minus $6w$
   (c) $18d$ more than $9e$
   (d) The product of $3s$, $6t$, and $5r$
   (e) a number divided by 5
   (f) The sum of $x$ and $y$ multiplied by $z$
   (g) Three added to the product of $x$ and $y$
   (h) 10 subtracted from the quotient of $p$ and $q$
   (i) One-fourth of a number plus the number
   (j) A number diminished by the sum of $s$ and $t$

   1. $9e + 18d$
   2. $x \div 5$
   3. $xy + 3$
   4. $2x + 5$
   5. $x - (s + t)$
   6. $10v - 6w$
   7. $\frac{x}{4} + x$
   8. $(3s)(6t)(5r)$
   9. $(x + y)z$
   10. $\frac{p}{q} - 10$

2. Use numbers and letters to express each statement in algebra.
   (a) $x$ plus 3
   (b) $x$ decreased by 6
   (c) $s$ divided by 7
   (d) 8 divided by $x$

    (e)  one-half of $y$             (f)  the product of $a$ and $b$

    (g)  3 more than $c$           (h)  $y$ subtracted from $x$

    (i)  4 less than $h$            (j)  $x$ diminished by $y$

3.  Express each of the statements in algebra.

    (a)  50 more than $x$           (b)  25 less than $x$

    (c)  2 more than twice $x$        (d)  $\frac{3}{4}$ of $x$

    (e)  half as much as $x$         (f)  5 less than four $x$

    (g)  increase twice $x$ by 9      (h)  3 divided by twice $x$

4.  If $s$ and $t$ represent two numbers, write expressions for:

    (a)  their sum.

    (b)  their product.

    (c)  twice $s$ divided by four $t$.

    (d)  twice the difference of $t$ and $s$.

    (e)  their sum divided by their product.

5.  The length of a board is $k$ inches. Represent, in terms of $k$, the length of a board 6 inches longer.

6.  The length of a rectangle is $L$ feet. The width is 5 feet more than the length. Write the width in terms of the letter $L$.

7.  A car travels $m$ miles per hour. A small plane travels 40 miles per hour faster. Write the speed of the airplane by using the letter $m$.

8.  Harvey is $x$ years old. Herman is three years older than twice Harvey's age. Express Herman's age by using the letter $x$.

9.  Three lock washers cost $z$ cents. What is the cost of one washer?

10.  If 4 stove bolts cost $x$ cents, what is the cost of 16 stove bolts?

11.  If 3 tires cost $m$ dollars, what is the cost of 5 tires?

12.  Sam bought items costing $x$ cents, $y$ cents, and $z$ cents. Write (in cents) the change he got from a $1 bill.

13.  Bob Akers borrowed $D$ dollars for a business investment at an annual rate of 8%.

    (a)  Show in terms of $D$ the amount of interest that Bob owed at the end of one year.

    (b)  Show in terms of $D$ the total amount that Bob paid at the end of one year when he paid both interest and principal.

14.  Jon rode his bike for $t$ hours. If he can ride at 25 miles per hour, how far did he ride?

15.  A casting weighs $w$ pounds. If $x$ pounds are removed in finishing the casting, what is the finished weight?

16.  There are 12 inches in 1 foot. How many inches are in $t$ feet?

17.  If a television set cost $k$ dollars, what is the cost of 5 sets?

18.  An engine tune-up class started with 35 students. At the end of the year $p$ students passed. How many students failed the course?

19.  If a gallon of gasoline cost $d$ cents, what is the cost of 22 gallons?

20. If a pint of oil cost *s* cents, what is the cost of 2 gallons?

21. Neil's new car cost $4,568.00. He received *k* dollars for his trade-in. How much is the balance?

22. Tony has *n* nickels and *d* dimes. If he changes them all to pennies, how many will he have?

23. A delivery truck gets *m* miles on a tank of gas. How many miles does it get on *t* tanks of gas?

**Problems 24 through 29.** Each statement contains the word "is" or the phrase "is equal to." Write the statement in algebra form. Use correct signs. Use *x* to replace "a number" or "a certain number."

**Example:** A number diminished by 7 is 14.

*Answer:* $x - 7 = 14$

24. A number added to 5 *is equal to* 18.

25. Ohm's law states, "*E is equal to* the product of *I* and *R*."

26. 17 times a number *is* 51.

27. A number *is equal to s* plus 15.

28. *w* minus 6 *equals* 8 more than twice *x*.

29. Profit (*p*) *is equal to* selling price (*s*) less cost (*c*).

30. The circumference (*c*) of a circle *is equal to* the product of $\pi$ and the diameter (*d*).

**Problems 31 through 41.** Write each of these expressions in word statements.

**Example:** $x + 4 = 9$

*Answer:* The sum of a number and 4 is equal to 9.

31. $x + 7 = 10$

32. $2x - 1 = 8$

33. $a + b = 3c$

34. $\frac{4}{5}t = 87$

35. $6t - 5 = -9$

36. $uv = 0$

37. $E = IR$

38. $A = .7854d^2$

39. $P = 2a + 2b$

40. $x + 6 - 3z = 10$

41. $\frac{a + b}{a - b} = 1$

## SEC. 2  FORMULAS

How do you find the number of inches in five feet? Can you find the diameter of a circle if you know the area? Do you know how to find the velocity of a falling body one, two, or ten seconds after it starts to fall? You can answer these questions and others like them by using certain rules.

Example of a rule:

> *Electrical power, expressed in watts, is equal to the product of the voltage and current.*

From this *rule*, you can find the power of an electrical device.

In science and technology you can state such rules in algebraic form. When you write a rule in terms of the *signs* and *symbols* of algebra, it is called a *formula*. To write a formula from a rule, follow these steps:

Step 1. Replace the quantities representing *general* numbers with *literal* symbols—that is, with letters. Replace *specific* numbers with arithmetic numerals.

Step 2. Replace words such as *plus*, *minus*, *product*, *divide*, *is equal to*, and others with algebraic signs.

> Replace *plus* with $+$
> Replace *minus* with $-$
> Replace *product* with "$\cdot$" or ( )
> Replace *divide* with $\div$
> Replace *is equal to* with $=$
> and so on

**Example 2:** Write the formula for the rule (Ohm's law):
The *voltage* is equal to the product of *current* and *resistance*.

*Solution:* The quantities representing general numbers are:

> *voltage, current*, and *resistance*

Replace these with letters: $E$ for voltage, $I$ for current, and $R$ for resistance. Replace the phrase "is equal to" with the sign $=$. Replace the word "product" with the multiplication sign ($\cdot$).

*Answer:* The formula is $E = I \cdot R$. The right-hand side may be written without the dot for multiplication because only letters are involved: $E = IR$.

There are many formulas in technology. Some of these formulas use only one or two letters and may be written as simply as Example 2. Others use several letters and numbers and are quite complicated.

## Evaluating Formulas

A formula is a statement of a rule in the language of algebra. The letters in a formula represent general numbers. If you replace all but one of the letters by known *specific* numbers, you can find the value of the letter representing the *unknown* number. This method of finding the value of the remaining letter is called *evaluating the formula for the unknown*.

To evaluate a formula for an unknown, substitute numerical values for all but one letter. Find the value of the remaining letter by using the methods and rules from arithmetic.

**Example 3:** Evaluate the formula relating interest ($I$), principal ($P$), rate ($r$), and time ($t$): $I = Prt$

for the unknown $I$ when $P = \$5,000$, $r = .05$, and $t = 9$ years.

*Solution:* Substitute the numbers for the letters in the formula.

$$I = (\$5,000) \cdot (.05) \cdot (9)$$

*Answer:*          $I = \$2,250.00$

**Example 4:** Evaluate the formula relating distance ($D$), speed ($r$), and time ($t$)

$$D = rt$$

for the unknown $D$ when $r = 65$ mph and $t = 7$ hours.

*Solution:*          $D = (65) \cdot (7)$

*Answer:*          $D = 455$ miles

**Example 5:** Evaluate the formula for the average ($A$) of two numbers, $a$ and $b$

$$A = \frac{a + b}{2}$$

for the unknown $A$ when $a = 10$ and $b = 6$.

*Solution:* Replace the letters with the given values:

$$A = \frac{10 + 6}{2}$$

*Answer:*          $A = 8$

**Example 6:** Find the value of $C$ in the formula relating Celsius temperature ($C$) and Fahrenheit temperature ($F$)

$$C = \frac{5}{9}(F - 32)$$

when $F = 212°$.

*Solution:*        $$C = \frac{5}{9}(212 - 32)$$

$$= \frac{5}{9}(180)$$

*Answer:*        $$C = 100°$$

### Practice Problems

In each practice problem a formula is given. Replace the letters with the given values. Then find the value of the remaining letter.

1. Formula: $D = RT$     $R = 45$ mph, $T = 2.5$ hours

2. Formula: $T = \dfrac{D}{R}$     $D = 450$ miles, $R = 9$ mph

3. Formula: $R = \dfrac{D}{T}$     $D = 600$ miles, $T = 12$ hours

4. Formula: $A = P(1 + rt)$     $P = \$800$, $r = .08$, $t = 4$ years

   *Answers:* 1. $D = 112.5$ miles
   2. $T = 50$ hours
   3. $R = 50$ mph
   4. $A = \$1,056$

It is important for you to be able to write a formula from a stated rule. It takes a lot of practice to write correct formulas. If a single letter is left out, or an incorrect sign of operation is used, the formula will not give the correct result.

Here are some suggestions for writing correct formulas from written rules:

1. Read the rule at least *twice*. Make sure you understand exactly what the rule states.
2. *Underline* the words that refer to general and specific numbers.
3. Place letters below the words that refer to general numbers. Place the correct numeral below the words that refer to specific numbers.

NOTE: It is not important what letter you choose for a general number. If you translate correctly, the resulting formula will be correct for any choice of letters. Some choices are better than others for specific rules. These choices will become clear as you work with formulas.

4. Replace words such as *addition, subtraction, multiplication, division, equality*, etc., with the correct sign ($+$, $-$, $\cdot$, $\div$, $=$, etc.).

Study these examples.

**Example 7:** Write the formula for the rule:

The area of a rectangle is equal to the product of the base and the height.

*Solution:* Underline the words *area*, *base*, and *height* since they refer to quantities represented by general numbers. Choose a letter for each word: $A$ for *area*, $b$ for *base*, and $h$ for *height*. Write these letters below the underlined words. Place the equality sign under the phrase "is equal to." A sign of multiplication is not needed because only letters are involved. The translation is:

The area of a rectangle is equal to
  $A$                            $=$

the product of the base and the height.
                        $b$          $h$

*Answer:* The formula for the rule is: $A = bh$

NOTE: You must supply any units of measure. In this case they are square units such as square feet, square inches, square millimeters, etc. The units depend upon the values that replace $b$ and $h$. The formula is true for all correct substitutions for $b$ and $h$.

**Example 8:** Write a formula for the rule:

The volume of a circular cylinder is equal to the product of $\pi$, the height, and the square of the radius.

*Solution:* Underline *volume*, $\pi$, *height*, and *square of radius* since they refer to numbers. Choose $V$ for *volume*, $h$ for *height*, and $r$ for *radius*. Put $V$ under volume, $h$ under height, and $r^2$ under the phrase "square of the radius." Put $=$ under "is equal to."

*Answer:* The formula for the rule is: $V = \pi h r^2$

NOTE: The sign of multiplication can be omitted because only one number is shown in the formula.

**Example 9:** Write a formula for the rule:

The electrical power in a direct current is equal to the product of the resistance and the square of the current.

*Solution:* Underline *power*, *resistance*, and *current*. Choose $W$ for power, $R$ for resistance, and $I$ for current. Put $W$ under *power*, $R$ under *resistance*, and $I^2$ under the square of the *current*. Put $=$ under *is equal to*.

*Answer:* $W = RI^2$

## PROBLEMS: Sec. 2 Formulas

1. Evaluate the formula $A = s^2$ for $A$ when $s = 15$.

2. Evaluate the formula $A = bh$ for $A$ when $b = 5$, $h = 8$.

3. Evaluate the formula $A = \frac{1}{2}bh$ when $b = 1.5$, $h = 4.6$.

4. Evaluate the formula $C = 3.14D$ when $D = 11.4$.

5. Evaluate the formula $A = .785d^2$ when $d = 4.6$.

6. Evaluate the formula $A = \frac{1}{2}rC$ when $r = 6$, $C = 14$.

7. Evaluate the formula $L = S \div h$ when $S = 18.5$ and $h = .5$.

8. Evaluate the formula $V = 3.14r^2h$ when $r = 3$, $h = 5$.

9. Evaluate the formula $V = \frac{1}{6}\pi d^3$ for $d = 3$. $\left(\pi = \frac{22}{7}\right)$

10. Evaluate the formula $c = \sqrt{a^2 + b^2}$ for $a = 3$, $b = 4$.

11. Evaluate the formula $A = \frac{1}{2}(b + B)h$ for $b = 2$, $B = 5$, $h = 6$.

12. Evaluate the formula $V = .785d^2h$ for $d = 3$, $h = 10$.

13. Write the formula for the rule:
    The *area* of a rectangle *is equal to* the product
    of the *length* and the *width*.

14. Write the formula for the rule:
    The volume of a box is equal to the
    product of the length, the width, and the height.

15. Write a formula for the rule:
    The resistance in an electrical circuit is equal
    to the voltage divided by the current.

16. Write a formula for the rule:
    The diameter of a circle is equal to two
    times the square root of the area divided by $\pi$.

17. Write a formula for the rule:
    The volume of a cube is equal to the
    cube of the length of a side.

18. Write a formula for the rule:
    The sine of an angle is equal to the
    ratio of the opposite leg to the hypotenuse.

19. Write a formula for the rule:
    The sum of the three interior angles
    of a triangle is equal to $180°$.

20. Write a formula for the rule:
    The square of the hypotenuse is equal to
    the sum of the squares of the other two sides.

21. Write a formula for the rule:
    The distance traveled is equal to the
    product of the rate of travel and the time traveled.

22. Write a formula for the rule:
    The power in an electrical circuit is equal to the
    product of the resistance and the square of the current.

23. Write a formula for the rule:
    The horsepower of an electrical motor is
    equal to the product of the voltage and the
    current divided by 746.

24. Write a formula for the rule:
    The interest rate on a loan is equal to the
    amount of interest paid divided by
    the product of the principal invested and
    the time.

25. Write a formula for the rule:
    The volume (in gallons) of a rectangular
    container is equal to the product of
    7.48, the length, the depth, and the
    height in feet.

## SEC. 3    SOLVING EQUATIONS IN ALGEBRA

One of the most important methods for problem solving is the *equation*. An *equation* is a statement in mathematical symbols that says:

*One algebraic expression is equal to another.*

For instance, the mathematical statement $2x + 1 = 6$ is an *equation*.

*An equation is a declaration of equality between two algebraic expressions.*

Most equations use one letter called a *variable*, which can change in value, and numbers or letters whose specific values are known. These *known* values are called *constants* (meaning without change). For instance, in the equation $5x - 4 = 17$ the *variable* is the letter $x$. The *constants* are the known values in the equation, and they are 5, 4, and 17. The main problem in algebra is to find the numerical value of the variable which will make the result a true statement. Whenever a number is found for the variable that makes both sides of an equatiin have the same value, the number is called a *solution* of the *equation*. A solution *solves* (or satisfies) the equation. That is:

*SOLUTIONS are numbers that make equations* true statements.

**Example 10:** Show that 2 is a solution to the equation
$$3x + 4 = 10.$$

*Solution:* Substitute 2 for $x$ in the equation:

$$3(2) + 4 = 6 + 4 = 10$$

Since the left-hand side equals 10, the number 2 satisfies the equation.

*Answer:* 2 is the solution of the equation $3x + 4 = 10$.

**Example 11:** Is 4 a solution to the equation:

$$x^2 + 2x = 16?$$

*Solution:* Substitute 4 for $x$ in the equation:

$$(4)^2 + 2(4) = 16?$$

The left-hand side has the value $16 + 8 = 24$ which is *not* equal to 16.

*Answer:* 4 is not a solution to the equation.

An equation is a statement that the left- and right-hand sides are *in balance* (equal). This idea of being in balance is similar to that of a scale or pan-balance. Figure 14-1 shows a scale that is in balance. The amount of weight in the two pans is the same.

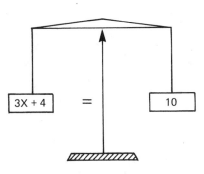

**Figure 14-1**

Some of the basic principles of equation solving are similar to the procedure of keeping the scale in balance while adding or removing weights from the two pans. Study the next examples to see how these principles are used to solve equations.

**Example 12:** Figure 14-2 shows a pan-balance. In the pan on the left are two objects. One of the objects weighs 4 ounces. The weight of the other object is unknown, and it is marked $x$ ounces. In the pan on the right there are nine 1-ounce weights. The scale is in bal-

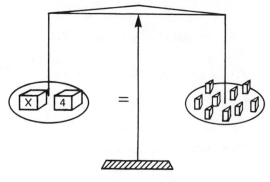

*Figure 14-2*

ance. Find the weight of the unknown weight labeled *x*. An algebraic equation for this equality is:

$$x + 4 = 9$$

*Solution:* If you remove 1 ounce from the right-hand pan, and nothing from the left, the scale is out of balance. If, however, you remove 4 ounces from BOTH sides, then the scale remains in balance. With the 4 ounces removed, 5 ounces remain in the right hand pan and the unknown weight *x* is in the left pan. BUT the pans are still *in balance*. Therefore, the unknown weight must equal 5 ounces. Write the algebraic result as:

$$x + 4 - 4 = 9 - 4 \text{ (take 4 from both sides)}$$

*Answer:* $x = 5$

From this example it is possible to state the:

### Rule of Subtraction in Equations

*To keep an equation in balance, subtract any quantity from one side when you subtract it from the other.*

This rule is often shortened to:

*Subtract equals from equals to get equals.*

**Example 13:** Solve the equation $x + 9 = 45$.

*Solution:* Use the rule of subtraction. Subtract 9 from BOTH sides of the equation:

$$x + 9 - 9 = 45 - 9$$

*Answer:* $x = 36$

*Check:* $36 + 9 = 45$

Instead of subtracting the same number from both sides of an equation, you may add a number to both sides. This is stated as the:

### Rule for Addition in Equations

*To keep an equation in balance add any quantity to one side when you add to the other.*

This rule is often shortened to:

*Add equals to equals to get equals.*

**Example 14:** Solve the equation $x - 5 = 45$.

*Solution:* Use the rule of addition. Add 5 to BOTH sides of the equation:

$$x - 5 + 5 = 45 + 5$$

*Answer:* $x = 50$

*Check:* Does $50 - 5 = 45$? Yes.

### Practice Problems

Use either the rule of addition or subtraction to find the solutions to these equations:

1. $x + 20 = 60$     $x = 40$       (Use rule of subtraction.)
2. $x - 20 = 60$     $x = 80$       (Use rule of addition.)
3. $50 + x = 65$     $x = 15$
4. $x - 10 = 100$     $x = 110$
5. $x + 15 = 10$     $x = -5$
6. $x - 20 = 0$     $x = 20$
7. $x + 2.6 = 4.7$     $x = 2.1$
8. $x - .8 = 5.9$     $x = 6.7$
9. $x + \dfrac{2}{3} = \dfrac{4}{5}$     $x = \dfrac{2}{15}$
10. $x - \dfrac{3}{8} = \dfrac{7}{16}$     $x = \dfrac{13}{16}$

Now look at a problem in which division is needed to solve an equation.

**Example 15:** One pan of a blanced scale contains three unknown weights of equal value. The other pan contains a 6-ounce weight. Find the weight of one of the unknown weights.

*Solution:* Call each of the unknown weights $x$. Because all three of the unknown weights are equal in value, the pan contains $3x$ ounces. That is:

$$x + x + x = 3x$$

The other pan contains 6 ounces. An equation for this is:

$$3x = 6$$

To find the value for the unknown $x$ (and solve the equation), divide both sides of the equation by 3.

$$\frac{3x}{3} = \frac{6}{3} = 2 \text{ (one-third of six equals two)}$$

When you divide $3x$ by 3 you get $\frac{3x}{3} = x$ (one-third of $3x$ is $x$).

*Answer:* $x = 2$ ounces

*Check:* Substitute 2 for $x$ in the equation $3(2) = 6$. Therefore, $x = 2$ is the solution.

From this example it is possible to state the

### Rule for Division in Equations

*Divide both sides of an equation by the same quantity to keep the equation in balance.*

Or, in short form:

*Divide equals by equals to get equals.*

NOTE: NEVER, NEVER DIVIDE BY ZERO!

**Example 16:** Solve each of the equations by applying the rule of division.
(a) Solve $4x = 16$

*Solution:* Divide both sides by 4.

$$\frac{4x}{4} = \frac{16}{4} = 4$$

*Answer* (a)*:* $x = 4$

*Check* (a)*:* $4(4) = 16$

**Example 16(b):** Solve $3x = 45$

*Solution* (b)*:* Divide both sides by the number 3.

$$\frac{3x}{3} = \frac{45}{3} = 15$$

*Answer* (b)*:* $x = 15$

*Check* (b)*:* $3(15) = 45$

**Example 16(c):** Solve $2.5x = 10$

*Solution* (c): Divide both sides by the number 2.5.

$$\frac{\cancel{2.5}x}{\cancel{2.5}} = \frac{10}{2.5} = 4$$

*Answer* (c): $x = 4$

*Check* (c): $2.5(4) = 10$

You need one additional rule to be able to solve more complicated equations. This is the

### Rule of Multiplication in Equations

*Multiply both sides of an equation by the same quantity to keep the equation in balance.*

This rule is often shortened to:

*Multiply equals by equals to get equals.*

**Example 17:** If one-half of an unknown quantity equals 10, find the amount of the unknown.

*Solution:* Call the unknown $x$. Write one-half of $x$ as $\frac{x}{2}$. Therefore, write the equation as:

$$\frac{x}{2} = 10$$

To solve this equation for $x$, ***multiply*** both sides by 2.

$$\cancel{2}\left(\frac{x}{\cancel{2}}\right) = 2(10) = 20$$

*Answer:* $x = 20$

*Check:* $\frac{20}{2} = 10$

Here are more examples in which the rule of multiplication is used to get the solutions.

**Example 18(a):** Solve $\frac{x}{3} = 5$

*Solution* (a): Multiply both sides by 3.

$$(\cancel{3})\frac{x}{\cancel{3}} = (3)5 = 15$$

*Answer* (a): $x = 15$

**Example 18(b):** Solve $\frac{x}{1.5} = 2.6$

*Solution* (b): Multiply both sides by 1.5.

$$(1.5)\frac{x}{(1.5)} = (1.5)(2.6) = 3.9$$

*Answer* (b): $x = 3.9$

**Example 18(c):** Solve $\frac{x}{10} = 7.9$

*Solution* (c): Multiply both sides by 10.

$$(10)\frac{x}{10} = (10)7.9 = 79$$

*Answer* (c): $x = 79$

NOTE: *Multiply* both sides of the equation by the number that *divides* the unknown number $x$.

Some equations use several of the rules for equation solving. When you need more than one rule to solve an equation, cautiously consider your plan of attack. A general rule of thumb is:

*Apply rules of addition or subtraction before applying the rules of multiplication or division.*

**Example 19:** Solve the equation $5x + 3 = 18$.

*Solution:* Isolate the term $(5x)$ that contains the unknown. Do this by subtracting 3 from both sides of the equation.

$$5x + 3 - 3 = 18 - 3 \text{ (rule of subtraction)}$$
$$5x = 15$$

Get the $x$ by dividing both sides by 5.

$$\frac{5x}{5} = \frac{15}{5} = 3 \qquad \text{(rule of division)}$$

*Answer:* $x = 3$

*Check:* $5(3) + 3 = 15 + 3 = 18$

**Example 20:** Solve the equation $\frac{x}{6} - 5 = 3$.

*Solution:* Add 5 to both sides to isolate the term $\frac{x}{6}$.

$$\frac{x}{6} - 5 + 5 = 3 + 5 \text{ (rule of addition)}$$

Solve for $x$ by multiplying both sides by 6.

$$(6)\frac{x}{6} = (6)8 = 48$$

*Answer:* $x = 48$

NOTE: Check the solution by substitution into the original equation.

**Example 21:** Solve the equation $\dfrac{3x + 6}{5} = 9$

*Solution:* The left-hand side of the equation is a fraction with 5 as the denominator. The unknown $x$ is in the numerator $3x + 6$. Multiply both sides by 5 as your first operation.

$$3x + 6 = 45 \qquad \text{(rule of multiplication)}$$

Subtract 6 from both sides, and then divide by 3.

$$3x + 6 - 6 = 45 - 6 \qquad \text{(subtract 6)}$$

$$3x = 39 \qquad \text{(simplify)}$$

$$\frac{\cancel{3}x}{\cancel{3}} = \frac{39}{3} = 13 \qquad \text{(divide by 3)}$$

*Answer:* $x = 13$

*Check:* Substitute $x = 13$ into original equation.

$$\frac{3(13) + 6}{5} = \frac{39 + 6}{5} = \frac{45}{5} = 9$$

The unknown in an equation always represents *some* number. The letter used for the unknown may be added, subtracted, or used as a factor or divisor just like any other number. The four rules of algebra apply to the unknown as well as to all other numbers that are in the equation.

**Example 22:** Solve the equation $5x - 3 = 2x + 12$

*Solution:* The basic idea in solving equations is to get the unknown letter on one side and the numbers on the other. To solve this equation, subtract $2x$ from both sides:

$$5x - 3 - 2x = 2x + 12 - 2x \qquad \text{(rule of subtraction)}$$

Regradless of the value of $x$:

$$5x - 2x = 3x \qquad \text{and} \qquad 2x - 2x = 0$$

The result is:     $3x - 3 = 12$

Add 3 to both sides:

$$3x - 3 + 3 = 12 + 3$$

$$3x = 15$$

Divide both sides by 3 to get the value for $x$:

$$\frac{\cancel{3}x}{\cancel{3}} = \frac{15}{3} = 5$$

*Answer:* $x = 5$

NOTE: *Check this solution!*

So far the letter $x$ has been used as the unknown variable to help you identify the unknown in the equation. When writing an equation, however, *any* letter may be used to represent the unknown variable. Although $x$ is the most common letter used in equations, other letters such as $s, t, u, v, w,$ $y, z$ are also used to represent unknowns.

Now you will use the four rules of algebra to solve equations. Before you begin the problems, take a minute or two and memorize:

### The Four Rules of Algebra

ADDITION:            *Add equals to equals to get equals.*
SUBTRACTION:         *Subtract equals from equals to get equals.*
MULTIPLICATION:      *Multiply equals by equals to get equals.*
DIVISION:            *Divide equals by equals to get equals.*

## PROBLEMS: Sec. 3 Solving Equations in Algebra

**Problems 1 through 8.** Check by substitution that the given number is a solution.

1. If $2x = 20$, then $x = 10$.
2. If $x + 2 = 8$, then $x = 6$.
3. If $x - 6 = 18$, then $x = 24$.
4. If $2x - 1 = 13$, then $x = 7$.
5. If $\frac{1}{2}x = 8$, then $x = 16$.
6. If $5 - x = 3$, then $x = 2$.
7. If $3x + 5 = -1$, then $x = -2$
8. If $\frac{y + 5}{6} = 3$, then $y = 13$

**Problems 9 through 16.** Find out if the number in parentheses is a solution to the equation.

9. $4x = 28$  (7)
10. $3x = 18$  (6)
11. $x + 3 = 9$  (6)
12. $2x - 3 = 21$  (9)
13. $4x + 2 = 26$  (7)
14. $3x + 5 = -4$  (−3)
15. $-m + 6 = 9$  (3)
16. $-2x - 3 = -11$  (4)

**Problems 17 through 26.** Use the rule of addition to solve for the variable.

17. $x - 3 = 7$
18. $s - 8 = 16$
19. $y - 2 = 10$
20. $14 = z - 10$
21. $x - 15 = 35$
22. $t - 7 = -14$
23. $t - .7 = 1.4$
24. $z - \frac{1}{3} = 5\frac{1}{3}$
25. $3 = h - .08$
26. $V - 1.09 = 0$

**Problems 27 through 36.** Use the rule of subtraction to solve for the variable.

27. $x + 10 = 20$
28. $b + 8 = 10$
29. $65 = k + 15$
30. $s + 1 = 11$

**31.**  $x + 10 = 10$

**32.**  $\dfrac{3}{4} = m + \dfrac{1}{2}$

**33.**  $y + .8 = 1.9$

**34.**  $x + 13 = -9$

**35.**  $20\dfrac{1}{3} = r + 1\dfrac{1}{6}$

**36.**  $V + 10 = -10$

**Problems 37 through 46.** Use the rule of division to solve for the variable.

**37.**  $2x = 20$

**38.**  $3x = 30$

**39.**  $18t = 36$

**40.**  $10a = 100$

**41.**  $4y = 4$

**42.**  $45z = 180$

**43.**  $4t = 1.6$

**44.**  $.5x = 10$

**45.**  $1.2x = 9.6$

**46.**  $\dfrac{1x}{3} = 9$

**Problems 47 through 54.** Use the rule of multiplication to solve for the variable.

**47.**  $\dfrac{x}{5} = 10$

**48.**  $\dfrac{t}{2} = 8$

**49.**  $\dfrac{m}{6} = 12$

**50.**  $\dfrac{v}{9} = 1$

**51.**  $\dfrac{x}{6} = \dfrac{2}{3}$

**52.**  $\dfrac{z}{.5} = 1.5$

**53.**  $\dfrac{a}{5} = \dfrac{1}{10}$

**54.**  $\dfrac{w}{10} = .01$

**Problems 55 through 100.** Solve each equation by using the proper rule.

**55.**  $3x + 25 = 50$

**56.**  $6x + 1x = 42$

**57.**  $10x - 4x = 24$

**58.**  $5x - x = 16$

**59.**  $x + 2x + 3x = 18$

**60.**  $3x - 4 = 5$

**61.**  $2a + 9 = 21$

**62.**  $3 = 2m + 3$

**63.**  $5x + \dfrac{1}{2} = 10\dfrac{1}{2}$

**64.**  $9t + 7 - 2 = 32$

**65.**  $\dfrac{2x + 5}{8} = 4$

**66.**  $4x - x + 4 = 31$

**67.**  $4a = 60 + 2a$

**68.**  $18 - z = 2z$

**69.**  $\dfrac{2x}{3} + 5 = 7$

**70.**  $\dfrac{x}{3} + 1\dfrac{1}{2} = 4\dfrac{1}{2}$

**71.**  $\dfrac{3b}{5} = -6$

**72.**  $-4n = -48$

**73.**  $4 - 2s = 5 - s$

**74.**  $5 + z = 1 - z$

**75.**  $m - 2 = 4 - 2m$

**76.**  $\dfrac{1x}{4} = 3$

**77.**  $\dfrac{5y}{8} = -10$

**78.**  $\dfrac{2s}{3} = 1 - \dfrac{s}{3}$

**79.** $\dfrac{4x}{3} - \dfrac{x}{3} - 1 = \dfrac{x}{3} - 7$      **80.** $3(x + 2) = 2x - 5$

**81.** $3(x - 2) = 2x + 3$      **82.** $3(4z - 2) = 2(3z + 6)$

**83.** $x + \dfrac{x}{3} = 20$      **84.** $x - 2 = \dfrac{x}{2} + 7$

**85.** $4 + k = 5 - k$      **86.** $3p - 5 = p - 5$

**87.** $x + \dfrac{x}{3} = 1 + \dfrac{1}{3}$      **88.** $x = \dfrac{x - 4}{3}$

**89.** $\dfrac{x}{5} = 10$      **90.** $\dfrac{x + 10}{x} = 3$

**91.** $\dfrac{2y}{3} - \dfrac{1}{2} = \dfrac{3y}{4} + \dfrac{1}{6}$      **92.** $\dfrac{3t}{4} = \dfrac{3}{8}$

**93.** $3 + g = -3 - g$      **94.** $3y = 6y$

**95.** $.5x + 10 = .6(x + 10)$      **96.** $.8x - 5 = 2x + 5$

**97.** $1.5x - 0.4x = 1.21$      **98.** $.50x + .25x + .10x = 1.90$

**99.** $\dfrac{1x}{2} + \dfrac{1x}{4} = \dfrac{3}{8}$      **100.** $.5x + 1\dfrac{1}{2} = 3.6x - \dfrac{2}{3}$

## SEC. 4   APPLYING ALGEBRA TO TECHNOLOGY

Algebra is the language of science and technology. It gives results that are not possible through arithmetic alone. It is used in all fields of science, such as engineering, electronics, medicine, atomic energy, business, and technology. Algebra is the basis for many of the formulas you have used, and without it you could not have found them. Algebra is a basic tool of applied mathematics, and it is as necessary to the technician as his other tools.

In this section you will learn to use algebra to solve practical problems. Study the methods in the examples. Be sure you understand each step and why you are doing it. Don't guess—guesswork is not the result of clear thinking.

Here are some general rules to use in analyzing and solving problems:

1. Read each problem carefully. Make sure you understand the meaning of each word and symbol that is in the problem. *Look before you leap*!
2. Make a neat sketch of any figures in the problem. Label all dimensions with correct numbers and units of measure.
3. Choose the correct formula or equation for solving the problem.
4. Identify the unknown in the formula. Use the rules of algebra to solve for the unknown.
5. Check your answer whenever possible. See if your answer is reasonable. A single wrong answer can be costly in time and money.

**Example 23:** The area of a rectangle is 450 square centimeters. The width is equal to 50 centimeters. What is the length? (See Fig. 14-3.)

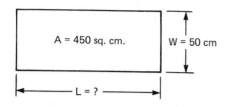

**Figure 14-3**

*Solution:* When solving problems, use known formulas and tables whenever possible. For this example, you know that the formula for the area of a rectangle is:

Area equals length times width,

which is written:

$$A = LW$$

The letter $L$ in Fig. 14-3 represents the unknown in the equation. Solve for the letter $L$ by *dividing both sides* by $W$, which represents the known value 50 centimeters.

$L = \dfrac{A}{W}$ (Equation is in the form: *Unknown = Known.*)

Substitute the known values for $A$ and $W$.

$$L = \frac{450 \text{ (square centimeters)}}{50 \text{ (centimeters)}} = 9 \text{ centimeters}$$

*Answer:* $L = 9$ centimeters

*Check:* $(50) \times (9) = 450$

**Example 24:** Temperature is measured by two different scales: the Fahrenheit scale, which is the most common, and the Celsius scale used by laboratory technicians, chemists, etc. A formula that relates the two scales is:

$$F = 1.8C + 32 \qquad (F = \text{Fahrenheit}; \ C = \text{Celsius})$$

What is the Celsius temperature (in degrees $C$) when the Fahrenheit temperature is 212°F?

*Solution:* The unknown in the equation is the letter $C$.
Evaluate the formula for the letter $C$.

$$212 = 1.8C + 32$$

*Subtract* 32 from both sides

$$212 - 32 = 1.8C + 32 - 32$$

$$180 = 1.8C$$

*Divide* both sides by 1.8

$$\frac{180}{1.8} = \frac{\cancel{1.8}C}{\cancel{1.8}} = C$$

$$100 = C.$$

*Answer:* 100°C = 212°F.

**Example 25:** Machinists use a formula to find the taper of a machined part. If a round bar $L$ inches long has a (large) diameter of $D$ inches and a (small) diameter of $d$ inches, the formula for the taper $T$, in inches per foot, is:

$$T = \frac{12(D - d)}{L}$$

Certain tapers are standard in the machine shop. One of these is the *Brown & Sharp* taper which is $\frac{1}{2}$ inch per foot. Figure 14-4 shows a tapered reamer. The small diameter is 0.750 inches. If the reamer has a Brown & Sharp taper, what is the diameter of the large end?

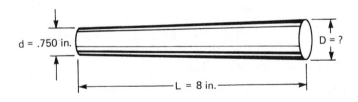

d = .750 in.     D = ?

L = 8 in.

**Figure 14-4**

*Solution:* A sketch of the reamer (Fig. 14-4) shows the given dimensions. Look at the formula $T = \frac{12(D - d)}{L}$. The letter $D$ is the unknown in the equation. All other letters $(T, L, d)$ represent known values. Evaluate the formula for $D$. Replace each letter by its given value.

$$\frac{1}{2} = \frac{12(D - .75)}{8}$$

*Multiply* both sides by 8.

$$8\left(\frac{1}{2}\right) = \cancel{8} \times \frac{12(D - .75)}{\cancel{8}}$$

$$4 = 12(D - .75)$$

Divide both sides by 12.

$$\frac{4}{12} = \frac{\cancel{12}(D - .75)}{\cancel{12}}$$

$$\frac{1}{3} = D - .75$$

Add $.75 = \dfrac{3}{4}$ to both sides.

$$D = \frac{1}{3} + \frac{3}{4} = \frac{13}{12}$$

$$D = 1.08 \text{ inches}$$

*Answer:* The large end has a diameter of 1.08 inches.

In the next examples there are no formulas. You must find an equation from the information given.

**Example 26:** Three boxes are weighed. They have a combined weight of 148 pounds (Fig. 14-5). The second box weighs 7 pounds more than the first. The third weighs 9 pounds less than the first. Find the weight of the three boxes.

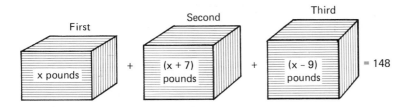

**Figure 14-5**

*Solution:* Let $x =$ weight of first box. From the information you know the second box weighs 7 pounds more than the first. The weight of the second box is: $(x + 7)$ pounds. The third box weighs 9 pounds *less* than the first box. The weight of the third box is: $(x - 9)$ pounds. The sum of the weights is:

$$\underset{\text{(first box)}}{x} + \underset{\text{(second box)}}{(x + 7)} + \underset{\text{(third box)}}{(x - 9)} = 148$$

To solve this problem, add 9 to both sides then subtract 7 from both sides:

$$x + x + x = 148 + 9 - 7$$

$$3x = 150 \quad \text{(Collect the } x\text{'s.)}$$

To solve for $x$, *divide both sides* by 3:

$$\frac{\cancel{3}x}{\cancel{3}} = \frac{150}{3} = 50$$

*Answer:* $x = 50$ lbs

$x + 7 = 50 + 7 = 57$ lbs.      (weight of second box)

$x - 9 = 50 - 9 = 41$ lbs.      (weight of third box)

*Check:* Sum of the three weights:

$$50 + 57 + 41 = 148$$

**Example 27:** A room that is in the shape of a rectangle has a perimeter of 96 feet. It is three times as long as it is wide. Find the length and width of this room.

*Solution:* Sketch a rectangle that is three times as long as it is wide. (See Fig. 14-6.)

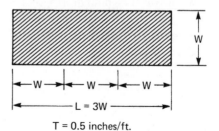

T = 0.5 inches/ft.

**Figure 14-6**

Both the length and width are unknown. The length is three times the width. Let $W$ equal the unknown width. The length is written as $3W$. From Chapter 11 you know that the perimeter of a rectangle is:

$$\text{Perimeter} = (\text{twice the length}) + (\text{twice the width})$$

$$96 \text{ feet} = 2(3W) + 2(W)$$

$$= 6W + 2W$$

$$96 = 8W$$

To solve for $W$, *divide both sides* by 8:

$$\frac{96}{8} = \frac{\cancel{8}W}{\cancel{8}} = W$$

*Answer:* Therefore,   (Width)        $W = 12$ feet

(Length)     $3W = 36$ feet

The room is 12 feet wide and 36 feet long.

*Check:* $2(36) + 2(12) = 72 + 24 = 96$

**Example 28:** A journeyman machinist is paid twice as much as an apprentice. In a 40-hour week their combined wages are $360.00.

(a) Find the amount each makes for a 40-hour week.

(b) Find the hourly wage of each.

*Solution* (a): Let $x =$ weekly wage of apprentice. Then:

$$2x = \text{weekly wage of journeyman}$$

since he is paid *twice* as much as the apprentice. From the statement of the problem:

$x + 2x = \$360.00$     (combined wages)

$3x = \$360.00$     (Add the $x$'s.)

$x = \$120.00$     (Divide both sides by 3.)

*Answer* (a): The apprentice makes $120.00 per week. The journeyman makes $2x = 2(\$120) = \$240$ per week.

*Solution* (b): To find the hourly wage of each, divide the weekly wage by the number of hours worked. In this case, divide by 40.

*Answer* (b): Hourly wage of apprentice $= \dfrac{\$120}{40} = \$3.00$

Hourly wage of machinist $= \dfrac{\$240}{40} = \$6.00$

**Example 29:** Mr. Brock, Mr. Weston, and Mr. Whitmore are journeymen mechanics. They decide to form a partnership and start the Westbrockmore Garage. To get started they need $25,000 for rent, supplies, shop tools, and general inventory. Mr. Brock agrees to put in a certain amount of money. Mr. Weston agrees to put in $1,000 *more* than *twice* Mr. Brock's amount. Mr. Whitmore agrees to put in $1,500 *less* than *three* times that put in by Mr. Brock.

(a) How much does each man put into the business?

(b) What percent of the business does each man own?

*Solution* (a): The amount each man puts into the business depends on the amount from Mr. Brock. Let his amount be $x$ dollars.

Amount from Mr. Brock = $x$
Amount from Mr. Weston = $2x + \$1,000$
Amount from Mr. Whitmore = $3x - \$1,500$
The sum of the three shares is $25,000.

Write this as an equation:

$$x \quad + (2x + 1,000) + (3x - 1,500) = 25,000$$
(Brock)      (Weston)      (Whitmore)

Collect $x$'s and numbers:

$$6x - 500 = 25,000$$
$$6x = 25,500 \quad \text{(Add 500 to both sides.)}$$
$$x = 4,250 \quad \text{(Divide both sides by 6.)}$$

*Answer* (a): Brock: $x = \$4,250$

Weston: $2x + 1,000 = 2(4,250) + 1,000 = \$9,500.$
Whitmore: $3x - 1,500 = 3(4,500) - 1,500$
$$= \$11,250.$$

*Check:* $\$4,250 + \$9,500 + \$11,250 = \$25,000.$

*Solution* (b): The *percent* of the business owned by each partner is found by dividing the total investment of $25,000. by the amount each man invested, and expressing this as a percent:

*Answer* (b): Percent owned by Brock $= \dfrac{\$4,250}{\$25,000} = 17\%$

Percent owned by Weston $= \dfrac{\$9,500}{\$25,000} = 38\%$

Percent owned by Whitmore $= \dfrac{\$11,250}{\$25,000} = 45\%$

*Check:*                                Sum $= \overline{100\%}$

**Example 30:** A service station attendant has this problem. A customer wants his car winterized. He specifies that the coolant in the radiator should be 30% antifreeze. The attendant checks the coolant and finds it is 10% antifreeze. How many quarts must be drained out and replaced by 100% antifreeze for the full radiator to test 30%?

*Solution:* Find the capacity of the cooling system by looking in the Owner's Manual or Motor's Auto Repair Manual. In this case the capacity is 20 quarts (5 gallons). Before draining the radiator, find the amount of antifreeze present:

$$10\% \text{ of } 20 \text{ qt.} = 2 \text{ quarts}$$

Let $x$ = number of quarts of mixture drained from the cooling system. Since the mixture is 10% antifreeze, after draining $x$ quarts, the cooling system has 10% of $(20 - x)$ quarts of antifreeze left (Fig. 14-7). Replace

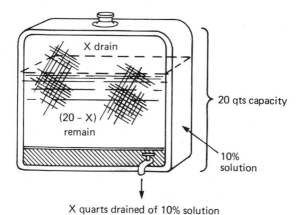

X quarts drained of 10% solution

**Figure 14-7**

the amount drained with 100% antifreeze. When this is done, the cooling system has 10% $(20 - x) + x$ quarts of antifreeze. If the correct amount is drained and re-placed, there are 6 quarts of antifreeze in the system. Note: 6 quarts = 30% of 20 = required amount (Fig. 14-8). The final equation for the amount $x$ is:

(amount replaced) + (amount of antifreeze drained)
$$= \text{(amount required)}$$

$x + 10\% \, (20 - x) = 30\%$ of 20

$x + .10(20) - .10x = 6$   (Use the Distributive Law, page 73 on left-hand side.)

$x + 2 - .10x = 6$

$.90x = 6 - 2$   (Collect $x$'s on left and subtract 2 from both sides.)

$.90x = 4$   Divide both sides by .90

$$x = \frac{4}{.90} = 4.44 \text{ quarts}$$

*Answer:* The attendant should drain 4.44 quarts of 10% solution and replace it with 100% antifreeze.

(Original mixture is 10% antifreeze)

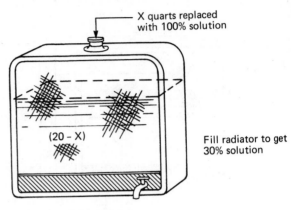

Drain off x quarts of solution of
which 10% of x is antifreeze

X quarts replaced
with 100% solution

(20 – X)

Fill radiator to get
30% solution

**Figure 14-8**

**Example 31:** A laboratory technician wants to dilute 25 cc (cubic
centimeters) of a 90% solution of acid and water with
a 20% solution to obtain a 40% solution. How many
cubic centimeters of 20% solution should he add?

*Solution:* Draw a sketch showing the amounts of solution present
and the amount of acid present. Figure 14-9 is a *picture
equation* of the problem. The amount of acid taken from
the first beaker is:

$$20\% \text{ of } x = .20x$$

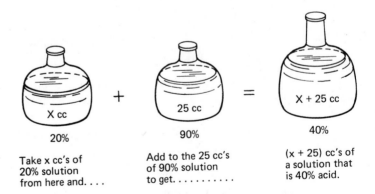

X cc

+

25 cc

=

X + 25 cc

20%

90%

40%

Take x cc's of
20% solution
from here and. . . .

Add to the 25 cc's
of 90% solution
to get. . . . . . . . . .

(x + 25) cc's of
a solution that
is 40% acid.

**Figure 14-9**

When this is added to the second container, there is a total of:

.20x + .90(25) = .20x + 22.5 cc of acid

There is now a total of (x + 25) cc of solution. If the right amount (x cc) has been added, the result is a 40% solution. Therefore,

.20x + 22.5 = .40(x + 25)

.20x + 22.5 = .40x + 10        (Use the Distributive Law)

12.5 = .20x        (Divide both sides by .20)

x = 62.5

*Answer:* 62.5 cubic centimeters of 20% solution must be added.

# PROBLEMS: *Sec. 4 Applying Algebra to Technology*

1. The perimeter of a rectangle is 48 inches. If the length is 10 inches, what is the width?

2. The area of a square is 64 square feet. What is the length of a side?

3. The perimeter of an equilateral triangle is 125 inches. How long is each side?

4. One angle of a right triangle is 54°. What is the size of the other acute angle?

5. Otis has 240 yards of wire with which to fence in a rectangular garden. The garden is to be three times as long as it is wide. What should be the length and width?

6. One angle of a triangle is 105°. Of the two remaining angles, one is twice as large as the other. How many degrees are there in the other two angles?

7. An 8-foot board is cut into three pieces. The second piece is 1 foot longer than the first, and the third is 1 foot longer than the second. What are the lengths of the three pieces?

8. Three oil tanks have a total capacity of 3,300 gallons. The first tank has a capacity of 100 gallons more than the second one, and the third has a capacity of twice the second one. What is the capacity of each tank?

9. How much money must be invested at 6% to receive $42 in interest in one year? (I = Prt)

10. The combined weight of three samples is 16 grams. The first sample is 2 grams more than the second, and the third is 1 gram less than the second. What are the weights of the samples?

11. A journeyman auto painter charges $20 per hour and $8 per hour for his assistant. Together they painted a fleet of delivery trucks for $1,040. If the journeyman worked 10 hours more than his helper, how many hours did each man work?

12. If there are 8 grams of alcohol in 70 grams of solution, how many grams are in 175 grams of the same solution? (Use direct proportion)

13. How many liters of pure alcohol (100%) must be added to 3 liters of a 25% solution to get a 50% solution?

14. How many cubic centimeters (cc) of water must be added to 100 cc of a 40% acid solution to dilute it to a 25% solution?

15. How many liters of an 80% solution of acid must be added to 10 liters of a 40% solution to obtain a 70% solution?

16. An automobile has a 12-quart cooling system. It is filled with a 30% anti-freeze solution. How many quarts must be drained off and replaced by pure alcohol to obtain a 70% solution?

17. Seventy grams of solder which is 40% tin is to be melted with solder that is 15% tin. How many grams of 15% solder should be melted to get a solder that is 25% tin?

18. A battery contains 2 quarts (64 fluid ounces) of a solution which is 25% acid. How many ounces must be drained off and replaced by 100% acid to get a 60% acid solution?

19. David Malcolm, a real estate broker, just sold 140 acres of land for $37,000. Part of the land is for cattle grazing, and it sold for $200 per acre. The rest of the 140 acres is for farming, and it sold for $300 per acre. How many acres were sold at each price?

20. When two electrical appliances are connected to the *same* current source, they are connected in *parallel* (Fig. 14-10). The total resistance of two ap-

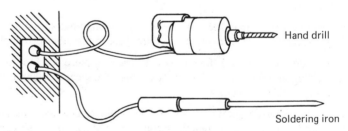

Hand drill

Soldering iron

**Figure 14-10**

pliances in parallel is not found by adding the separate resistances as can be done in a series connection. Instead, the total resistance $R_{total}$ is found by using the formula:

$$R_{total} = \frac{R_1 \cdot R_2}{R_1 + R_2}$$

where $R_1$ and $R_2$ represent the resistances of the two appliances.

(a) What is the total resistance when a hand drill (500 ohms) and a soldering iron (750 ohms) are connected in parallel?

(b) What is the total resistance when a drop light (150 ohms) and a metal saw (1,000 ohms) are connected in parallel?

**21.** A pair of pliers has 5-inch handles and 1-inch jaws. If a force of 75 pounds is applied to the end of the handles, what force is at the jaws? (This is an indirect proportion problem.)

**22.** A piece of wire 64 inches long is bent into the shape of a rectangle. The *ratio* of the length to the width is to be 3:2. What are the dimensions of the rectangle?

**23.** Two cars are 540 miles apart. They start toward each other at the same time. The first car travels an average of 32 mph. The second car travels an average of 40 mph. How long will it take them to meet?

**24.** On two technical mathematics tests Baron scored 75 and 82. What score must he make on a *third* test to have an average of 85?

**25.** A length of rope is hanging from the top of a power pole to the ground. Ten feet of rope lies coiled on the ground. When the free end of the rope is brought out 30 feet, it just touches the ground. How high is the power pole?

**26.** A hydrographic research vessel has just set off a small explosive charge a certain distance away. The underwater hydrophone picked up the sound of the explosion 12 seconds before the explosion was heard on the surface. If sound travels 4,800 fps in water and 1,125 fps through the air, how far away was the explosion?

**27.** In morning traffic Mr. Piper averages 24 mph going to work. In the evening he averages 30 mph going home. If the total driving time is 1.5 hours, how far is it from where he lives to where he works?

**28.** By using both trigonometry and algebra, find the radius of the circle in Fig. 14-11.

Note: $\dfrac{r}{r + \sqrt{8}} = \sin 45°$

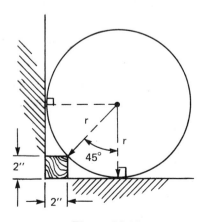

*Figure 14-11*

## Practice Test for Chapter 14

1.  Use letters and numbers to write each of the statements in algebra.
    (a) Twice a certain number decreased by five.  *Answer:* _____
    (b) The product of $s$ and $t$ divided by their sum.
    
    *Answer:* _____

2.  One automatic screw machine can produce $x$ number of parts in 1 hour. Another faster machine can produce 100 more per hour than the first machine. Express in terms of $x$ the number of parts produced by both machines in 2 hours.    *Answer:* _____

3.  The length of a rectangular solid (box) is $L$ inches. The width is 1 inch less than the length, and the height is 2 inches more than the length. Express the volume in terms of the letter $L$.    *Answer:* _____

4.  Write a formula for the rule:
    The kinetic energy ($K$) of a moving object is equal to one-half the product of the mass ($m$) and the square of the velocity ($v$).
    
    *Answer:* _____

5.  Solve each of the equations for the letter $x$.
    (a) $2x - 3 = x + 4$    *Answer :* _____
    (b) $\dfrac{5x + 4}{3} = 12$    *Answer :* _____

6.  A steel bar 18 feet long is cut into three pieces. The second piece is to be twice as long as the first. The third piece is three feet shorter than the first. What is the length of the longest piece?    *Answer:* _____

7.  During their annual tire sale, Mid-Nite Auto Supply sold 500 tires for $20,000. Some of the tires sold for $55 and the others sold for $30 each. How many of each type were sold during the sale?
    
    *Answer:* _____

8.  An automobile has a 20-quart cooling system. It is filled with a 15% solution of antifreeze. How many quarts must be drained and replaced by 100% antifreeze to get a 75% solution?    *Answer:* _____

# Appendix

**TABLE A-1** Tables of Interrelation of Units of Measurement
U.S. Customary and Metric Systems
Source: National Bureau of Standards, U.S. Department of Commerce

### Units of Length

| Units | Inches | Feet | Yards | Rods | Miles | Cm. | Meters |
|---|---|---|---|---|---|---|---|
| 1 inch = | 1 | 0.083 333 | 0.027 778 | 0.005 051 | 0.000 016 | 2.54 | 0.025 4 |
| 1 foot = | 12 | 1 | 0.333 333 | 0.060 606 | 0.000 189 | 30.48 | 0.304 8 |
| 1 yard = | 36 | 3 | 1 | 0.181 818 | 0.000 568 | 91.44 | 0.914 4 |
| 1 rod = | 198 | 16.5 | 5.5 | 1 | 0.003 125 | 502.92 | 5.029 2 |
| 1 mile = | 63,360 | 5,280 | 1,760 | 320 | 1 | 160,934.4 | 1,609.344 |
| 1 cm. = | 0.3937 | 0.032 808 | 0.010 936 | 0.001 988 | 0.000 006 | 1 | 0.01 |
| 1 meter = | 39.37 | 3.280 840 | 1.093 613 | 0.198 839 | 0.000 621 | 100 | 1 |

### Units of Area

| Units | Square Inches | Square Feet | Square Yards | Square Rods |
|---|---|---|---|---|
| 1 sq. inch = | 1 | 0.006 944 | 0.000 772 | 0.000 026 |
| 1 sq. foot = | 144 | 1 | 0.111 111 | 0.003 673 |
| 1 sq. yard = | 1,296 | 9 | 1 | 0.033 058 |
| 1 sq. rod = | 39,204 | 272.25 | 30.25 | 1 |
| 1 acre = | 6,272,640 | 43,560 | 4,840 | 160 |
| 1 sq. mile = | 4,014,489,600 | 27,878,400 | 3,097,600 | 102,400 |
| 1 sq. cm. = | 0.155 000 | 0.001 076 | 0.000 120 | 0.000 004 |
| 1 sq. meter = | 1,550.003 | 10.763 91 | 1.195 990 | 0.039 537 |

| Units | Acres | Square Miles | Square Cm. | Square Meters |
|---|---|---|---|---|
| 1 sq. inch = | 0.000 000 159 | 0.000 000 000 2 | 6.451 6 | 0.000 645 |
| 1 sq. foot = | 0.000 023 | 0.000 000 036 | 929.030 4 | 0.092 903 |
| 1 sq. yard = | 0.000 207 | 0.000 000 323 | 8,361.273 6 | 0.836 127 |
| 1 sq. rod = | 0.006 25 | 0.000 009 766 | 252,928.526 4 | 25.292 85 |
| 1 acre = | 1 | 0.001 563 | 40,468,564 | 4,046.856 |
| 1 sq. mile = | 640 | 1 | 25,899,881,103 | 2,589,988 |
| 1 sq. cm. = | 0.000 000 025 | 0.000 000 000 04 | 1 | 0.000 1 |
| 1 sq. meter = | 0.000 247 | 0.000 000 386 | 10,000 | 1 |

## Units of Volume

| Units | Cubic Inches | Cubic Feet | Cubic Yards | Cubic Cm. | Cubic Meters |
|---|---|---|---|---|---|
| 1 cubic inch = | 1 | 0.000 579 | 0.000 021 | 16.387 064 | 0.000 016 |
| 1 cubic foot = | 1,728 | 1 | 0.037 037 | 28,316.847 | 0.028 317 |
| 1 cubic yard = | 46,656 | 27 | 1 | 764,554.858 | 0.764 555 |
| 1 cubic cm. = | 0.061 024 | 0.000 035 | 0.000 001 | 1 | 0.000 001 |
| 1 cubic meter = | 61,023.74 | 35.314 67 | 1.307 951 | 1,000,000 | 1 |

## Units of Capacity (*Liquid Measure*)

| Units | Liquid Ounces | Liquid Pints | Liquid Quarts |
|---|---|---|---|
| 1 liquid ounce = | 1 | 0.062 5 | 0.031 25 |
| 1 liquid pint = | 16 | 1 | 0.5 |
| 1 liquid quart = | 32 | 2 | 1 |
| 1 gallon = | 128 | 8 | 4 |
| 1 cubic inch = | 0.554 113 | 0.034 632 | 0.017 316 |
| 1 cubic foot = | 957.507 | 59.844 16 | 29.922 08 |
| 1 liter = | 33.814 02 | 2.113 376 | 1.056 688 |

| Units | Gallons | Cubic Inches | Cubic Feet | Liters |
|---|---|---|---|---|
| 1 liquid ounce = | 0.007 813 | 1.804 688 | 0.001 044 | 0.029 574 |
| 1 liquid pint = | 0.125 | 28.875 | 0.016 710 | 0.473 176 |
| 1 liquid quart = | 0.25 | 57.75 | 0.033 420 | 0.946 353 |
| 1 gallon = | 1 | 231 | 0.133 681 | 3.785 412 |
| 1 cubic inch = | 0.004 329 | 1 | 0.000 579 | 0.016 387 |
| 1 cubic foot = | 7.480 519 | 1,728 | 1 | 28.316 847 |
| 1 liter = | 0.264 172 | 61.023 74 | 0.035 315 | 1 |

## Units of Capacity (*Dry Measure*)

| Units | Dry Pints | Dry Quarts | Pecks | Bushels | Cubic Inches |
|---|---|---|---|---|---|
| 1 dry pint = | 1 | 0.5 | 0.0625 | 0.015 625 | 33.600 313 |
| 1 dry quart = | 2 | 1 | 0.125 | 0.031 25 | 67.200 625 |
| 1 peck = | 16 | 8 | 1 | 0.25 | 537.605 |
| 1 bushel = | 64 | 32 | 4 | 1 | 2,150.42 |
| 1 cubic inch = | 0.029 762 | 0.014 881 | 0.001 860 | 0.000 465 | 1 |
| 1 liter = | 1.816 166 | 0.908 083 | 0.113 510 | 0.028 378 | 61.023 74 |

## Units of Capacity (*Dry Measure*) (*cont'd.*)

| Units | Liters |
|---|---|
| 1 dry pint = | 0.550 610 |
| 1 dry quart = | 1.101 221 |
| 1 peck = | 8.809 768 |
| 1 bushel = | 35.239 07 |
| 1 cubic inch = | 0.016 387 |
| 1 liter = | 1 |

## Units of Mass and Weight

| Units | Kilograms |
|---|---|
| 1 grain = | 0.000 065 |
| 1 ounce (avdp) = | 0.028 350 |
| 1 pound (avdp) = | 0.453 593 |
| 1 milligram = | 0.000 001 |
| 1 gram = | 0.001 |
| 1 kilogram = | 1 |

## Units of Mass and Weight

| Units | | Grains | Avdp.<br>Ounces | Avdp.<br>Pounds | Milligrams | Grams |
|-------|---|--------|-------|-------|------------|-------|
| 1 grain | = | 1 | 0.002 286 | 0.000 143 | 64.799 | 0.064 799 |
| 1 ounce (avdp.) | = | 437.5 | 1 | 0.062 5 | 28,349.523 | 28.349 523 |
| 1 pound (avdp.) | = | 7,000 | 16 | 1 | 453,592.37 | 453.592 37 |
| 1 milligram | = | 0.015 432 | 0.000 035 | 0.000 002 | 1 | 0.001 |
| 1 gram | = | 15.432 36 | 0.035 274 | 0.002 205 | 1,000 | 1 |
| 1 kilogram | = | 15,432.36 | 35.274 | 2.204 623 | 1,000,000 | 1,000 |

**TABLE A-2**   TABLES OF IMPORTANT EQUIVALENTS
(Miscellaneous Values)

### Units of Length

| | |
|---|---|
| 1 fathom | = 6 feet |
| 1 furlong | = $\frac{1}{8}$ mile = 220 yards = 201.168 meters |
| 1 inch | = 2.54 centimeters (By law) |
| 1 league | = 3 miles = 2.6069 nautical miles |
| 1 mil | = 0.001 inch |
| 1 millimeter | = 0.039 370 inch |
| 1 mile | = 1.609 34 kilometers = 0.868 98 nautical mile |
| 1 nautical mile | = 1.150 78 statute mile = 6,076.1155 feet |

### Units of Area

| | |
|---|---|
| 1 circular mil | = 0.7854 sq. mil (Definition: area of a circle with diameter of 1 mil) |
| 1 hectare | = 10,000 sq. meters = 2.471 05 acres = 107,637 sq. ft. |
| 1 section | = 1 mile square = 640 acres |
| 1 township | = 36 sections = 6 miles square = 36 square miles |

### Units of Volume

| | |
|---|---|
| 1 acre-foot | = 43,560 cu. ft. = 325,851 gallons |
| 1 board-foot | = 144 cubic inches |
| 1 cord | = 128 cubic feet (4 × 4 × 8 feet) |
| 1 cubic centimeter | = 0.061 024 cu. in. = 0.033 814 ounces = 0.999 972 ml.<br>(For all practical purposes use 1 cubic centimeter = 1 milliliter so that 1 liter = 1,000 cubic centimeters.) |
| 1 cubic foot | = 7.481 gallons |
| 1 cubic yard | = 46,656 cu. in. = 202.0 gallons |
| 1 gallon | = 231 cu. in. = 8.3452 lb. of water |
| 1 liter | = 0.2642 gallon = 61.02 cu. in. = 0.03531 cu. ft. |
| 1 ⠿ | = 1 cubic meter |
| 1 barrel (cement) | = 376 pounds |
| 1 barrel (oil) | = 42 gallons |

## Units of Mass and Weight

```
1 carat         = 0.200 gram = 3.086 grains = 0.007 oz.
1 cu. ft. of water = 62.4 lbs. = 28.3 kg.
1 ton = 2,000 lbs.
The British system uses:
1 stone          = 14 pounds
1 hundredweight = 112 pounds
1 long ton       = 20 hundredweight = 2,240 pounds
1 metric ton     = 2.205 tons
```

## Units of Pressure

```
1 atmosphere (std.)  = 29.92 in. mercury = 14.7 psi = 10,329 kg./sq. m.
1 kg./sq. meter      = 0.2048 psf = 0.001 422 psi
1 pound/sq. ft. (psf) = 0.000 473 atm.
1 pound/sq. in. (psi) = 0.068 04 atm. = 703.1 kg/sq. m. = 70.31 gram/sq. cm.
```

## Units of Moment or Torque

```
1 pound-feet = 0.1383 kg.-m.
1 kg.-meter   = 7.233 lb.-ft.
```

## Units of Linear Velocity

```
1 centimeter/sec. = 1.939 fpm = 0.032 81 fps = 0.036 km./hr. = 0.02237 mph
                  = 3.728 miles/min. = 0.019 43 knot
1 foot/min. (fpm) = 0.5080 cm./sec. = 0.018 29 km./hr. = 0.011 36 mph = 0.009 868 knot
1 foot/sec. (fps)  = 30.48 cm./sec. = 1.097 km./hr. = 0.6818 mph = 0.5921 knot
1 kilometer/hour  = 27.78 cm./sec. = 54.68 fpm = 0.9113 fps = 0.6214 mph = 0.539 96 knot
1 knot            = 5144 cm./sec. = 101.3 fpm = 1.688 fps = 1.852 km./hr. = 1.150 78 mph
                  = 1 nautical mile per hour.
1 mile/hour (mph) = 44.70 cm./sec = 88 fpm = 1.467 fps = 1.609 km./hr. = 0.868 98 knot
```

## Units of Flow or Discharge

```
1 acre-foot/hr.        = 726 cfm = 5,432 gpm = 45,320 lb./min. = 342.7 liters/sec.
1 cubic foot/sec. (cfs)  = 7.481 gpm = 62.43 lb./min. = 0.472 liter/sec.
1 cubic foot/min. (cfm) = 448.8 gpm = 3,746 lb./min. = 28.32 liter/sec. = 1.9834 acre-ft./day
                       = 0.646 35 million gal./day
1 cubic meter/min.     = 1 stere/min. = 35.32 cfm = 0.5886 cfs = 264.2 gpm
1 gallon/min.          = 0.1337 cfm = 8.345 lb./min. = 0.063 09 liter/sec. = 0.001 44 mgd
                       = 0.004 419 acre-ft./day.
1 liter/sec.           = 2.119 cfm = 1,585 gpm = 0.06 cu. m./min.
1 million gallon/day (mgd) = 1.547 cfs = 43.806 liter/sec. = 694.44 gpm
```

| | |
|---|---|
| 1 Btu/min. | = 778.0 ft.-lb./min. = 0.023 57 hp. = 0.017 58 kw. |
| 1 foot-pound/min. | = 0.001 285 Btu/min. = 0.0000303 hp. = 2.260 kw. |
| 1 horsepower (hp) | = 550 ft.-lb./sec. = 42.418 Btu/min. = 33,000 ft.-lb./min. = 0.746 kw. |
| 1 kilowatt (kW.) | = 56.89 Btu/min. = 44,254 ft.-lb./min. = 737.6 ft.-lb./sec. = 1.341 hp. |
| 1 metric horsepower | = 41.83 Btu/min. = 32,550 ft.-lb./min. = 0.9863 hp. = 0.7355 kw. |
| | = 75 kg.-m./sec. |

*Units of Time*

1 minute = 60 seconds
1 hour  = 60 minutes = 3,600 seconds
1 day   = 24 hour = 1,440 minutes
1 week  = 7 days = 168 hours
1 year  = 52 weeks = 365 days (360 days to get interest)
1 decade = 10 years
1 century = 100 years

*Units of Angular Measure*

1 minute (′) = 60 seconds (″)
1 degree (°) = 60 min. = 3,600 sec.
1 quadrant  = 90 degrees
1 circle    = 4 quadrants = 360 degrees

*Weights of Certain Common Materials*

1 cu. in. aluminum = 0.0963 lb.
1 cu. in. cast iron  = 0.2604 lb.
1 cu. in. mild steel = 0.2833 lb.
1 cu. in. copper    = 0.3215 lb.
1 cu. in. lead      = 0.4106 lb.
1 cu. ft. aluminum = 167 lb.
1 cu. ft. cast iron  = 464 lb.
1 cu. ft. mild steel = 490 lb.
1 cu. ft. copper    = 556 lb.
1 cu. ft. lead      = 710 lb.

*Units of Fuel Consumption*

1 kilometer per liter = 2.3527 miles per gallon
1 mile per gallon    = 0.4249 kilometers per liter

**TABLE A-3**  DECIMAL EQUIVALENTS OF PARTS OF AN INCH BY 64THS

| Fraction (inches) | | | Decimal (inches) | Millimeters (mm.) | Fraction (inches) | | | Decimal (inches) | Millimeters (mm.) |
|---|---|---|---|---|---|---|---|---|---|
| | | $\frac{1}{64}$ | 0.015625 | 0.3968 | | | $\frac{33}{64}$ | 0.515625 | 13.0966 |
| | $\frac{1}{32}$ | | 0.03125 | 0.7937 | | $\frac{17}{32}$ | | 0.53125 | 13.4934 |
| | | $\frac{3}{64}$ | 0.046875 | 1.1906 | | | $\frac{35}{64}$ | 0.546875 | 13.8903 |
| $\frac{1}{16}$ | | | 0.0625 | 1.5875 | $\frac{9}{16}$ | | | 0.5625 | 14.2872 |
| | | $\frac{5}{64}$ | 0.078125 | 1.9843 | | | $\frac{37}{64}$ | 0.578125 | 14.6841 |
| | $\frac{3}{32}$ | | 0.09375 | 2.3812 | | $\frac{19}{32}$ | | 0.59375 | 15.0809 |
| | | $\frac{7}{64}$ | 0.109375 | 2.7780 | | | $\frac{39}{64}$ | 0.609375 | 15.4778 |
| $\frac{1}{8}$ | | | 0.125 | 3.1749 | $\frac{5}{8}$ | | | 0.625 | 15.8747 |
| | | $\frac{9}{64}$ | 0.140625 | 3.5718 | | | $\frac{41}{64}$ | 0.640625 | 16.2715 |
| | $\frac{5}{32}$ | | 0.15625 | 3.9686 | | $\frac{21}{32}$ | | 0.65625 | 16.6684 |
| | | $\frac{11}{64}$ | 0.171875 | 4.3655 | | | $\frac{43}{64}$ | 0.671875 | 17.0653 |
| $\frac{3}{16}$ | | | 0.1875 | 4.7625 | $\frac{11}{16}$ | | | 0.6875 | 17.4625 |
| | | $\frac{13}{64}$ | 0.203125 | 5.1592 | | | $\frac{45}{64}$ | 0.703125 | 17.8590 |
| | $\frac{7}{32}$ | | 0.21875 | 5.5561 | | $\frac{23}{32}$ | | 0.71875 | 18.2559 |
| | | $\frac{15}{64}$ | 0.234375 | 5.9530 | | | $\frac{47}{64}$ | 0.734375 | 18.6527 |
| $\frac{1}{4}$ | | | 0.25 | 6.3498 | $\frac{3}{4}$ | | | 0.75 | 19.0496 |
| | | $\frac{17}{64}$ | 0.265625 | 6.7467 | | | $\frac{49}{64}$ | 0.765625 | 19.4465 |
| | $\frac{9}{32}$ | | 0.28125 | 7.1436 | | $\frac{25}{32}$ | | 0.78125 | 19.8433 |
| | | $\frac{19}{64}$ | 0.296875 | 7.5404 | | | $\frac{51}{64}$ | 0.796875 | 20.2402 |
| $\frac{5}{16}$ | | | 0.3125 | 7.9373 | $\frac{13}{16}$ | | | 0.8125 | 20.6371 |
| | | $\frac{21}{64}$ | 0.328125 | 8.3342 | | | $\frac{53}{64}$ | 0.828125 | 21.0339 |
| | $\frac{11}{32}$ | | 0.34375 | 8.7310 | | $\frac{27}{32}$ | | 0.84375 | 21.4308 |
| | | $\frac{23}{64}$ | 0.359375 | 9.1279 | | | $\frac{55}{64}$ | 0.859375 | 21.8277 |
| $\frac{3}{8}$ | | | 0.375 | 9.5248 | $\frac{7}{8}$ | | | 0.875 | 22.2245 |
| | | $\frac{25}{64}$ | 0.390625 | 9.9216 | | | $\frac{57}{64}$ | 0.890625 | 22.6214 |
| | $\frac{13}{32}$ | | 0.40625 | 10.3185 | | $\frac{29}{32}$ | | 0.90625 | 23.0183 |
| | | $\frac{27}{64}$ | 0.421875 | 10.7154 | | | $\frac{59}{64}$ | 0.921875 | 23.4151 |
| $\frac{7}{16}$ | | | 0.4375 | 11.1125 | $\frac{15}{16}$ | | | 0.9375 | 23.8120 |
| | | $\frac{29}{64}$ | 0.453125 | 11.5091 | | | $\frac{61}{64}$ | 0.953125 | 24.2089 |
| | $\frac{15}{32}$ | | 0.46875 | 11.9060 | | $\frac{31}{32}$ | | 0.96875 | 24.6057 |
| | | $\frac{31}{64}$ | 0.484375 | 12.3029 | | | $\frac{63}{64}$ | 0.984375 | 25.0026 |
| $\frac{1}{2}$ | | | 0.5 | 12.6997 | 1 | | | 1. | 25.4000 |

## TABLE A-4  APPROXIMATE SQUARE ROOTS AND CUBE ROOTS

| $n$ | $\sqrt{n}$ | $\sqrt{10n}$ | $\sqrt[3]{n}$ | $\sqrt[3]{10n}$ |
|---|---|---|---|---|
| 1 | 1.000 | 3.162 | 1.000 | 2.154 |
| 2 | 1.414 | 4.472 | 1.260 | 2.714 |
| 3 | 1.732 | 5.477 | 1.442 | 3.107 |
| 4 | 2.000 | 6.325 | 1.587 | 3.420 |
| 5 | 2.236 | 7.071 | 1.710 | 3.684 |
| 6 | 2.449 | 7.746 | 1.817 | 3.915 |
| 7 | 2.646 | 8.367 | 1.913 | 4.121 |
| 8 | 2.828 | 8.944 | 2.000 | 4.309 |
| 9 | 3.000 | 9.487 | 2.080 | 4.481 |
| 10 | 3.162 | 10.000 | 2.154 | 4.642 |
| 11 | 3.317 | 10.488 | 2.224 | 4.791 |
| 12 | 3.464 | 10.954 | 2.289 | 4.932 |
| 13 | 3.606 | 11.402 | 2.351 | 5.066 |
| 14 | 3.742 | 11.832 | 2.410 | 5.192 |
| 15 | 3.873 | 12.247 | 2.466 | 5.313 |
| 16 | 4.000 | 12.649 | 2.520 | 5.429 |
| 17 | 4.123 | 13.038 | 2.571 | 5.540 |
| 18 | 4.243 | 13.416 | 2.621 | 5.646 |
| 19 | 4.359 | 13.784 | 2.668 | 5.749 |
| 20 | 4.472 | 14.142 | 2.714 | 5.848 |
| 21 | 4.583 | 14.491 | 2.759 | 5.944 |
| 22 | 4.690 | 14.832 | 2.802 | 6.037 |
| 23 | 4.796 | 15.166 | 2.844 | 6.127 |
| 24 | 4.899 | 15.492 | 2.884 | 6.214 |
| 25 | 5.000 | 15.811 | 2.924 | 6.300 |
| 26 | 5.099 | 16.125 | 2.962 | 6.383 |
| 27 | 5.196 | 16.432 | 3.000 | 6.463 |
| 28 | 5.292 | 16.733 | 3.037 | 6.542 |
| 29 | 5.385 | 17.029 | 3.072 | 6.619 |
| 30 | 5.477 | 17.321 | 3.107 | 6.694 |
| 31 | 5.568 | 17.607 | 3.141 | 6.768 |
| 32 | 5.657 | 17.889 | 3.175 | 6.840 |
| 33 | 5.745 | 18.166 | 3.208 | 6.910 |
| 34 | 5.831 | 18.439 | 3.240 | 6.980 |
| 35 | 5.916 | 18.708 | 3.271 | 7.047 |
| 36 | 6.000 | 18.974 | 3.302 | 7.114 |
| 37 | 6.083 | 19.235 | 3.332 | 7.179 |
| 38 | 6.164 | 19.494 | 3.362 | 7.243 |
| 39 | 6.245 | 19.748 | 3.391 | 7.306 |
| 40 | 6.325 | 20.000 | 3.420 | 7.368 |
| 41 | 6.403 | 20.248 | 3.448 | 7.429 |
| 42 | 6.481 | 20.494 | 3.476 | 7.489 |
| 43 | 6.557 | 20.736 | 3.503 | 7.548 |
| 44 | 6.633 | 20.976 | 3.530 | 7.606 |
| 45 | 6.708 | 21.213 | 3.557 | 7.663 |
| 46 | 6.782 | 21.448 | 3.583 | 7.719 |
| 47 | 6.856 | 21.679 | 3.609 | 7.775 |
| 48 | 6.928 | 21.909 | 3.634 | 7.830 |
| 49 | 7.000 | 22.136 | 3.659 | 7.884 |
| 50 | 7.071 | 22.361 | 3.684 | 7.937 |

# Answers to Odd-Numbered Problems

*Chapter 1, Section 1, Page 6*

**1.** (a) 364 (b) 21,421 (c) 62,302 (d) 15

*Chapter 1, Section 2, Page 9*

**1.** (a) $3 \times 100 + 8 \times 10 + 9$ (b) $5 \times 100 + 7 \times 10 + 6$
(c) $7 \times 100 + 6 \times 10 + 5$ (d) $1 \times 1,000 + 6 \times 100 + 5 \times 10 + 2$
(e) $1 \times 1,000 + 5 \times 100$
(f) $1 \times 10,000 + 7 \times 1,000 + 4 \times 100 + 8 \times 10 + 8$
(g) $4 \times 10,000 + 5 \times 1,000 + 8 \times 100 + 9 \times 10$
(h) $4 \times 10,000 + 7 \times 100 + 8 \times 10 + 5$
(i) $1 \times 10,000 + 6 \times 1,000 + 5$
(j) $3 \times 100,000 + 4 \times 10,000 + 5 \times 1,000 + 8 \times 100 + 8 \times 10 + 4$
(k) $9 \times 100,000 + 7 \times 10,000 + 7 \times 100 + 5 \times 10$
(l) $1 \times 1,000,000 + 6 \times 100,000 + 5 \times 10,000 + 7 \times 1,000 + 7 \times 100$
$+ 5 \times 10 + 4$
(m) $5 \times 1,000,000 + 8 \times 100,000 + 6 \times 100 + 5 \times 10$
(n) $3 \times 10,000,000 + 4 \times 1,000,000 + 6 \times 100,000 + 5 \times 10,000$
$+ 8 \times 100 + 9 \times 10 + 5$

**3.** (a) 1,000 (b) 100,000 (c) 1 (units) (d) 100

**5.** (a) 726,354 cubic feet (b) 146,059 cubic feet
(c) 8,234 kilowatt hours (d) 73,452 gallons

*Chapter 1, Section 3, Page 14*

**1.** (a)

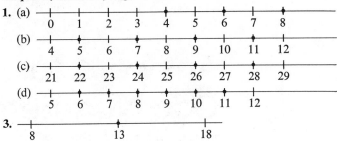

(b)

(c)

(d)

**3.**

**5.** $r = 25$    $s = 75$    $t = 175$    $u = 225$

*Chapter 1, Practice Test, Page 15*

**1.** 560,124    **2.**

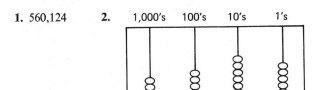

**3.**

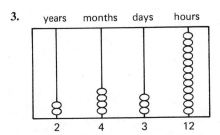

**4.** $2 \times 1,000,000 + 4 \times 100,000 + 5 \times 10,000 + 6 \times 1,000 + 6 \times 100 + 5 \times 10 + 4$

**5.** $4 \times 1,000,000 + 4 \times 10$    **6.** 7, 6, 5    **7.** 5,341

**8.** 1,257    **9.** Twenty-four thousand, five hundred nine.

**10.** Four hundred fifty-six thousand, five hundred ninety-seven

**11.** 35,469    **12.** 853,058    **13.** 64,107

**14.**

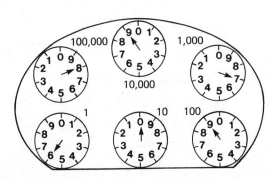

*Chapter 2, Section 1, Page 22*

1. (a) 480    350    650       930         50     80    180    580
   (b)   4,570     2,190     4,660       5,100         4,010   6,700
   (c)  14,750    16,980   20,080     15,010      19,730  22,800
   (d) 234,660   695,790   1,789,460   25,899,710

3. (a)    2,000      4,000      5,000       6,000
   (b)   55,000     45,000     22,000     66,000
   (c)  146,000    236,000    561,000    642,000
   (d) 1,477,000   2,565,000  16,760,000  12,121,000

5. (a) 400    (b) 2,000    (c) 2,000    (d) 10,000    (e) 7,400
   (f) 16,000    (g) 3,460    (h) 5,600    (i) 2,000,000    (j) 4,000,000

7. 14,200,000    15,500,000; 11,100,000

9. (a)   5,499     4,501    (b)   16,500     15,500
   (c)  45,499    44,501    (d)   40,500     39,500
   (e) 165,499   164,501    (f)  1,470,500   1,469,500

11. Greatest 454. Least 446.

13. (a) 6,080    (b) 6,100

15. (a) 63,800,000    (b) 31,830,000    (c) 28,000,000

17. (a) 239,000    (b) 240,000

19. (a) 3,600,000    (b) 78,000    (c) 3,540,000

*Chapter 2, Section 2, Page 28*

1. (a) 4; 2,500    (b) 6; 340,400    (c) 5; 36,100    (d) 3; 600
   (e) 5; 35,500    (f) 7; 1,908,900

3. (a) 3; 1    (b) 4; 1    (c) 2; 10    (d) 3; 10    (e) 2; 1,000
   (f) 4; 1,000    (g) 2; 100    (h) 2; 1

5. 56,000 gal.

7. (a) 8,900,000    (b) $61,355,000,000

9. (a) 5,200    (b) 5,200

11. 650,000 to 1

13. 6,270,000 cu. in.

15. 213,000 Japanese; 170,000 Chinese; 139,000 Filipino

*Chapter 2, Practice Test, Page 30*

1. (a)  23,460       23,500       23,000
   (b)   1,100        1,100       1,000
   (c)   2,500        2,500       3,000
   (d) 236,650,010  236,650,000  236,650,000

2. (a) 5; 24,000    (b) 7; 2,400,000    (c) 4; 350,000    (d) 3; 250

3.

          800             880  900            1,000
                           (c)  (b)           (a)

**4.** (a) 44,501          45,499
   (b)    226             234
   (c)  1,550           1,650
   (d) 1,999,950      2,000,050

**5.** (a) 2$\bar{0}$0   200     (b) 2,000   2,490     (c) 234,580   230,000

**6.** 92,500,001      93,499,999

**7.** (a) 2,0$\bar{0}$0,000     (b) 3,400,$\bar{0}$00     (c) 3,$\bar{0}$00     (d) 10$\bar{0}$

**8.** 27,000,000

**9.** (a) 46,000,000   28,000,000     (b) 45,800,000   28,400,000

**10.** (a) 1,400     (b) 1,500     (c) 1,$\bar{0}$00     (d) 1,400

### *Chapter 3, Section 1, Page 39*

**1.** (a) gallons     (b) gallon, pounds     (c) square feet
   (d) years     (e) degrees     (f) atmosphere, pounds per square inch
   (g) miles per hour, kilometers per hour     (h) ton     (i) board-feet

**3.** (a) 90 square feet     (b) 26,400 yards     (c) 748 gallons
   (d) 6 miles (approximate)     (e) 30,000 pounds     (f) 244 cubic inches
   (g) 7 pounds (approximate)     (h) 62 miles per hour (approximate)

### *Chapter 3, Section 2, Page 44*

**1.** (a) ten; 10; $\pm$5 meters     (b) ten; 10; $\pm$5 cm.
   (c) ten; 10; $\pm$5 ft.
   (d) million; 1,000,000; $\pm$500,000 cubic miles
   (e) hundred; 100; $\pm$50°C

**3.** (a) Inexact: 45 psi, 47 psi, 46 psi. 48 psi   Exact: four tires
   (b) Inexact: 2,750 sq. ft., $35,900. Exact: six-room     1067 Ivy Drive
   (c) Inexact: 247 pounds. Exact: six kegs, 23,500 bolts.
   (d) Inexact: 10 years old, 85 pounds, 13 years old, 126 pounds, 5 ft. 2 in. tall
       Exact: Two children, second of May
   (e) Inexact: 88 psi, 98 psi, 116 psi, 190 psi.
       Exact: 3 cylinder

### *Chapter 3, Practice Test, Page 46*

**1.** (a) MACH   1   miles per hour   (b) rod   rod   feet
   (c) hour   minute     (d) inch     (e) pecks   quarts
   (f) cord   cord   feet   cubic feet     (g) gallon     square feet

**2.** Weight, inside diameter, outside diameter, tread wear, tread width, circumference, etc.

**3.** (a) 2     10          10 oz.         $\pm$5 oz.
   (b) 1     1,000       1,000 bu.      $\pm$500 bu.
   (c) 3     1           1 acre         $\pm\frac{1}{2}$ acre
   (d) 1     100         100 ton        $\pm$50 ton
   (e) 3     10,000      10,000 yr.     $\pm$5,000 yr.
   (f) 1     1           1°F            $\pm\frac{1}{2}$°F

**4.** (a) 10 pounds    (b) 10 cubic centimeters    (c) 1 gram    (d) 100 (exact)

**5.** (a) left to right   right to left     (b) precise     (c) precision (d) 100

**6.** (a) 144     (b) 63,360     (c) 160,000

**7.** (a) 10 $\pm$5 miles     (b) 100 $\pm$50 miles     1 $\pm\frac{1}{2}$ miles
(c) 1 gal. $\pm$2 qt. 10 $\pm$5 miles
(d) 1 $\pm\frac{1}{2}$ cu. in.   10 $\pm$5 hp.   100 $\pm$50 rpm

**8.** (a) V8 (exact)   4-barrel (exact)   4.36 $\times$ 3.59 (inexact)   429 cu. in. (inexact)
10.5 to 1 (inexact)   360 hp. at 4,600 rpm (inexact)   35 to 60 psi (inexact)
(b) 396 people (exact)   45,000 ft. (inexact)   650 mph (inexact)
(c) 1$\frac{1}{2}$ sec. (inexact)   19,200 cars (exact)

### *Chapter 4, Section 1, Page 54*

**1.** (a) 26    38    58    79    39    21    94
(b) 34    88    105    61    108    94    66
(c) 52    50    30    70    101    77    108
(d) 31    100    41    130    36    32    61

**3.** (a) 1,408    (b) 988    (c) 3,806

**5.** 33,349

**7.** 2,779,399

**9.** 21,533,918

**11.** 2,110,490

**13.** 40,608,282

**15.** (a) Sum = 72    (b) Sum = 181    (c) Sum = 706    (d) Sum = 3,060

**17.** 26,310 gallons

**19.** (a) 2,510 sq. ft.    (b) 15,050 sq. ft.    (c) 19,780 sq. ft.

**21.** 13 ft. 10 in.

**23.** 7 yd. 1 ft. 0 in.

**25.** 7 qt. 1 pt. 3 oz.

**27.** 38 hr. 18 min.

**29.** 11 yr. 2 mo. 20 day

**31.** 22 mi. 2,790 ft. 5 in.

**33.** 4 hr. 21 min. 23 sec.

**35.** $1,847

**37.** 274 inches

**39.** 14 hr. 40 min. 16 sec.

### *Chopter 4, Section 2, Page 62*

**1.** (a)    11    13    11    9    10    16    18
(b)    22    13    31    2    2    13    9
(c)    122    141    109    387    101    185    9
(d)    332    18,892    88,892    78,328

**3.** (a)   4,884        1,198        8,829        1,891
    (b)   30,979       97,900       69,780       56,449
    (c)   101,191      20,145       8,889
    (d)   909,290      579,920

**5.** 5,816

**7.** 14,133

**9.** 1,200 k w h.

**11.** $29,008

**13.** 92,658,557 miles

**15.** 90,520 days

**17.** (a)  0 yd. 1 ft. 9 in.          (b)  1 gal. 3 qt. 1 pt.
    (c)  1 m. 88 cm. 2 mm.        (d)  0 mi. 4,280 ft. 1 in.
    (e)  1 lb. 14 oz.               (f)  9° 49 min. 36 sec.
    (g)  1 hr. 48 min. 55 sec.      (h)  4 mi. 3 furlong 193 yd.

**19.** 7 : 51 : 06 A.M.

## Chapter 4, Problems, Sections 1 and 2, Page 64

**1.** 11,679

**3.** 6 hr. 26 min.

**5.** (a)  M $112     Tu $9     W $71     Th $9     F $19     Sa $84     Su 67
    (b) $953
    (c) $1,324
    (d) $371

**7.** (a)  4 hr. 30 min. 20 sec.       (b)  5 hr. 47 min. 6 sec. (c)  4 : 06 : 08 P.M.

**9.** $9,280

**11.** (a) depth = 25 ft.       (b) width = 15 ft.

**13.** $740

## Chapter 4, Practice Test, Page 67

**1.** (a)  542,203      (b)  2,481,870
**2.** (a)  4,292        (b)  254,123
**3.** 29,807
**4.** Total = 544
**5.** (a)  31 yd. 0 ft. 1 in.       (b)  11 gal. 0 qt. 0 pt.

**6.** (a)  152,348,000   and   176,480,000
    (b)  24,131,867

**7.** (a)  128,612      (b)  $268,012

**8.** (a)  119,791      (b)  $327,200

**9.** (a)  8,821        (b)  $59,188

**10.** May ($22,993)

## Chapter 5, Section 1, Page 76

**1.** (a) 36    35    48    35    48    36    56
   (b) 121    84    108    90    72   66    48
   (c) 42    48    60    64    300    75    54
   (d) 12    56    108    84
   (e) 24    36    96

**3.** (a) 56     (b) 108     (c) 91     (d) 44

**5.** (a) 12     (b) 6     (c) 6     (d) 0

**7.** (a) 1,638    1,835    2,316   1,970   2,779   2,976   6,672
   (b) 4,701    23,202    38,096   47,856   48,700   27,000
   (c) 94,296    96,588    132,335   71,100   112,640
   (d) 3,792    53,925    156,832

**9.** (a) 48    360    378    60
   (b) 560    270    276    1,800
   (c) 1,360    2,016    4,704    9,345
   (d) 17,388    304,200    1,171,584    17,109,360

**11.** (a) 17 wk. 1 day 4 hr.     (b) 45 wk. 6 day 12 hr.
   (c) 17 ft. 8 in.     (d) 82 yd. 1 ft. 4 in.
   (e) 341 yd. 1 ft. 0 in.     (f) 11 yr. 10 mo. 15 day
   (g) 79 yr. 5 mo. 18 day     (h) 50 m. 25 cm.
   (i) 2,225 m. 18 cm.     (j) 312 m. 93 cm.

**13.** 870 miles        **15.** 135 cu. ft.

**17.** 108 doz.        **19.** 2,920 hr.

**21.** $5,076        **23.** (a) 1,160     (b) 5,800

**25.** 44 hr. 17 min. 30 sec.     **27.** (a) 11,340   (b) 680,400

**29.** 52 lb.        **31.** $792,000

**33.** 7,200        **35.** $1,260

## Chapter 5, Section 2, Page 90

**1.** (a) 13     (b) 12     (c) 6     (d) 12     (e) 2

**3.** (a) 8r2     (b) 6r1     (c) 10r1     (d) 10r2     (e) 11

**5.** (a) 15     (b) 20r18     (c) 58r12     (d) 41r14

**7.** (a) 53r25     (b) 284r5     (c) 310r24

**9.** (a) 1r29     (b) 2     (c) 1r22     (d) 4

**11.** (a) 30r26     (b) 26r12     (c) 21r4     (d) 27r5

**13.** (a) 283r15     (b) 173r15     (c) 87r7

**15.** (a) 12,484r38     (b) 540r48     (c) 668r13

**17.** (a) 201     (b) 2,000r1     (c) 2,453r18

**19.** (a) 329r59     (b) 1,002r88     (c) 34r8

**21.** (a) 1,578r9     (b) 236r58     (c) 2,028r44

**23.** (a) 1,361r400    (b) 652r587    (c) 1,516r200

**25.** (a) 8r3,650    (b) 16r1,272    (c) 47r9,188

**27.** (a) 54,991r1,994    (b) 3,905r3,534    (c) 16,088r1,605

**29.** (a) 3 ft. 2 in.    (b) 1 yd. 1 ft. 2 in.

**31.** (a) 2 yr. 2 mo. 2 wk.    (b) 1 min. 7 sec.

**33.** 1,073r21    **35.** Q = 15    r = 0

**37.** 86    **39.** 8

**41.** 78    **43.** 375

**45.** $1,250    **47.** (a) 186    (b) 23    (c) 30    (d) 26

**49.** 48 lb. (Troy)    **51.** 2 amps

**53.** 38 mph    **55.** 19 in.

**57.** 4,280 mph    **59.** 88

**61.** 247

### Chapter 5, Sections 1 and 2, Page 94

**1.** (a) 24    48    45    63    20    56    81    56    42    44
   (b) 35    72    48    55    96    121    63    64    49
   (c) 99    24    36    132    16    84    54    21    36    45
   (d) 48    84    72    132    216    280    504    1,728

**3.** (a) 9    4    5    9    3    (b) 9    5    12    6    8
   (c) 6    9    7    12    8

**5.** (a) 168    (b) 432    (c) 72    (d) 240    (e) 105    (f) 288

**7.** (a) 5,117,112    (b) 6,986,616    (c) 78,328,800    (d) 199,847,700

**9.** (a) 739    (b) 985

**11.** (a) 36    (b) 34

**13.** (a) 231    (b) 121    (c) 61    (d) 21    (e) 122

**15.** (a) 1,234    (b) 2,121    (c) 1,568    (d) 3,246    (e) 764

**17.** (a) 4    (b) 21    (c) 314    (d) 847

**19.** (a) 6    (b) 43    (c) 37

**21.** 1,536

**23.** 108 miles

**25.** 28,648 pounds

**27.** 6,564 feet    **29.** 886 feet

**31.** (a) 6,552 loads    (b) 728

**33.** 4,128    **35.** 4 hr. 26 min.

### Chapter 5, Practice Test, Page 97

**1.** (a) 138,014    (b) 803,451

**2.** (a) 14r17    (b) 17    · **3.** 129r33

**4.** 482     **5.** (a) $40     (b) $200     (c) $920

**6.** 53     **7.** 1,053     **8.** 66 fps

**9.** 64 hr. 39 min. 24 sec.     **10.** 1 hr. 31 min. 50 sec.

**11.** $7,622

### Chapter 6, Section 1, Page 100

**1.** a = 3, b = −2, c = 1, d = −5

**3.** (a) −10     (b) +2     (c) −5     (d) −13

**5.** (a) −7     (b) +3     (c) +1     (d) +1

**7.** (a) +$56     (b) −$23     (c) −$45     (d) −$10

### Chapter 6, Section 2, Page 109

**1.** (a) +8     (b) +9     (c) +14     (d) −8     (e) −13     (f) +2
(g) −2     (h) +2     (i) +4     (j) −10

**3.** (a) 0     (b) −1     (c) 0     (d) 98     (e) 26     (f) −5,627
(g) −879,928     (h) −1

**5.** +14 V.     **7.** A = +10 V.   B = +20 V.

**9.** 1-75-reject, 2-69-reject, 3-44-reject, 4-41-reject,
5-46-reject, 6-67-accept, 7-39-reject, 8-68-reject,
9-34-reject, 10-36-reject

**11.** −28°F     **13.** 108°F

### Chapter 6, Section 3, Page 116

**1.** (a) 2     (b) 1     (c) −13     (d) −14     (e) −29     (f) 95

**3.** (a) −2°     (b) 6°     (c) −7°     (d) −20°

**5.** (a) −30 ft.     (b) +29 ft.     (c) 1,538 ft.

### Chapter 6, Section 4, Page 123

**1.** (a) 3,535     (b) −38,610     (c) 297,216     (d) 432     (e) −5,880
(f) −2,795     (g) 2,006     (h) −31,185

**3.** (a) −30     (b) 15     (c) −28

**5.** +9°F     **7.** 600 lb.     **9.** 21,132 lb.-in.

### Chapter 6, Section 5, Page 130

**1.** (a) +5     (b) −5     (c) −7     (d) +4     (e) −27     (f) +11
(g) −2     (h) +3

**3.** (a) 15     (b) 150     (c) −5     (d) −30

**5.** forward     **7.** −8° per min.

**9.** 250 ft. per sec.     **11.** F = 60 lb.

**13.** −2,000 fpm

*Chapter 6, Sections 1 to 5, Page 133*

**1.** (a) $-2$    (b) $-20$    (c) 10    (d) $-35$    (e) 3    (f) $-4$

**3.** (a) 40    (b) $-13$    (c) 27

**5.** $+21$ V    **7.** No. Total $= +1°$

**9.** 750 lb.    **11.** 9,000 lb.-in.

*Chapter 6, Practice Test, Page 137*

**1.** (a) 19    (b) $-26$

**2.** (a) $-10$    (b) 13    **3.** $-92°$F

**4.** 100 lb.    **5.** Sum $= 0$    **6.** 541°F

**7.** 31,458 lb-in. decrease

**8.** 810 lb.-in.    **9.** 1,275 lb.

*Chapter 7, Section 1, Page 146*

**1.** Table 1 (a) 5,8    (b) 7,16    (c) 15,32
Table 2 (a) 15,64    (b) 7    (c) 11/64    (d) 25/128    25

**3.** (a) 14/48    (b) 3/6    (c) 4/8

**5.** One part $= 1/12$, A $= 2/12$, B $= 3/12$, C $= 4/12$

**7.** (a) 9/20    (b) 3/4    (c) 7/16    (d) 13/32    (e) 1/2    (f) 9/16
(g) 1/5    (h) $2/1 = 2$    (i) 1    (j) 1/8

**9.** (a) All fractions equal 1/2.
(b) All fractions equal 1/3.
(c) All fractions equal 1/4.
(d) 1/8  1/4  1/2  9/16  7/16  3/8  5/16
(e) 1/4  1/2  7/8  1  5/8  1/8  3/4
(f) 3/4  1  1/2  1/4  5/4  7/4
(g) 4/5  2/5  5/9  3/10  2/5  1/2  7/9
(h) 3/4  5/8  3/4  1/4  1/2  3/8  3/4

**11.** (a) 1/2 ft.    (b) 1/2 ft.    (c) 1/4 ft.    (d) 1/2 yd.    (e) 2/3 yd.

**13.** (a) 3/8 lb.    (b) 1/2 lb.    (c) 7/8 lb.    (d) 5/8 qt.

**15.** (a) 3/20 cm.    (b) 3/20 km.    (c) 17/20 liter

**17.** 1/3 day    **19.** 4/9 yd.    **21.** 1/4 yr.

**23.** 3/4 lb.    **25.** (a) 4/7 (b) 3/7

**27.** 1/3 sleeping, 1/12 eating, 1/8 relaxing

**29.** 3/5 won, 2/5 lost

**31.** Auto mechanics 8/15, 2/15 diesel mechanics, 1/5 machine shop, 1/15 metal trades, 1/15 other.

*Chapter 7, Section 2, Page 159*

**1.** (a) 4    (b) 2/4, 1/4    **3.** (a) 6    (b) 4/6, 5/6

**5.** (a) 18    (b) 15/18, 2/18    **7.** (a) 12    (b) 21/12, 14/12

9. (a) 24    (b) 14/24, 3/24      11. (a) 30    (b) 18/30, 25/30
13. (a) 48    (b) 32/48, 15/48     15. (a) 21    (b) 7/21, 6/21
17. 2/3 < 5/6      19. 4/10 < 41/100
21. 1/4 < 1/3 < 5/12     23. 12/25 < 3/5 < 7/10
25. 7/16 < 1/2 < 3/5     27. (a) 5 1/4    (b) 3 3/4    (c) 4 3/4
29. (a) 3 1/8    (b) 5 7/8    (c) 2 1/8
31. (a) 1 33/64    (b) 2 11/64    (c) 2 29/64
33. (a) 5 1/2    (b) 4 1/2    (c) 2 3/4
35. (a) 1 7/10    (b) 1 89/100    (c) 15 13/15
37. (a) 5/2    (b) 27/8    (c) 83/16    (d) 37/8
39. (a) 101/8    (b) 497/32    (c) 89/4    (d) 75/4
41. (a) 3,457/100    (b) 923/100    (c) 16,147/1,000    (d) 2,450,897/10,000
43. (a) 22/3 ft.    (b) 23/4 ft.    (c) 223/12 ft. 551/12 ft.
45. (a) 43/8 gal.    (b) 254/100 cm.    (c) 231/24 day    (d) 766/365 yr.

*Chapter 7, Section 3, Page 169*

1. (a) 2/3    (b) 3/4    (c) 1/4    (d) 1/4    (e) 3/10    (f) 3/5
   (g) 1/4    (h) 1/5
3. (a) 1/4    (b) 1/4    (c) 3/5    (d) 5/1    (e) 3/4    (f) 4/9
   (g) 1/5    (h) 1/26
5. (a) 5/1    (b) 10/3    (c) 20/1    (d) 6/1
7. 5/6      9. 18 : 1
11. (a) 1/4    (b) 1/4    (c) 1/6    (d) 1/8    (e) 1/8    (f) 1/2
13. 4 : 1      15. 50 : 1      17. 9 : 1      19. 1 : 5

*Chapter 7, Section 4, Page 173*

1. (a) proportional    (b) not proportional    (c) not proportional
   (d) not proportional    (e) proportional    (f) not proportional
   (g) proportional
3. 24 teeth      5. 350 hp.      7. 300 lb.
9. 18 pairs      11. 18

*Chapter 7, Section 5, Page 183*

1. 240 rpm      3. 5 in.      5. D = 25 in.
7. (12 in.) 500 rpm (6 in.) 2,000 rpm, (3 in.) 7,000 rpm
9. 4 in.
11. (4 in.) 8,400 rpm, (6 in.) 5,600 rpm, (10 in.) 3,360 rpm, (8 in.) 4,200 rpm
13. (a) 350 rpm    (b) 6,400 rpm    (c) 150 rpm    (d) 176 rpm
15. 2      17. 256 teeth      19. 224 teeth

**21.** 156 rpm

**23.** (13T) 240 rpm, (16T) 195 rpm, (19T) 164 rpm, (22T) 142 rpm, (25T) 125 rpm

### Chapter 7, Section 6, Page 190

**1.** (a) 7    (b) 69    (c) 351    (d) 1,650

**3.** 12 lb. 11 oz.    **5.** 1 hr. 33 min.

**7.** 140 psi    **9.** (a) 1 hr. 28 min.    (b) 2 : 53 P.M.

**11.** (a) 222 lb.    (b) 222 lb.    (c) 222 lb.

### Chapter 7, Practice Test, Page 192

**1.** (a) 1/2    (b) 6/16    **2.** (a) equal    (b) unequal

**3.** (a) 9/4    (b) 29/8    **4.** (a) 5 3/4    (b) 3 5/16

**5.** (a) 3/4    (b) 7/8    **6.** (a) 1/4    (b) 3/4    (c) 4/5

**7.** (a) 3/8    (b) 4/3    **8.** (a) 4 : 3    (b) 4 : 1

**9.** 17,500 lb.    **10.** 9 : 1

**11.** $B = 45$ rpm, $C = 45$ rpm, $D = 24$ rpm, $E = 24$ rpm, $F = 20$ rpm

### Chapter 8, Section 1, Page 200

**1.** (a)  3/5    (b) 7/8    (c) 1/2    (d) 1/2    (e) 4/5    (f) 1 1/12
(g)  15/16    (h) 4/5    (i) 21/32    (j) 11/32    (k) 1 5/16    (l) 2 1/2
(m) 23/32    (n) 1    (o) 1

**3.** (a) 15 1/2 in.    (b) 12 3/4 lb.    (c) 18 7/8 in.    (d) 4 1/12 ft.
(e) 14 5/6 yd.    (f) 14 3/4 mi.    (g) 29 77/100 cm.

**5.** (a) 4/8, 2/8    (b) 3/6, 2/6    (c) 9/12, 2/12    (d) 12/16, 10/16
(e) 3/12, 4/12    (f) 4/24, 15/24    (g) 4/20, 5/20    (h) 9/15, 5/15
(i) 22/32, 12/32, 8/32    (j) 24/64, 20 /64, 7/64    (k) 4/12, 6/12, 9/12
(l) 16/20, 6/20, 15/20    (m) 33/48, 30/48, 32/48
(n) 4/24, 9/24, 10/24    (o) 30/36, 16/36, 21/36
(p) 27/48, 30/48, 28/48

**7.** (a) 7 5/8    (b) 24 1/16    (c) 15 9/16    (d) 19 19/32    (e) 16 1/32

**9.** (a) 4 31/64    (b) 10 41/64    (c) 31 5/12    (d) 46 5/6    (e) 7 1/12

**11.** (a) 13 1/25    (b) 10 39/50    (c) 14 79/100    (d) 11 1/4
(e) 11 13/18

**13.** (a) 9 1/8    (b) 14 9/10    (c) 16 29/64

**15.** (a) 20 7/20    (b) 17 5/24    **17.** 23 3/16

**19.** 1 5/24    **21.** 1 3/8 in.

**23.** 43 9/32 in.    **25.** 2 31/32 in.

**27.** Yes (total = 49 in.)

**29.** $A = 1$ 17/32, $B = 2$ 23/32, $C = 3$ 31/32

**31.** 11 11/12 ohms    **33.** 9 7/12 hr.

*Chapter 8, Section 2, Page 208*

**1.** (a) 4     (b) 4     (c) 16

**3.** (a) 5     (b) 13     (c) 21     **5.** 8 7/4 = 8 28/16

**7.** (a) 3/2     (b) 1 20/16     (c) 3 14/10     (d) 7 15/10

**9.** (a) 1 10/7     (b) 4 46/24     (c) 1 12/9     (d) 5 52/20

**11.** (a) 1/8     (b) 3/16     (c) 1/8     (d) 1/32

**13.** (a) 1/8     (b) 11/64     (c) 1/3     (d) 11/32

**15.** (a) 5/8     (b) 1     (c) 1/2     (d) 7/36

**17.** (a) 2 3/4 in.     (b) 3-3/10 cm.

**19.** (a) 3 1/12 yd.     (b) 1 2/7 week

**21.** (a) 12 11/32     (b) 5 1/10     (c) 2 19/32     (d) 1/4

**23.** (a) 1 1/6     (b) 1 3/10     (c) 3 7/12     (d) 5/12

**25.** (a) 1/4     (b) 3 3/8     (c) 1 63/100     (d) 9 9/10

**27.** (a) 2 15/16     (b) 11 3/4     (c) 5 2/3

**29.** (a) 2 13/20     (b) 2 5/6     (c) 1 7/12

**31.** (a) 5 11/15     (b) 1 3/4     (c) 1 19/20

**33.** 3 1/6     **35.** 5 1/8     **37.** 1 9/10

**39.** 6 31/32     **41.** 6 7/10     **43.** 2 7/16

**45.** 6 1/16     **47.** 11/64 in.     **49.** 1 7/64 in.

**51.** 29/32 in.     **53.** 7/8 lb.     **55.** 2 11/12 hr.

**57.** 5 3/8 lb.     **59.** 5/8     **61.** 2 3/4 ft.

**63.** A = 2 31/32 in.     **65.** X = 5/8 in.

*Chapter 8, First Practice Test, Pages 214–218*

**1.** (a) 1 1/2     (b) 2     (c) 1 1/2     (d) 1 5/16

**2.** (a) 12 3/4     (b) 23 3/4     (c) 22 1/6     (d) 39 9/64

**3.** (a) 10 ft. 3 7/8 in.     (b) 13 lb. 6 1/2 oz.
(c) 14 yd. 1/12 ft.     (d) 92 m. 38 2/5 cm.

**4.** (a) 15 13/32 in.     (b) 16 11/16 lb.     (c) 13 9/100 cm.

**5.** No (need 16 1/6 ft.)     **6.** 39/64 in.

**7.** 89 7/50 gm.     **8.** 2 63/64 in.

**9.** 5 3/32 in.     **10.** 3 7/16 in.     **11.** 4 3/64 in.

**12.** 2 3/4 in.     **13.** 23/32 in.

**14.** (a) 2 1/4     (b) 11 1/4     (c) 36 3/16     (d) 3 3/64

**15.** (a) 1 5/8     (b) 3 1/4     (c) 3 25/32     (d) 22 1/32

**16.** 17/64 in.     **17.** 31/64 in.     **18.** 3/64 in.

**19.** 3/128 in.     **20.** Y = 7/16 in.     **21.** 29 3/8 in.

**22.** 5/8 in.     **23.** 1 107/128 in.

## Chapter 8, Section 3, Page 223

**1.** (a) 1/16     (b) 1/4     (c) 1/32     (d) 1/64     (e) 1/20

**3.** (a) (a) 1/5     (b) 1/5     (c) 1/10     (d) 1/3     (e) 3/10

**5.** (a) 1/6     (b) 1/2     (c) 1/5     (d) 1/6     (e) 1/3

**7.** (a) 21/160     (b) 7/10     (c) 4/3     (d) 1     (e) 3/40

**9.** (a) 1     (b) 1     (c) 1     (d) 1     (e) 1

**11.** (a) 4     (b) 6     (c) 7/3     (d) 33/4     (e) 34

**13.** (a) 2 2/3     (b) 3 3/4     (c) 5 1/4     (d) 5 5/8     (e) 4 1/8

**15.** (a) 5 5/32     (b) 6 3/32     (c) 7 11/64     (d) 94 1/4     (e) 40

**17.** (a) 1     (b) 1 2/3     (c) 9/10     (d) 11/16     (e) 9/32

**19.** (a) 6 3/8     (b) 5 1/4     (c) 33 1/4     (d) 31 1/2     (e) 33 1/2

**21.** (a) 29 1/3     (b) 12     (c) 146 2/3     (d) 3 3/5     (e) 3 1/5

**23.** (a) 18 in.     (b) 3 ft.     (c) 8 in.     (d) 12 oz.

**25.** (a) 7/8 ft.     (b) 3 3/5 mi.     (c) 3 1/2 ft.     (d) 5 2/5 qt.

**27.** 48 3/8 in.     **29.** 312 3/4 in.

**31.** 483 3/4 min. = 8 hr. 3 3/4 min.

**33.** $124     **35.** 60 gal.     **37.** 1 29/64 gal.

**39.** 102 in.

**41.** (a) $192     (b) $336     (c) $288     (d) $696     (e) $220 4/5
    (f) $67 1/5

## Chapter 8, Section 4, Page 232

**1.** (a) 2/1     (b) 4/3     (c) 5/4     (d) 8/3     (e) 16/5     (f) 8/7
    (g) 10/9     (h) 16/15

**3.** (a) 2/3     (b) 4/15     (c) 8/43     (d) 4/33     (e) 16/151     (f) 3/16
    (g) 12/55

**5.** (a) 3/4     (b) 6     (c) 7/12     (d) 3/2     (e) 10/3

**7.** (a) 10/3     (b) 28/5     (c) 35/18     (d) 42/23

**9.** (a) 15/2     (b) 35/48     (c) 125/78     (d) 189/8     (e) 9/16

**11.** (a) 3/16     (b) 1/12     (c) 7/40     (d) 7/24     (e) 47/120

**13.** (a) 1 1/2 in.     (b) 4/3 ft.     (c) 3/4 lb.

**15.** (a) 1 7/8 in.     (b) 11/12 in.     (c) 8 8/9 ft.

**17.** (a) 3 7/9 qt.     (b) 9 1/25°F     (c) 16 mi.

**19.** (a) 5 1/9     (b) 5 2/7     (c) 1 2/13     (d) 9/22

**21.** 8     **23.** 40 mph     **25.** (a) 12     (b) $159

**27.** 1 ft. 9 in.     **29.** 5 37/320     **31.** 96

**33.** 25     **35.** 2,112

**Chapter 8, Second Practice Test, Sections 3 and 4, Pages 236–237**

**1.** (a) 15/128     (b) 23/24     **2.** (a) 57/16     (b) 76

**3.** 3 15/16     **4.** 11 7/8     **5.** 4 ft. 7 in.

**6.** 112 gal.     **7.** $132     **8.** (a) 1 1/8     (b) 1 1/3

**9.** (a) 1 3/5     (b) 1 29/40     **10.** 3 105/128 in.

**11.** 24     **12.** 1,120 nails     **13.** 159 screws

**Chapter 9, Section 1, Page 244**

**1.** (a) five tenths     (b) eight tenths     (c) seven tenths
(d) four tenths     (e) six tenths     (f) nine tenths
(g) one tenth     (h) three tenths

**3.** (a) twenty-five hundredths     (b) thirty-eight hundredths
(c) fifty-seven hundredths     (d) sixty-four hundredths
(e) twenty-six hundredths     (f) forty-nine hundredths
(g) seventy-three hundredths

**5.** (a) two and five hundredths     (b) three and four hundredths
(c) five and seven hundredths     (d) four and nine hundredths
(e) three and eight hundredths     (f) eight and eight hundredths

**7.** (a) twenty-five and one hundredth     (b) forty-five and five hundredths
(c) fifty and two hundredths     (d) thirty-eight and seven hundredths
(e) forty-four and six hundredths     (f) ninty-nine and three hundredths

**9.** (a) .3     (b) .8     (c) .4     (d) .9

**11.** (a) 25.9     (b) 50.5

**13.** (a) 2.35     (b) 6.49     (c) 12.18     (d) 30.03

**15.** (a) four and four tenths     (b) four hundredths
(c) forty and four tenths     (d) forty and four hundredths
(e) forty-four and forty hundredths
(f) forty and forty-four hundredths
(g) fourteen and forty-one hundredths

**17.** (a) five and five hundredths     (b) five and fifty hundredths
(c) five and five tenths     (d) fifty and five hundredths
(e) fifty-five and fifty hundredths
(f) fifty and fifty hundredths
(g) fifteen and fifty-one hundredths

**19.** (a) seventy and seven tenths     (b) seventy and seven hundredths
(c) seven hundred and seven tenths
(d) seven hundred and seventy hundredths
(e) seven hundred seventy and seven tenths
(f) seven hundred seven and seven tenths
(g) seven hundred seven and seven hundredths

**21.** (a) 44.4     (b) 60.66     (c) 30.33     (d) 55.5     (e) 303.33
(f) 550.55     (g) 100.10     (h) 101.01     (i) 222.02     (j) 69.96

**23.** (a) one and three hundred forty-five thousandths
(b) two and four hundred thirty-eight thousandths
(c) six and seven hundred eighty-four thousandths
(d) nineteen and five hundred fifty-six thousandths
(e) twenty-three and eight hundred fifty-six thousandths
(f) thirty-four and eight hundred fifty-five thousandths

**25.** (a) one hundred forty-five and six hundred forty-six thousandths
(b) four hundred fifty-six and two hundred forty-three thousandths
(c) seven hundred eighty-nine and five hundred fifty-four thousandths
(d) three hundred forty-four and ninety-eight thousandths
(e) two hundred seventy-five and seven hundred fifty thousandths

**27.** (a) forty and forty thousandths
(b) thirty-five and fifty-three thousandths
(c) sixty-four and six hundred forty thousandths
(d) four hundred seventy-five and four hundred seventy-five thousandths
(e) thirty-three and three hundred thirty-three thousandths

**29.** (a) eight thousand, seven hundred sixty-four ten-thousandths
(b) five thousand, six hundred seventy-eight ten-thousandths
(c) three thousand, five hundred sixty-four ten-thousandths
(d) one thousand, nine hundred eighty-two ten-thousandths
(e) two thousand, three hundred seventy-five ten-thousandths
(f) one thousand, eight hundred seventy-five ten-thousandths

**31.** (a) twenty-three and five thousand fifty-six ten-thousandths
(b) fifty-six and seven thousand seven hundred eighty ten-thousandths
(c) forty-five and eight thousand seventy ten-thousandths
(d) one hundred forty and six thousand sixty-five ten-thousandths
(e) two hundred thirty-four and four thousand three hundred twenty ten-thousandths.

**33.** (a) ten thousand and nine hundred five ten-thousandths
(b) ten thousand ten and one hundred one ten-thousandths
(c) two thousand and two ten-thousandths
(d) one and one ten-thousandth

**35.** (a) tenths     (b) hundredths     (c) thousandths     (d) ten-thousandths
(e) ones     (f) ones     (g) hundredths     (h) hundreds
(i) hundreds     (j) thousandths     (k) thousands     (l) ten thousands

**37.** 581.2495

**39.** (a) 0.05, 0.55     (b) 0.098, 0.98     (c) 1.12, 2.11     (d) 0.035, 5.3
(e) 1.01, 1.110

**41.** (a) 2.5     (b) 5.8     (c) 1.1     (d) 2.0     (e) 0.9     (f) 34.1

**43.** (a) 3.11     (b) 5.66     (c) 8.23     (d) 0.01     (e) 45.15

**45.** (a) 0.565     (b) 0.670     (c) 1.008     (d) 35.123     (e) 1.998     (f) 3.000

**47.** (a) 0.6     (b) 2.6     (c) 3.4     (d) 6.7     (e) 10.1     (f) 23.0

**49.** (a) 3.07     (b) 11.68     (c) 14.01     (d) 23.46     (e) 9.98     (f) 2.00

**51.** (a) 1.305   (b) 3.156   (c) 2.544   (d) 8.118   (e) 5.007
(f) 4.900

**53.** (a) 1.09 yd.   (b) 2.21 lb.   (c) 0.39 in.   (d) 1.15 mi.
(e) 6.44 cm.

**55.** (a) 1.1 yd.   (b) 1.09 yd.   (c) 1.094 yd.

**57.** (a) 56.8 ft.   (b) 57 ft.   (c) 60 ft.   (d) 60 ft.

**59.** (a) 16.454 gm.   (b) 16.45 gm.   (c) 16 gm.   (d) 16.5 gm.

**61.** (a) $87.50   (b) $87   (c) $90

*Chapter 9, Section 2, Page 254*

**1.** (a) 1.8   (b) 1.2   (c) 1.2   (d) 1.8   (e) 3.0

**3.** (a) 1.00   (b) 1.82   (c) 1.85   (d) 2.91   (e) 2.25

**5.** (a) 7.92   (b) 14.21   (c) 10.65   (d) 14.08   (e) 24.71

**7.** (a) 702.19   (b) 493.41   (c) 129.20   (d) 1,304.21   (e) 5,519.63

**9.** (a) 8.105   (b) 8.685   (c) 12.335   (d) 95.546   (e) 9.885

**11.** (a) 46.787   (b) 12.824   (c) 367.23   (d) 5.973   (e) 191.293

**13.** (a) 210.105 gm.   (b) 1.7987 in.   (c) 71.36 yd.   (d) 192.2 km.

**15.** (a) 50.25 in.   (b) 82.537 cm.   (c) 82.86 m.   (d) $30.47
(e) 120.415 gm.

**17.** 2.625 in.   **19.** 1.6221 in.   **21.** 1,382.3 mi.

**23.** 11.875

**25.** (a) 1.0   (b) 3.09   (c) 4.98   (d) 3.85   (e) 0.79

**27.** (a) 2.899   (b) 4.120   (c) 121.783   (d) 0.0929   (e) 1.2078

**29.** (a) 2.343   (b) 10.6524   (c) 17.7813   (d) 34.933   (e) 3.935

**31.** (a) 3.225   (b) 13.594   (c) 32.225   (d) 1.94   (e) 10.999

**33.** 27.88   **35.** 33.07   **37.** 21.105

**39.** 16.91   **41.** 2.89   **43.** .666 in.

**45.** L = 0.799 cm.

*Chapter 9, Section 3, Page 262*

**1.** (a) 4.8   (b) 6.3   (c) 2.1   (d) 1.8   (e) 6.4

**3.** (a) 4.42   (b) 16.20   (c) 26.52   (d) 38.25   (e) 15.13

**5.** (a) 0.4005   (b) 1.232   (c) 29.24   (d) 79.20   (e) 3.2296

**7.** (a) 13.0525   (b) 59.3009   (c) 36.673   (d) 0.551375
(e) 58.0145

**9.** (a) $36.48   (b) $4.74   (c) $31.94   (d) $80.27   (e) $0.06

**11.** (a) 2.75 cm.   (b) 4.06 m.   (c) 5.69 in.   (d) 0.04 ft.
(e) 407.20 liters

**13.** $151.20      **15.** 69.144 lb.      **17.** 2.625 in.

**19.** 4.33 lb.      **21.** $9.45      **23.** $724.73

**25.** 1 hr. 7.5 min.      **27.** No. (Cost = $615.82)

**29.** (a) $420.64      (b) $849.70      (c) $93.60      (d) $217.80
     (e) $82.00      (f) 357 hr.      (g) $1,663.74

### Chapter 9, Section 4, Page 268

**1.** (a) 1.4      (b) 1.5      (c) 0.4      (d) 0.09      (e) 11.21

**3.** (a) 0.06      (b) 0.121      (c) 0.32      (d) 0.95      (e) 7.5

**5.** (a) 2.3      (b) 2.2      (c) 7.4      (d) 4.7      (e) 1.3

**7.** (a) 5.62      (b) 6.71      (c) 2.92      (d) 6.53      (e) 2.29

**9.** (a) 0.125      (b) 0.250      (c) 0.200      (d) 0.750      (e) 0.875

**11.** (a) 0.4375      (b) 0.6562      (c) 0.3125      (d) 0.8438      (e) 0.2812

**13.** (a) 35      (b) 112      (c) 7.9      (d) 13.4      (e) 2.2

**15.** (a) 0.002      (b) 0.0012      (c) 0.002      (d) 0.0005      (e) 10.03

**17.** (a) 201      (b) 54.075      (c) 8.715      (d) 2,836      (e) 20.02

**19.** (a) 94.9      (b) 3480.018      (c) 367.1      (d) 62.5

**21.** (a) 2.5      (b) 25.6      (c) 2.3      (d) 502.4

**23.** (a) 0.03      (b) 236.25      (c) 0.44      (d) 0.75

**25.** (a) 0.175      (b) 0.306      (c) 1.291      (d) 1.180

**27.** (a) 0.0067      (b) 39.444      (c) 0.0200      (d) 0.1646

**29.** 0.0049 lb.      **31.** 0.025 in.

**33.** 36.48 mph      **35.** 1,280

**37.** (a) No      (b) 8.1 mpg      **39.** $0.36

**41.** 0.872 inches per hour

### Chapter 9, Section 5, Page 275

**1.** (a) 0.2      (b) 0.5      (c) 0.9      (d) 0.1      (e) 0.8

**3.** (a) 0.95      (b) 0.90      (c) 0.28      (d) 0.8      (e) 0.75

**5.** (a) 2.8      (b) 5.6      (c) 5.28      (d) 6.09

**7.** (a) 0.125      (b) 0.375      (c) 0.062      (d) 0.438      (e) 0.875

**9.** (a) 1.625      (b) 4.219      (c) 8.688      (d) 9.172      (e) 10.188

**11.** (a) 0.086      (b) 0.563      (c) 0.0013      (d) 0.598      (e) 0.0165
     (f) 0.1987      (g) 0.0946      (h) 0.0019      (i) 0.0002

**13.** (a) 0.78      (b) 0.35      (c) 0.08      (d) 0.49      (e) 0.8129      (f) 7/8

**15.** (a) 46/64      (b) 26/64      (c) 23/64      (d) 37/64      (e) 15/64
     (f) 1 38/64      (g) 5 34/64      (h) 50/64      (i) 5 43/64

**17.** (a) 0.875      (b) 1.375      (c) 0.688      (d) 2.188      (e) 2.312
     (f) 3.047

**19.** 0.14      **21.** 0.08      **23.** 0.43

**25.** 0.87

**27.** Bill (.333), Homer (.379), Jake (.280), LeRoy (.421), Noel (.423), "Fingers" (.267), Carl (.095), Jim (.458), Charlie (.308)

**29.** Carl

### Chapter 9, Section 6, Page 286

**1.** (a) 25.4 cm.    (b) 63.5 cm.    (c) 114.3 cm.    (d) 62.23 cm.
(e) 226.06 cm.    (f) 190.5 cm.

**3.** (a) 1.82 m.    (b) 9.1 m.    (c) 31.85 m.    (d) 15.925 m.
(e) 409.5 m.    (f) 1,601.6 m.

**5.** (a) 8.05 km.    (b) 16.1 km.    (c) 24.15 km.    (d) 161 km.
(e) 73.255 km.    (f) 11,249.07 km.

**7.** (a) 3.9 in.    (b) 9.75 in.    (c) 19.5 in.    (d) 39 in.    (e) 175.5 in.
(f) 1 in.

**9.** (a) 6.2 mi.    (b) 31 mi.    (c) 46.5 mi.    (d) 9.98 mi.    (e) 102.3 mi.
(f) 0.31 mi.

**11.** (a) 25.4 cm.    (b) 76.2 cm.    (c) 457.2 cm.    (d) 91.4 cm.
(e) 635 cm.

**13.** (a) 0.622 m.    (b) 11.43 m.    (c) 2.74 m.    (d) 1,609.34 m.
(e) 7,081.11 m.

**15.** (a) 82 ft.    (b) 1.48 ft.    (c) 6.22 ft.    (d) 13,120 ft.
(e) 1.29 ft.    (f) 0.49 ft.

**17.** (a) 38.86 cm.    (b) 43.18 cm.    (c) 132.08 cm.    (d) 1,706.9 ft.

**19.** (a) 2.25 kg.    (b) 4.5 kg.    (c) 22.5 kg.    (d) 121.5 kg.
(e) 900 kg.

**21.** (a) 11 lb.    (b) 22 lb.    (c) 220 lb.    (d) 59.2 lb.    (e) 2,200 lb.

**23.** (a) 1 31/32    (b) 3 5/32    (c) 4 23/32    (d) 9 53/64    (e) 1
(f) 4 21/64

**25.** (a) 29 mm.    (b) 24 mm.    (c) 19 mm.    (d) 22 mm.    (e) 17 mm.

**27.** (a) 2.84    (b) 14.19    (c) 23.18

**29.** (a) 3.79    (b) 7.1    **31.** (a) 4.23    (b) 3.62    (c) 13.21

**33.** 1.4 in.    **35.** 169.2 liters    **37.** 0.28 in.

**39.** 6.21 mi.

**41.** (a) 11.8 mpg    (b) 15.3 mpg    (c) 23.5 mpg    (d) 19.3 mpg

**43.** 109.8 cu. in.    **45.** (a) 5,359 cc.    (b) 5.359 liters

**47.** 395.54 cc.    **49.** 335.5 mps

**51.** (a) 11,040 m. 6,031 fath.    (b) 35,693 ft. 5949 fath.
(c) 29,988 ft. 9,146 m.    (d) 8,415 m. 4,598 fath.
(e) 26,568 ft. 4,428 fath.    (f) 27,534 ft. 8,398 m.
(g) 24,715 ft. 4,119 fath.    (h) 6,123 m. 3,346 fath.

### Chapter 9, Practice Test, Page 291

**1.** 0.045     **2.** 0.0065     **3.** five and seventy-eight thousandths

**4.** 50.765     **5.** 0.8     **6.** 0.05, 0.45, 0.5, 0.54

**7.** 0.7     **8.** 0.898     **9.** 13.52     **10.** 4.8985     **11.** 1.899

**12.** 17.82     **13.** 1.7     **14.** $120.60     **15.** 0.9375     **16.** 50/64

**17.** 14.61 cm.     **18.** 21.54 in.     **19.** 0.03785 ohm/m.     **20.** 93.21 mph

**21.** Import (28.23 mpg)     **22.** 1,247.3     **23.** 12.13¢

**24.** 157.5 kg./sq. m.     **25.** 7.06 cm.

### Chapter 10, Section 1, Page 299

**1.** (a) 0.25   25%     (b) 65/100   65%     (c) 75/100   0.75
(d) 0.58   58%     (e) 80/100   80%     (f) 15/1000   0.015
(g) 1.15   115%     (h) 1 45/100   145%     (i) 1 25/100   1.25

**3.** (a) 55%     (b) 78%     (c) 23%     (d) 156%     (e) 0.5%
(f) 3.25%

**5.** (a) 25%     (b) 75%     (c) 12.5%     (d) 37.5%     (e) 62.5%
(f) 80%     (g) 6.3%     (h) 18.8%     (i) 43.8%     (j) 46.9%
(k) 33.3%     (l) 66.7%     (m) 1.6%     (n) 39%

**7.** (a) 44%     (b) 51%     **9.** 37.5%     **11.** (a) 75%     (b) 25%

**13.** 37.5%     **15.** 86%     **17.** 22.8%     **19.** 16.7%

**21.** 3/4 copper, 1/4 nickel     **23.** 60% tin, 40% lead

**25.** (a) 52 lb.     (b) 65.4%     **27.** 2,750%     **29.** (a) 9/10     (b) 90%

**31.** 13.3%     **33.** 5.4%     **35.** 14.9%     **37.** 1.67%

**39.** 5.3%     **41.** 20%

**43.** (a) 10.2%     (b) 17.0%     (c) 9.2%     (d) 17.3%     (e) 25.1%
(f) 21.2%

### Chapter 10, Section 2, Page 313

**1.** (a) 12.5     (b) 18     (c) 81     (d) 38.4     (e) 11.2     (f) 16.3
(g) 26.25     (h) 108.75     (i) 96

**3.** (a) 1.72     (b) 0.27     (c) 1.84     (d) 9.41     (e) 0.188     (f) 25.42
(g) 0.83     (h) 0.18     (i) 1.42

**5.** (a) $70     (b) 35.2 ft.     (c) 26.1 in.     (d) 157.9 lb.     (e) 105.6 in.
(f) 35 cm.     (g) 210 yd.     (h) 108.1 gal.     (i) 360 gm.

**7.** (a) $11.41     (b) $.51     (c) $5.89     (d) $75.38     (e) $30.70
(f) $.07     (g) $.18     (h) $.82     (i) $.59

**9.** 15.8 ft.     **11.** 17     **13.** $438     **15.** 26     **17.** 6

**19.** 0.11 gal.     **21.** $9.00

**23.** (a) $18.90   $8.50   $161.60     (b) $2.45   $1.10   $20.95
(c) $ .58   $ .26   $4.91     (d) $3.58   $1.61   $30.56
(e) $4.78   $2.15   $40.92

**25.** (a) $20.93    (b) $252.91    (c) $86.12    (d) $55.60    (e) $39.24
(f) $69.77    (g) $250.73

**27.** $101.74 (triple), $98 (straight), the 30% discount

**29.** $331.88    **31.** $7.23    **33.** Honest Abe ($9.25 saved per unit)

**35.** $3,772.75

**37.** (a) $300    (b) $90    (c) $203    (d) $350    (e) $73.34
(f) $157.58    (g) $225.23    (h) $580.80

**39.** 51 amp.    **41.** $128.43    **43.** (a) $1,230    (b) 21.7%

**45.** lead 18.75, ant. 4.75 lb., tin 1.25 lb., copper 0.25 lb.

**47.** $175.40    **49.** 7.2%    **51.** 4.74%

**53.** Cost $61,309.60, expense $56,930.34, profit $27,735.29

*Chapter 10, Practice Test, Page 319*

**1.** 76%    **2.** 185%    **3.** 0.015    **4.** 21.5 gal.    **5.** 83.7%

**6.** $77.80    **7.** 7%    **8.** $354.02    **9.** $56.25

**10.** $2,133.33

*Chapter 11, Section 1, Page 325*

**1.** (a) 6 ft., 2 sq. ft.    (b) 14 in., 12 sq. in.    (c) 32 cm., 63 sq. cm.
(d) 24 yd., 36 sq. yd.    (e) 62 m., 234 sq. m.

**3.** 9.5 ft., 16.6 sq. ft.    **5.** $90    **7.** $186

**9.** (a) 32 ft., 44 sq. ft.    (b) 26 ft., 29 sq. ft.

**11.** 755 ft.    **13.** 236,425 sq. ft.    **15.** 490,750 sq. ft.

**17.** 51.8%    **19.** $15.97

*Chapter 11, Section 2, Page 335*

**1.** (a) 31.4 in.    (b) 65.94 in.    (c) 43.96 cm.    (d) 8.48 ft.
(e) 45.53 ft.    (f) 489.84 ft.

**3.** 152.8 ft.    **5.** 103.98 in.    **7.** 175.93 in.

**9.** 132 in.    **11.** 6,283 fpm

**13.** (a) 3.14    (b) 28.26    (c) 1.77    (d) 14,200.85    (e) 0.0000338

**15.** (a) 0.157 sq. in.    (b) 0.05%

**17.** 108.4 sq. cm.    **19.** 71.83 sq. ft.

**21.** (a) 9,425 ft.    (b) 37.7 fps    (c) 25.7 mph

**23.** (a) 583,412,000 mi.    (b) 938,860,000 km.

*Chapter 11, Section 3, Page 345*

**1.** (a) 12 in.    (b) 15 cm.    (c) 9 ft. 9 in.    (d) 12 m. 39 cm.
(e) 9.05 in.    (f) 3.58 in.

**3.** No. Total = 179°

**5.** (a) 14 sq. in.     (b) 21 sq. cm.     (c) 1.69 sq. ft.     (d) 3.68 sq. in.
(e) 53,868.5 sq. cm.     (f) 0.0028 sq. in.

**7.** 78.6 lb.     **9.** $4,485

### Chapter 11, Section 4, Page 350

**1.** (a) 4.79     (b) 7.48     (c) 13.34     (d) 16.58     (e) 1.26

**3.** (a) 12.49     (b) 1.57     (c) 0.088     (d) 0.059

**5.** (b)     **7.** (a) 60.39     (b) 1.21

### Chapter 11, Section 5, Page 358

**1.** (a) 5 in.     (b) 13 cm.     (c) 12.53 ft.     (d) 8.75 in.
(e) 3.12 ft.     (f) 11.53 in.     (g) 1.003 cm.     (h) 10.69 ft.
(i) 1.18 cm.

**3.** 596 ft.     **5.** 2.92 lb.

**7.** (a) 82.65     (b) 16.25 sq. cm.     (c) 15.59     (d) 20.78
(e) 29.94 sq. in.

**9.** 12.32 sq. in.     **11.** 492 sq. in.     **13.** 56.3 naut. miles

**15.** (a) 0     (b) 0%     (c) Yes

### Chapter 11, Section 6, Page 370

**1.** Surface Area = 52 sq. ft., Vol. = 24 cu. ft.     **3.** 584 sq. ft.

**5.** 27, 8, 12, 6, 3, 9, 3, 9     **7.** 2.78 cu. yd.

**9.** (a) 18,144 cu. in.     (b) 10.5 cu. ft.

**11.** (a) 6,480 cu. in.     (b) 28 gal.     **13.** $166.14

**15.** (a) 47,719 cu. ft.     (b) 6,363 gal.     **17.** 3.22 liters

**19.** 99.28 lb.     **21.** 503 cu. yd.     **23.** 785.5 sq. in.

### Chapter 11, Sections 1 to 6, Pages 373–374

**1.** P = 20 ft. 6 in., A = 26.12 sq. ft.

**3.** 7 in.     **5.** 69%     **7.** 15.7 sq. in.

**9.** (a) 54,671 gal.     (b) 400,922 lb.

**11.** 26.7 in.     **13.** 16,800 cu. ft.     **15.** $112.50

**17.** 2,262 gal.     **19.** 0.798 cu. in.

### Chapter 11, Practice Test, Page 377

**1.** $550     **2.** 24 ft.     **3.** 120 sq. cm.     **4.** 1.47 cu. ft.

**5.** 196,111 cu. ft.     **6.** 355 sq. in.     **7.** 227 cu. in.     **8.** 269,280 lb.

### Chapter 12, Section 1, Page 386

**1.** (a) 0.3420, 0.3640      (b) 0.4226, 0.4663
     (c) 0.5000, 0.5774      (d) 0.6691, 0.9004
     (e) 0.7071, 1.000      (f) 0.8387, 1.5399
     (g) 0.8660, 1.7321      (h) 0.9816, 5.6713
     (i) 0.9998, 57.2900

**3.** (a) $x = 50°$    (b) $x = 45°$    (c) $x = 55°$    (d) $x = 28°$
     (e) $x = 34°$    (f) $x = 50°$

**5.** (a) $x = 6.53$    (b) $x = 31.47$    (c) $x = 2.31$
     (d) $x = 2.33$    (e) $x = 6.88$    (f) $x = 9.77$

**7.** AB $= 4.2$, BC $= 7.27$, CD $= 8.4$, AD $= 5.94$, DE $= 13.24$, EA $= 6.62$

### Chapter 12, Section 2, Page 394

**1.** 56 ft. 8 in.     **3.** 14 ft. 9 in.     **5.** 691.5 ft.     **7.** 254 ft.

**9.** 3°     **11.** 1,410.5 ft.     **13.** 16,710 ft.     **15.** 17/64 in.

**17.** 283.1 ft.

**19.** (a) 19.15 sq. ft.     (b) 123.9 sq. in.     (c) 244.3 sq. in.
     (d) 598.2 sq. cm.     (e) .60 sq. mm.     (f) 1.00 sq. cm.
     (g) 0.89 sq. mi.

### Chapter 12, Practice Test, Page 397

**1.** (a) 0.1736, 0.1763     (b) 0.4226, 0.4663

**2.** (a) 21°    (b) 30°     **3.** (a) 10°    (b) 64°

**4.** 11.33 and 8.23 inches

**5.** (a) 707.6 ft.     (b) 863.7 ft.     (c) 156.1 ft.

### Chapter 13, Section 1, Page 409

**1.**

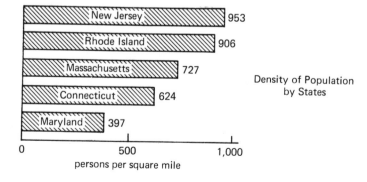

Density of Population
by States

**3.**

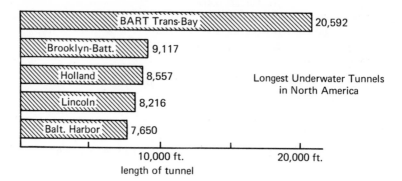

BART Trans-Bay   20,592
Brooklyn-Batt.   9,117
Holland   8,557
Lincoln   8,216
Balt. Harbor   7,650

Longest Underwater Tunnels
in North America

10,000 ft.          20,000 ft.
length of tunnel

**5.**

Composition of Air

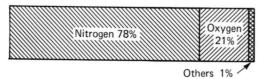

Nitrogen 78%     Oxygen 21%

Others 1%

**7.**

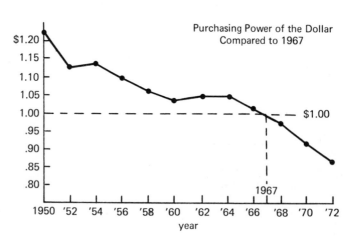

Purchasing Power of the Dollar
Compared to 1967

$1.20
1.15
1.10
1.05
1.00 — — — — — — — — — $1.00
.95
.90
.85
.80

1967

1950  '52  '54  '56  '58  '60  '62  '64  '66  '68  '70  '72
year

**9.**

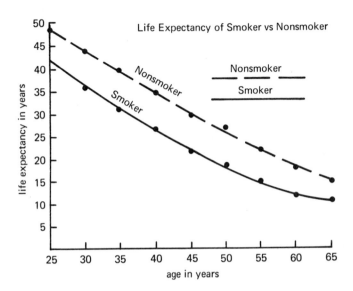

**11.**

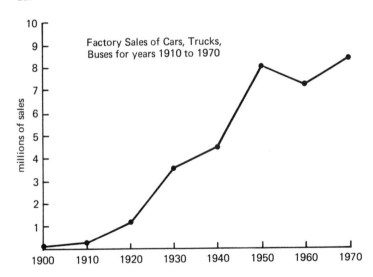

*Chapter 13, Section 2, Page 418*

**1.**

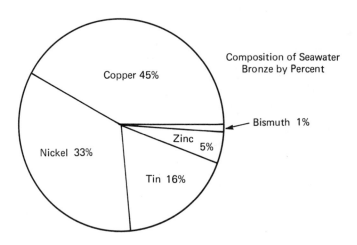

Composition of Seawater
Bronze by Percent

Copper 45%

Bismuth 1%

Zinc 5%

Nickel 33%

Tin 16%

**3.**

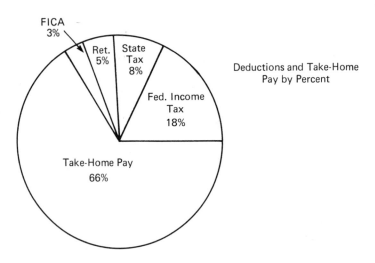

FICA 3%

Ret. 5%

State Tax 8%

Fed. Income Tax 18%

Deductions and Take-Home
Pay by Percent

Take-Home Pay
66%

**5.**

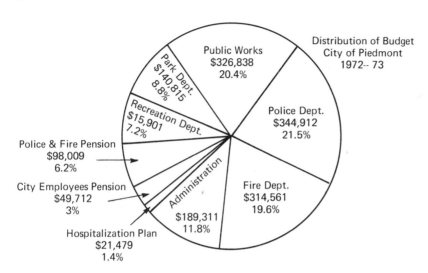

*Chapter 13, Section 3, Page 424*

**1.**

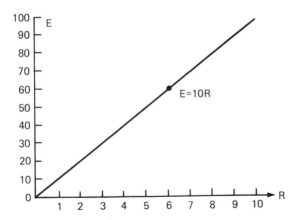

**3.**

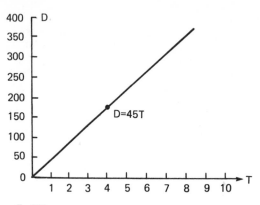

**5.** 570 rpm

**7.**

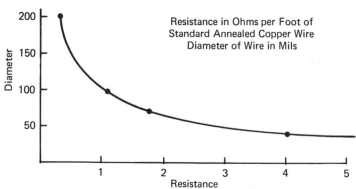

Resistance in Ohms per Foot of
Standard Annealed Copper Wire
Diameter of Wire in Mils

**9.**

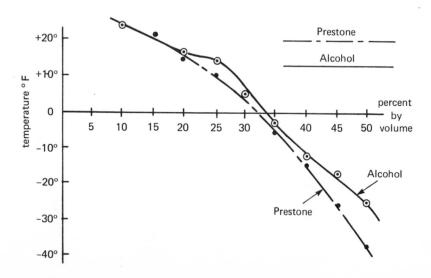

**11.**

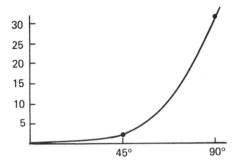

**13.**

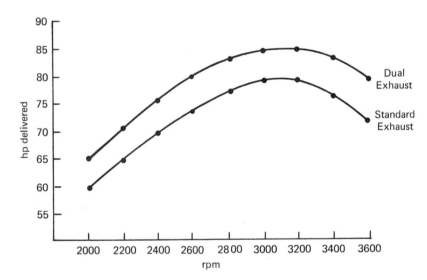

*Chapter 13, Practice Test, Page 427*

**1.**

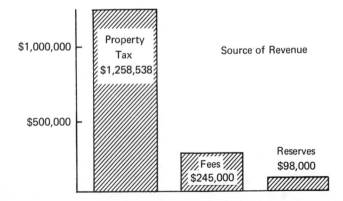

**2.**

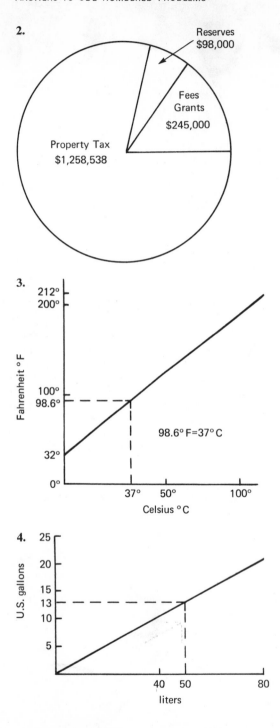

**5.** 50 liters = 13 gallons (approximately)

### Chapter 14, Section 1, Page 434

**1.** 1-c, 2-e, 3-g, 4-a, 5-j, 6-b, 7-i, 8-d, 9-f, 10-h

**3.** (a) $x + 50$     (b) $x - 25$     (c) $2x + 2$     (d) $3x/4$
(e) $x/2$     (f) $4x - 5$     (g) $2x + 9$     (h) $3/2x$

**5.** $k + 6$     **7.** $m + 40$

**9.** $z/3$     **11.** $(5/3)m$     **13.** (a) $0.08D$     (b) $D + .08D$

**15.** $w - x$ lb.     **17.** $\$5k$     **19.** $22d¢$     **21.** $\$4568 - k$

**23.** $mt$ miles     **25.** $E = IR$     **27.** $x = s + 15$

**29.** $P = S - C$     **31.** A number plus seven equals ten.

**33.** An amount $a$ plus an amount $b$ equals three times an amount $c$.

**35.** Five less than six $t$ equals negative nine.

**37.** $E$ equals $I$ times $R$.

**39.** An amount $P$ equals twice $a$ plus twice $b$.

**41.** The sum of $a$ and $b$ divided by their difference equals one.

### Chapter 14, Section 2, Page 441

**1.** $A = 225$     **3.** $A = 3.5$     **5.** $A = 16.61$

**7.** $L = 37$     **9.** $V = 14\ 1/7$     **11.** $A = 21$

**13.** $A = LW$     **15.** $R = E/I$     **17.** $V = L^3$

**19.** $A + B + C = 180°$     **21.** $d = rt$

**23.** $H = EI/746$     **25.** $V = 7.48LWH$

### Chapter 14, Section 3, Page 450

**9.** Yes     **11.** Yes     **13.** No     **15.** No     **17.** $x = 10$

**19.** $y = 12$     **21.** $x = 50$     **23.** $t = 2.1$     **25.** $h = 3.08$

**27.** $x = 10$     **29.** $k = 50$     **31.** $x = 0$     **33.** $y = 1.1$

**35.** $r = 19\ 1/6$     **37.** $x = 10$     **39.** $t = 2$     **41.** $y = 1$

**43.** $t = 0.4$     **45.** $x = 8$     **47.** $x = 50$     **49.** $m = 72$

**51.** $x = 4$     **53.** $a = 0.5$     **55.** $x = 8\ 1/3$     **57.** $x = 4$

**59.** $x = 3$     **61.** $a = 6$     **63.** $x = 2$     **65.** $x = 13.5$

**67.** $a = 30$     **69.** $x = 3$     **71.** $b = -10$     **73.** $s = -1$

**75.** $m = 2$     **77.** $y = -16$     **79.** $x = -9$     **81.** $x = 9$

**83.** $x = 15$     **85.** $k = 1/2$     **87.** $x = 1$     **89.** $x = 50$

**91.** $y = -8$     **93.** $g = -3$     **95.** $x = 40$     **97.** $x = 1.1$

**99.** $x = 1/2$

### Chapter 14, Section 4, Page 461

**1.** 14 inches    **3.** 41 2/3 inches    **5.** 90 by 30 yd.

**7.** 1 ft. 8 in., 2 ft. 8 in., 3 ft. 8 in.

**9.** $700    **11.** Journeyman 40 hr., helper 30 hr.

**13.** 1 1/2 liters    **15.** 30 liters    **17.** 105 gm.

**19.** 50 acres at $200 and 90 acres at $300

**21.** 375 lb.    **23.** 7 1/2 hr.    **25.** 40 ft.    **27.** 20 miles

### Chapter 14, Practice Test, Page 464

**1.** (a) $2x - 5$    (b) $st/(s + t)$

**2.** $2(2x + 100)$    **3.** Volume $= L(L - 1)(L + 2)$

**4.** $K = 1/2\,mv^2$    **5.** (a) $x = 7$    (b) $x = 6\,2/5$

**6.** 10 1/2 ft.    **7.** 200 at $55 and 300 at $30

**8.** 10.2 quarts

# Index